# Fluid Mechanics and Its Applications

Volume 111

## Aims and Scope of the Series

The purpose of this series is to focus on subjects in which fluid mechanics plays a fundamental role.

As well as the more traditional applications of aeronautics, hydraulics, heat and mass transfer etc., books will be published dealing with topics which are currently in a state of rapid development, such as turbulence, suspensions and multiphase fluids, super and hypersonic flows and numerical modeling techniques.

It is a widely held view that it is the interdisciplinary subjects that will receive intense scientific attention, bringing them to the forefront of technological advancement. Fluids have the ability to transport matter and its properties as well as to transmit force, therefore fluid mechanics is a subject that is particularly open to cross fertilization with other sciences and disciplines of engineering. The subject of fluid mechanics will be highly relevant in domains such as chemical, metallurgical, biological and ecological engineering. This series is particularly open to such new multidisciplinary domains.

The median level of presentation is the first year graduate student. Some texts are monographs defining the current state of a field; others are accessible to final year undergraduates; but essentially the emphasis is on readability and clarity.

More information about this series at http://www.springer.com/series/5980

Daniel T.H. New · Simon C.M. Yu
Editors

# Vortex Rings and Jets

## Recent Developments in Near-Field Dynamics

Springer

*Editors*
Daniel T.H. New
Division of Aerospace Engineering,
School of Mechanical
and Aerospace Engineering
Nanyang Technological University
Singapore
Singapore

Simon C.M. Yu
Singapore Institute of Technology
Singapore
Singapore

ISSN 0926-5112 ISSN 2215-0056 (electronic)
Fluid Mechanics and Its Applications
ISBN 978-981-10-1167-2 ISBN 978-981-287-396-5 (eBook)
DOI 10.1007/978-981-287-396-5

Springer Singapore Heidelberg New York Dordrecht London

Softcover reprint of the hardcover 1st edition 2015


Printed on acid-free paper

Springer Science+Business Media Singapore Pte Ltd. is part of Springer Science+Business Media (www.springer.com)

*This book is dedicated to our families,*
*as well as past and present graduate students.*

# Preface

Vortex rings and jets have always fascinated and captivated researchers and engineers alike, partly due to the surprisingly rich flow phenomena underpinning such seemingly simple flow scenarios, and partly due to their immediate relevance towards a significant number of industrial applications. Without going into the details, smoke rings exhaled from the mouth, squid propulsion, industrial waste issuing out of chimneys and exhaust from jet engines are just some of the immense numbers of real-world instances that illustrate the unique characteristics of vortex rings and jets. More interestingly, from the vortical structures and behaviour standpoint, vortex rings can often be treated, albeit simplistically, as the fundamental building blocks of more complex jet flows. From the extensive pool of experimental, numerical and theoretical studies conducted up to this point in time, both phenomena share many similarities in their underlying behaviour. In fact, many of the flow characteristics and vortex flow models postulated to better explain jet flow behaviour are based on or associated with vortex rings.

Despite their long history of being at the heart of many studies performed in the past, interest in vortex rings and jets has not waned. In reality, unique demands from both well-established and new emerging engineering applications, such as flow control, renewable energy and noise emissions, just to name a few, ensure that interest in them remains as high as before, if not greater. The main discernible difference between most present and past investigations is that the present focus is now on more complex flow configurations surrounding the use of vortex rings and jets, and how to better exploit them for useful purposes in an efficient and robust manner. And coupled with significant advances made in measurement and numerical tools such as particle image velocimetry, large-eddy simulation, and not to mention data analysis in the past two decades, more exacting details can now be extracted from the flow fields of vortex rings and jets for an unprecedented level of understanding.

In view of these new developments, it will be timely to provide some fresh updates to our collective understanding on vortex ring and jet flow phenomena through the present book, while not overwhelming the readers at the same time.

There have been many excellent seminal work in this area, including *The theory of turbulent jets* by G.N. Abramovich, *Turbulent jets* by N. Rajaratnam, *Vortex dynamics* by P.G. Saffman and *Fluid vortices: Fluid mechanics and its applications* edited by S.I. Green, just to name a selected few, and the authors hope the present contribution will provide new perspectives and inspirations to readers who desire to find out some of the latest studies in related areas. In particular, this book attempts to relate vortex rings with jets in a systematic manner through several key areas, namely free vortex rings, vortex ring-structure interactions, jets formed by vortex-ring trains, jets issuing from unconventional nozzles and jet-structure interactions. The editors believe that this book will serve the readers well, either as an update to some of the emerging knowledge on vortex rings and jets, or as a guide on how to tap some of their unique flow behaviour for new engineering applications. Lastly, the editors are especially thankful to all the authors for their contributions towards this book, without which this book would not have been possible.

Singapore
October 2014

Daniel T.H. New
Simon C.M. Yu

# Contents

# Starting Jets and Vortex Ring Pinch-Off

**L. Gao and S.C.M. Yu**

**Abstract** Starting jet is commonly defined as the transient motion produced when a viscous incompressible fluid is forced from an initial state of rest Cantwell (Journal of Fluid Mechanics, 173, 159–189 [6]). The applied force can be time dependent in the form of an impulsive, step or ramp function acting at a point or along a line. Starting jet can be used in fundamental study of vortex ring dynamics, synthetic jets, mixing enhancement, and vortex-enhanced unsteady propulsion systems. Researches related to starting jets have been carried out broadly in two directions. The first direction is focused on the underlying mechanism for the vortex ring pinch-off, which is defined as the process whereby a forming vortex ring is no longer able to absorb vorticity flux from the jet source via the separated shear layer. Several theoretical models are proposed to predict a critical time scale for the pinch-off process, dubbed as the formation number $F$, for different flow conditions. The second direction is focused on its practical applications in entrainment enhancement as well as pulsed-jet propulsion systems. Specifically, due to the restricted vortex ring formation in starting jet, the propulsive efficiency can be effectively improved over the steady jet propulsion by increasing the generated thrust via the vortex over-pressure in the near-wake and by decreasing the kinetic energy loss in the wake via vortex entrainment. In this chapter, we intend to provide the readers with some basic ideas on the dynamic process of vortex ring formation in a starting jet, and its practical application in nature and engineering fields. This chapter is divided into four parts. The first part provides a brief introduction of the starting jet and the phenomenon of vortex ring pinch-off. The discussion of the underlying mechanisms and its theoretical explanation are provided in part two. In the third part, the

L. Gao (✉)
School of Aeronautics and Astronautics, Sichuan University, Chengdu 610065, People's Republic of China
e-mail: lei.gao@scu.edu.cn

S.C.M. Yu
Academic Programme Division, Singapore Institute of Technology, Singapore 179104, Singapore
e-mail: Simon.yu@singaporetech.edu.sg

 
D.T.H. New and S.C.M. Yu (eds.), *Vortex Rings and Jets*,
Fluid Mechanics and Its Applications 111, DOI 10.1007/978-981-287-396-5_1

practical application of starting jets in engineering systems will be explained and discussed. This chapter ends with a summary and an outlook for future study on the starting jet.

**Keywords** Vortex ring pinch-off · Entrainment · Unsteady-jet propulsion

# 1 Introduction

Before going into the details of the above-mentioned areas, the background information about a starting jet will first be introduced. As a relatively new topic of research, its introduction will be made in relation to two classical topics, i.e., the steady jets and the vortex ring dynamics.

## *1.1 General Description of the Starting Jet*

It is well known that jet flow is referred to as an efflux of fluid being injected into the surrounding medium from a nozzle, or orifice. From a classical point of view on jet flow, a starting jet is commonly defined as a single-pulsed jet or an initial development of a continuous (steady) jet, depending on the duration of the fluid ejection. It also corresponds to the transient motion produced when a viscous fluid is forced from an initial state of rest.

If the region of interest is in the far field where the source is relatively small, the mass entrained by the flow eventually overwhelms those issued from the source, so that the jet is mostly described by their source momentum. Another feature of steady jets in the far field is that its characteristics succumb to similarity solutions with an appropriate virtual origin correction. In theoretical analysis, a jet coming out of a point source can be obtained in the limit of a practical jet when the size $D$ of the nozzle tends to be zero while the discharge velocity $U_0$ increases such that the initial momentum flux $\pi U_0^2 D^2/4$ remains constant. The volume flux at the exit $\pi U_0 D^2/4$, on the other hand, will become negligible. This is a good approximation on the properties of a jet in the far field. But the idealization ignores completely the near-field flow details of the jet from a nozzle with finite size, by replacing it with a point source. Unfortunately, in the study of starting jet, the focus lies mainly on the unsteady near-field development. The finite size of the jet source can be important especially when the flow in the vicinity of the source is considered. Therefore, it requires a totally different way to investigate this kind of starting flows.

For the near field of a steady axisymmetric jet, the basic sequence of its evolution was summarized by List [31] as: "*in the immediate neighborhood of the orifice, the high-speed jet flow causes a laminar shear layer to be produced. The shear layer is unstable and grows very rapidly, forming ring-vortices that enhance the mixing between the irrotational ambient fluid and the ejected fluid*". The separated shear layer, its instability growth, roll-up of a series of vortices and their interactions thus determine the near field behavior of a steady jet. The motion induced in the fluid by each vortex affects other vortices in such a way that adjacent vortices pair off. The vortex motion develops a secondary circumferential instability that causes the eventual breakdown of the each vortex [32]. For steady plane jets, the basic difference is the existence of two modes of large-scale vortex formation, on alternate sides, or simultaneously on both sides. These have been called the "flapping" and "puffing" modes of oscillation. In addition, a secondary instability of vortex pairs in two-dimensional jets was found by Leweke and Williamson [28], and it is termed as, elliptic instability.

The second perspective to look at the starting jet is related to the fundamental study of vortex ring. In laboratory, vortex rings can be generated by the piston-cylinder apparatus, in which the motion of a piston pushes a column of fluid of length $L$ through an orifice or nozzle of diameter $D$. The boundary layer at the edge of the orifice or nozzle separates from the orifice or nozzle edge, and subsequently rolls up into spiral, as sketched in Fig. 1.

By dimensional analysis, the starting jet flow is mainly governed by two dimensionless parameters: (1) Formation time: $t^* = U_0 t/D$ and (2) Reynolds number $Re = U_0 D/\nu$, where $\nu$ is the kinematic viscosity. It is noted that other parameters, such as piston velocity programs, geometry of the nozzle or orifice exit, may also play important roles in the dynamics of a starting jet.

The analysis of Mohseni and Gharib [34] suggested that the properties of the leading vortex ring formed in a starting jet are the final outcome of a relaxation process, dependent only on three integrals of the motion, namely the kinetic energy $E$, impulse $I$, and circulation $\Gamma$. These three quantities are invariant under the assumption of inviscid flow and no external forces. In the viscous case, total energy and circulation of the flow will be diffused with time by the viscosity. For the axisymmetric jet flow with no swirl (namely $u_\theta = 0$), the kinetic energy, axial component of the hydrodynamic impulse and circulation can be obtained by integration over the plane of symmetry as

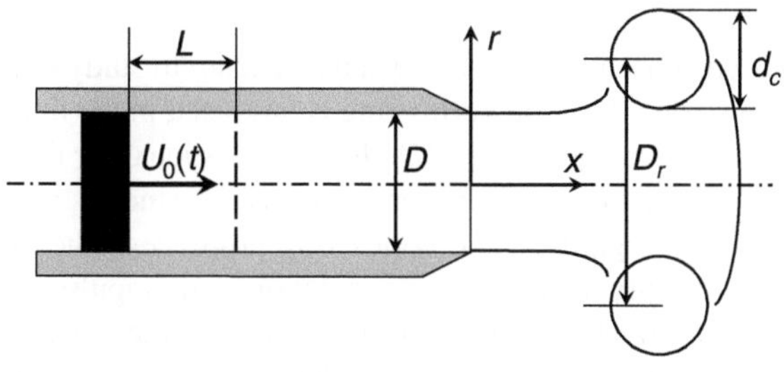


**Fig. 1** Illustration of a starting jet generated by the piston-cylinder apparatus and the formation of a leading vortex ring

$$E = \pi\rho \int_{0}^{\infty} \int_{-\infty}^{\infty} \psi\omega dx\, dr \qquad (1)$$

$$I = \pi\rho \int_{0}^{\infty} \int_{-\infty}^{\infty} \omega r^2 dx\, dr \qquad (2)$$

$$\Gamma = \int_{0}^{\infty} \int_{-\infty}^{\infty} \omega dx\, dr \qquad (3)$$

where $\rho$ is the fluid density and $\psi$ is the Stokes stream function.

For the properties of vortex rings, readers can refer to the detailed review of Shariff and Leonard [53] and Lim and Nickels [29] on the topic. Here, we only focus on the initial formation of the vortex ring from a starting jet and its kinematics. In order to understand better the following discussion in this chapter, the definitions of several kinematic concepts related to the vortex ring in starting jets are given below (which are also illustrated in Fig. 1):

1. Core center position ($x_c$, $r_c$): The vortex center can be identified through detecting the maximum of the stream function. The second method is to locate the position of minimum static pressure [19]. The time history of the vortex core center position is termed as the vortex trajectory. Vortex ring diameter $D_r$, defined by the distance between the spiral centers.
2. Vortex core diameter $d_c$: Based on the experimental results, Weigand and Gharib [56] estimated the vortex ring core size by two ways. The core diameter was determined by the distance between the minima and maxima of the axial velocity profiles along the core. Another method was to measure the core diameter in terms of the vorticity distribution. Larger core diameter by vorticity profile resulted from that the vorticity is non-zero at locations where the gradient $\partial u/\partial y = 0$. It suggested that the term $\partial v/\partial x$ contributed to the magnitude of the vorticity.

## 1.2 Vortex Ring Formation in a Starting Jet: Early Studies

A starting jet has been used extensively as a mechanism to study the properties of a vortex ring. Because only a single vortex ring is produced in the final flow field of a starting jet with small stroke ratio $L_{max}/D$, this set-up is ideal for the study of vortex rings. By using the piston-cylinder apparatus, early studies generated a large amount of results with regards to the vortex ring properties. They mainly focused on the structure, vorticity distribution, circulation and impulse of vortex rings. However, we are not going to repeat here the details of the vortex ring dynamics. Instead, we want to examine the effect of the flow parameters (e.g. geometry, initial conditions, Reynolds number, etc.) of the starting jet on the sequence of events leading to the creation of a vortex ring. We review the features of vortex ring formation only to highlight their effects on the general properties of a starting jet.

When a piston ejects fluid through a nozzle, boundary layer is generated on the inner wall of the nozzle, and separates from the sharp edge in order to satisfy the Kutta condition. Then the separated shear layer rolls-up and the vortex ring formation starts. There is strong evidence in support of the notion that the formation of vortex rings is mainly an inviscid process [38, 39]. The effect of viscosity over much of the formation process is small (except at very small times). The relaxation time toward a steady vortex ring is much shorter than the viscous diffusion time at high Reynolds number. The main effect of viscosity is to remove the singular, non-analytical behavior of the velocity the center of the spiral [36]. Immediately after the flow initiation, the dynamics of the starting jet is solely dominated by the characteristics of a large-scale vortex in the near field, and then gradually approaches to a steady jet in the far field.

It is well known that the roll-up of an inviscid plane vortex sheet, which separates at the edge of a body, is a self-similar process. The analysis stems originally from the work of Kaden [22]. It was then applied to the roll-up of a cylindrical vortex sheet by Saffman [49]. Recently, Hettel et al. [19] revisited the similarity law for the formation of free viscid vortex rings by using numerical simulation.

Kaden [22] showed that the analytical solution of the roll-up of a semi-infinite inviscid planar discontinuity sheet could be obtained using the conformal transformation of a flow along a plane wall. The velocity at position of constant angle $\varphi$ in the polar coordinate system ($r'$, $\varphi$) centered at the spiral center was found to be proportional to $1/\sqrt{r'}$. As a result of a self-similar process, the length scale $l_1$ and $l_2$ of the similar flow patterns at two points in the flow field can be related to their characteristic times $t_1$ and $t_2$ as

$$\frac{l_1}{l_2} = \left(\frac{t_1}{t_2}\right)^{\frac{2}{3}} \tag{4}$$

Similar to the planar flow around a semi-infinite flat plate, there is no length scale at the early stage of the axisymmetric flow when the spiral size $d_c$ is much smaller than the nozzle diameter $D$. It implies the initial axisymmetric vortex sheet roll-up is also self-similar and the scaling behavior can be predicted from its planar counterpart. Saffman [49] applied similarity theory of the formation of a two-dimensional vortex to obtain the location of the center of the axisymmetric vortex spiral in an impulsively started jet. Therefore, the similarity laws for the time dependent axial and radial position of the vortex center, $x_c$ and $r_c$, respectively, as well as the diameter of the vortex spiral $d_c$ are given in dimensionless form:

$$x_c \sim (t^*)^{2/3}, r_c \sim (t^*)^{2/3}, d_c \sim (t^*)^{2/3} \tag{5}$$

All quantities are normalized in terms of the characteristic velocity $U_c$, i.e., $U_0$ and characteristic length scale $L_c$, i.e., $D$. These scaling laws are only valid as long as the dimension $d_c$ of the vortex spiral is much smaller than the characteristic length scale $L_c$ (i.e. nozzle diameter, channel height, length of plate) of the appropriate geometry. According to the experimental data of Didden [12] (see Fig. 7 therein), the self-similar development of cylindrical vortex sheet roll-up should be valid for $d_c < 0.1\,D$, or equivalently for $t^* < 0.6$.

Subsequent studies, however, have revealed contradictory results on the time-dependence of the axial position $x_c$ of the cylindrical vortex sheet roll-up to the prediction of similarity laws. Experimental [12] and numerical [18, 21, 39] data showed $x_c \sim t^{*3/2}$. However, the similarity laws for plane vortices predict $x_c \sim t^{*2/3}$. Hettel et al. [19] suggested that the discrepancy could be explained by the fact that in each system (vortex formation behind a circular nozzle and roll-up of a semi-infinite free vortex sheet), a different frame of reference is used. The plane vortex investigated by Kaden [22] moves in the negative $x$-direction, while vortex ring moves in the positive $x$-direction during its formation process. The relative position between jet front and vortex center, i.e., $x_f - x_c$, exhibits a time dependence which can be predicted by the similarity law $(t^*)^{2/3}$.

As concluded in Hettel et al. [19], the similarity law for the self-similar roll-up of the axisymmetric vortex sheet at the early stage can be expressed as,

$$\begin{cases} \dfrac{x_c}{D} = 0.28t^{*3/2} \\ \dfrac{r_c}{D} = 0.16t^{*2/3} \\ \dfrac{d_c}{D} = 0.25t^{*2/3} \end{cases} \tag{6}$$

The development of vortex ring roll-up during this initial stage plays a key role in determining the characteristics of the vortex ring later on.

For $t^* > 0.6$, the size of the vortex core $d_c$ increases to become comparable to the diameter of the nozzle $D$, leading to the failure of the similarity law. Afterwards, the self-induced velocity of the vortex ring starts to play a more important role in its downstream translation. Thus, during the self-induced translation stage, more attention should be paid to the dynamic properties of the forming vortex ring.

A simple slug flow model has been used by Didden [12], Glezer [17], and Maxworthy [33] to predict the circulation of the total starting jet with piston-cylinder apparatus. The model assumes a uniform velocity equal to the piston velocity across the exit plane of the nozzle or orifice type of generators. The increasing rate of the total starting jet circulation, i.e., the flux of vorticity is calculated as

$$\frac{d\Gamma_{jet}}{dt} = \int \omega u_x dr \approx \int -\frac{\partial u_x}{\partial r} u_x dr \approx \frac{1}{2} U_0^2 \tag{7}$$

It should be noted that the slug model for the circulation is independent of the geometry of the vortex generator (in other words, its effects are not considered). By comparing with the experimental and numerical data, the slug model was found to generally underestimate the total jet circulation. James and Madnia [21] attributed the over shoot of nozzle exit velocity profile near the edge to the higher circulation in their simulation data. Krueger [25] indicated that the under-predicted circulation by the slug model is caused by its ignorance of the effect of a rapid pressure rise at the nozzle exit during the initial vortex ring formation, which is termed as the over-pressure effect.

As a convenient tool of estimating the dynamic properties of the jet flow, the slug model can also be applied to obtain the kinetic energy and impulse of a starting jet as

$$E_{jet} = \frac{1}{8} \pi D^2 \rho L U_0^2 \tag{8}$$

$$I_{jet} = \frac{1}{4} \pi D^2 \rho L U_0 \tag{9}$$

$$\Gamma_{jet} = \frac{1}{2} L U_0 \tag{10}$$

The above estimation of the kinetic energy, impulse and circulation will be an important tool in the theoretical study of the vortex ring formation and pinch-off as discussed in the next section.

## 2 Vortex Pinch-Off and the Formation Number

It is obvious that at sufficiently high Reynolds number the stroke ratio *L*/*D* is an important controlling parameter for the development of starting jet. In the early stage of the starting jet, the size and circulation of the vortex ring will continue to grow as *L*/*D* increases. Gharib et al. [16] addressed the question of whether there is an upper limit in the circulation that a vortex ring can acquire by continuing to increase the stroke ratio. The investigation on this question leads to the finding of vortex ring pinch-off process, as well as the critical dimensional parameter in starting jet, i.e., the formation number.

## 2.1 The Process of Vortex Ring Pinch-Off

Gharib et al. [16] examined the vortex formation process for larger maximum stroke ratio, $L_{max}/D > 4$, and observed a limit on the maximum growth of vortex rings formed using a piston-cylinder apparatus. Unlike the production of a single vortex ring in starting jet with small stroke ratio, the flow field generated by larger maximum stroke ratio consists of a leading vortex ring followed by a trailing jet. The vorticity field of the leading vortex ring gradually disconnects from that of the trailing jet. This dynamical process of the vorticity field disconnection was dubbed by Gharib et al. [16] as vortex ring "pinch-off". An equivalent definition of "pinch-off" was given by Dabiri [8] as a process whereby a forming vortex ring is no longer able to entrain additional vorticity from the vortex generator. The pinch-off process is characterized by a critical dimensionless time, i.e., vortex formation number $F$, at which the total circulation supplied by the vortex generator is equal to the pinched-off vortex ring circulation. According to its definition, formation number should correspond to the onset of the vortex pinch-off process.

For more thorough understanding of this phenomenon, the pinch-off process is described from two perspectives. First, it is about the final state of the starting jet determined by the maximum stroke ratio $L_{max}/D$. Gharib et al. [16] showed that if the piston was forced to stop at $L(t)/D = 4$ (i.e., $L_{max}/D = 4$), the resulting vortex ring (in its long term development) leaves no trailing jet. Otherwise, the final state of the starting flow consists of a leading vortex ring as well a trailing jet stem. This description of the vortex ring pinch-off process can be clearly illustrated by the vorticity contours shown in Fig. 2.

Second, the dynamics of the pinch-off process can be observed as transient development of the flow field as the instantaneous stroke ratio $L(t)/D$ (or the equivalent dimensionless formation time $t^*$) increases. As shown in Fig. 3a, b, the leading vortex ring remains connected with the trailing shear layer near the formation number, i.e., the onset of the pinch-off process. After the critical formation number, the vorticity flux from the trailing shear layer gradually decreases, leading to the final disconnection between the vorticity fields between the leading vortex ring and the trailing jet at the end of the pinch-off process. The almost complete separation of the leading vortex ring after the pinch-off can be clearly observed in Fig. 3c. Therefore, the pinch-off process is not a sudden event and its completion will usually take up to several formation time units to finish.

According to Dabiri [8], it is should also be noted that the dimensionless formation time $t^*$ is basically the normalized total circulation. Therefore, the normalization of the formation number in the starting jet indicates that the maximum circulation of the vortex ring is determined mainly by the length and velocity scale of the vortex-shedding flow.

In order to determine the formation number, several methods have been proposed based on its definition and physical implications. The most straightforward method to determine the formation number is by comparison of the total circulation

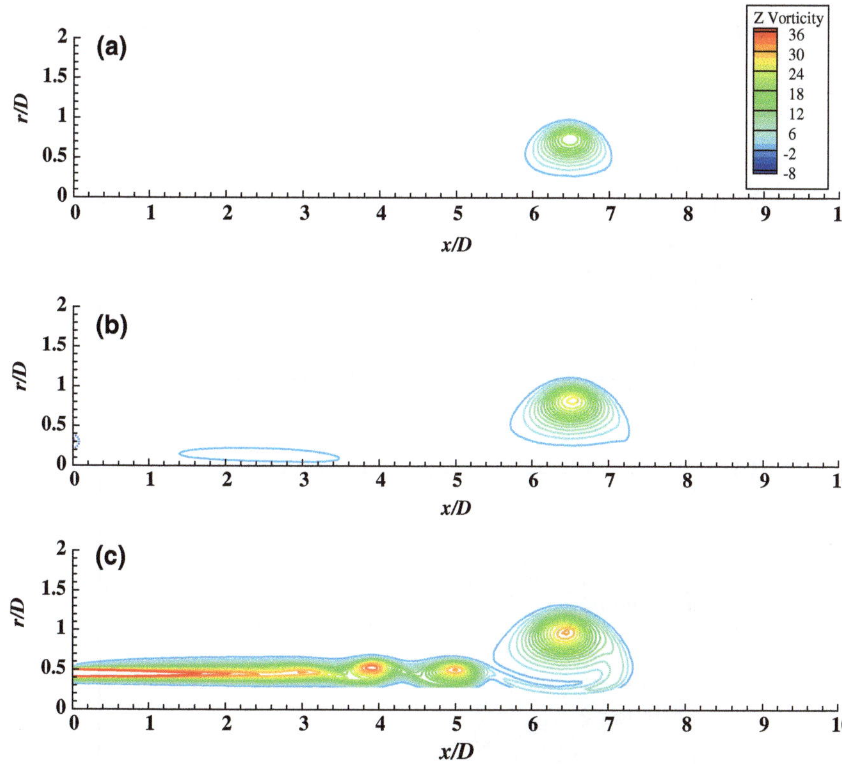


**Fig. 2** Vorticity contours of flow field for the starting jets with maximum stroke ratio: **a** $L_{max}/D = 2$, **b** $L_{max}/D = 4$, and **c** $L_{max}/D = +\infty$, when vortex rings translate at downstream location $x/D \approx 6.5$. The minimum absolute vorticity value shown is 2 $s^{-1}$

discharged by the starting jet and the circulation of the vortex ring after pinch-off [16]. This procedure is illustrated by the dash line and the arrow in Fig. 4. This method has been extensively used in experimental and numerical studies of the vortex ring pinch-off for a variety of flow conditions using the piston-cylinder apparatus.

However, the method of Gharib et al. [16] poses several challenges for complex flow conditions, as pointed out by O'farrell and Dabiri [41]. First, determining the final circulation of the vortex ring requires the ring to be indistinguishable from its trailing jet. Second, it is essential that the vorticity field does not diffuse and the vortex does not distort by interaction with other structures in the flow. To overcome these difficulties, O'farrell and Dabiri [41] proposed an alternative method for identifying the formation number using the Lagrangian coherent structures (LCSs). It is found that the appearance of a new disconnected LCS and the termination of the original LCS is indicative of the initiation of vortex pinch-off. Using the Lagrangian criterion, the formation number is found to be consistent with the results found by the circulation criterion. It is noted that the appearance of a new LCS in

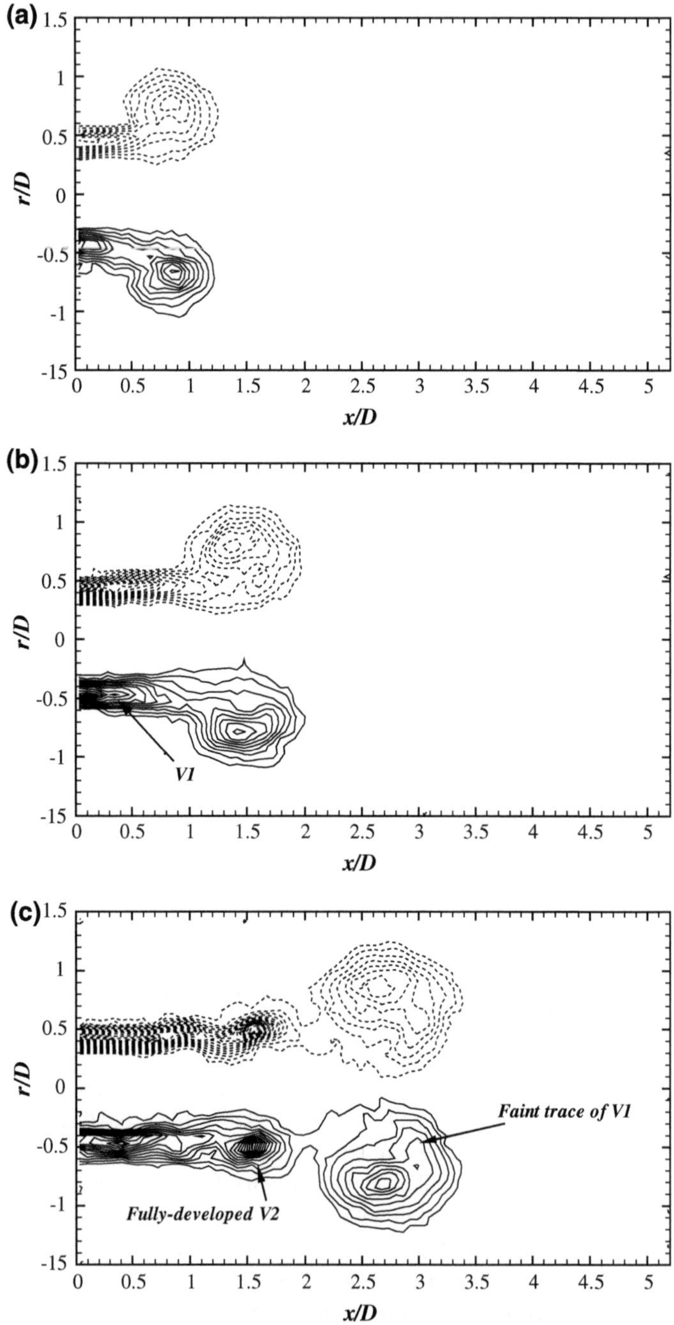


**Fig. 3** The evolution of vorticity contours of a starting jet at: **a** $t^* = 3.1$, **b** $t^* = 4.9$, **c** $t^* = 8.1$. The formation number for this case is $F = 4.5 \pm 0.1$. *V1* and *V2* indicate the first and second trailing vortices, respectively. Graphs are reproduced from Gao and Yu [15]

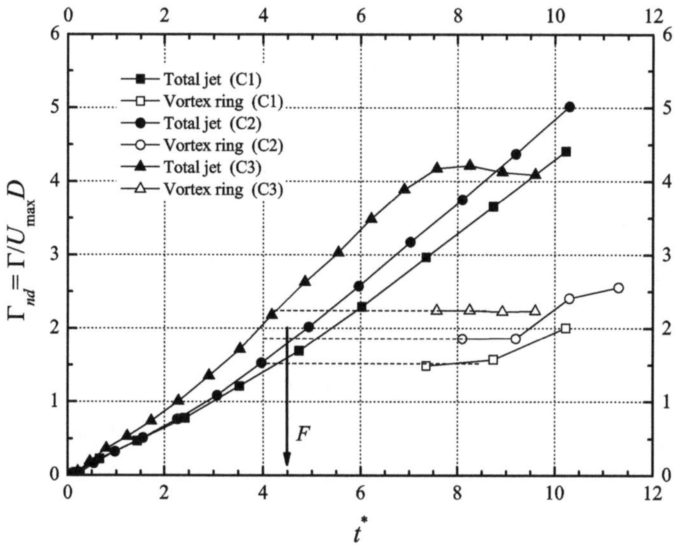


**Fig. 4** The circulation method for determining the formation number $F$ (reproduced from Gao and Yu [15])

the starting jet corresponding to the development of shear layer instability in the trailing jet.

In a recent experimental study of Lawson and Dawson [27], the Trailing pressure maximum (TPM) formed behind the leading vortex ring was used to identify the formation number. Their results showed that the formation number found by TPM was in agreement with those reported in previous studies.

## 2.2 Theoretical Analysis on the Vortex Ring Pinch-Off

The existence of the formation number was first explained by Gharib et al. [16] based on Kelvin-Benjamin variational principle for steady axis-touching vortex ring. They stated that the limiting stroke ratio $(L/D)_{\text{lim}}$, i.e., the formation number, occurs when "*the apparatus is no longer able to deliver energy at a rate compatible with the requirement that a steadily translating vortex ring have maximum energy with respect to impulsive-preserving iso-vortical perturbations.*" During the vortex formation process, the ejected fluid, carrying a specific amount of kinetic energy, impulse and circulation, is fed into the leading vortex ring and serves as continuous perturbations to the "quasi-steady" vortex ring. Therefore, in that sense, the vortex ring pinch-off process was considered as a relaxation process of the leading vortex ring to an equilibrium state. To apply the Kelvin-Benjamin variational principle, a normalized energy $\alpha$, which is defined as the kinetic energy normalized by impulse and circulation as,

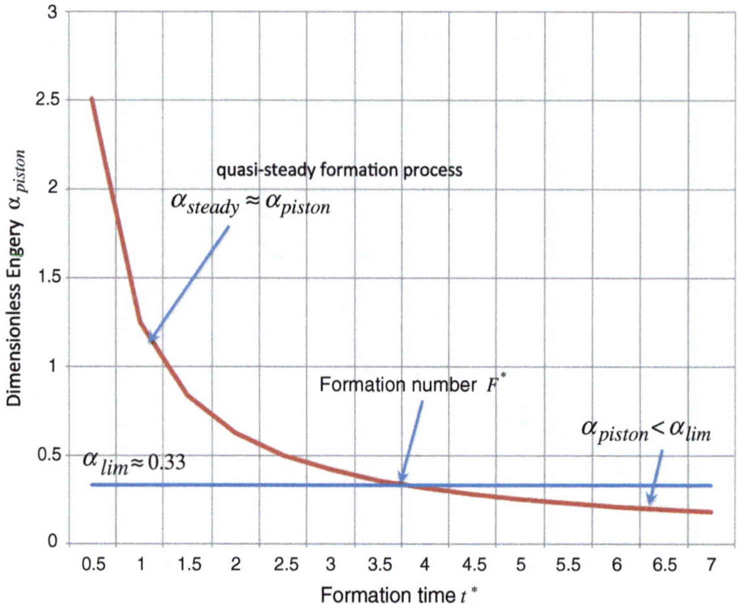


**Fig. 5** Explanation of the vortex ring pinch-off process based on the Kelvin-Benjamin variational principle

$$\alpha = \frac{E}{\sqrt{\rho I \Gamma^3}} \tag{11}$$

was introduced to describe the energy restriction. The definition of the normalized energy $\alpha$ in terms of its impulse and circulation is consistent with the requirement of the Kelvin-Benjamin variational principle that an alternative arrangement of the vortex ring vorticity should preserve the total impulse. The theory of Gharib et al. [16] is illustrated in Fig. 5.

As shown in Fig. 5, normalized energy of the total jet $\alpha_{piston}$, which is calculated by substituting Eqs. (8)–(10) into Eq. (11), decreases monotonically with the stroke ratio *L*/*D*. By assuming that all ejected fluids (with their *E*, *I* and Γ) are absorbed by the resulting vortex ring, one can deduce that $\alpha_{steady} \approx \alpha_{piston}$. Here $\alpha_{steady}$ is the value of $\alpha$ for the final steady vortex ring which is accessible from the vorticity field at stroke ratio *L*/*D* via an iso-vortical impulse-preserving rearrangement. This fact indicates that the Kelvin-Benjamin variational principle is not about the instantaneous properties of the generation process, but rather focuses on matching the quantities provided by the apparatus and those of the final vortex ring. This is actually the key point to understand the explanation proposed by Gharib et al. [16]. Due to the properties of the steady vortex rings in theoretical analysis, lower value of the $\alpha$ corresponds to thicker vortex ring. It means the vortex ring becomes thicker and thicker as the stroke ratio increases.

However, it is realized that the vortex ring cannot be infinitely thick. We expect $\alpha_{steady}$ to diminish to a limiting value, $\alpha_{lim}$, as the core thickens because the existence of a $\alpha_{lim}$ for every family of steady vortex rings is an unproven generalization based on the Norbury-Fraenkel vortex family [13, 40]. For the Norbury-Fraenkel family of vortex rings which have uniform vorticity density ($\omega/r = const$), Hill's spherical vortex [20] is the theoretical limit for the thickest member with $\alpha = 0.16$. Nevertheless, the vortex rings generated in a piston-cylinder apparatus cannot approach this theoretical limit. It should have a slightly higher limiting normalized energy. Gharib et al. [16] suggested that, a limiting state of the resulting vortex ring, characterized in terms of the dimensionless mean core radius $\varepsilon$ [40], is then restricted by the axis-touching state of the vortex ring. They found in their experimental results that for axis-touching vortex rings, $\alpha_{lim} = 0.33 \pm 0.01$ for a wide variety of flow conditions. They also pointed out that the specific value of $\alpha_{lim}$ actually depends on the shape of the vorticity profile, which in turn depends on piston history and Reynolds number. It may vary among different starting flow configurations. For example, by temporally increasing the nozzle exit diameter while keeping the volume flux constant, Dabiri and Gharib [11] found $\alpha_{lim} = 0.5$ for the final vortex ring, implying the smaller core size than that for $\alpha_{lim} = 0.33 \pm 0.01$. As a result, this critical value of $\alpha_{lim}$ can only be determined from experiment. In short, the total kinetic energy, impulse and circulation from the generator at a specific stroke ratio can reach the final vortex ring as long as $\alpha_{piston} > \alpha_{lim}$.

For longer stroke ratio of the starting jet, $\alpha_{piston}$ will become less than $\alpha_{lim}$. If the assumption $\alpha_{steady} \approx \alpha_{piston}$ is still valid, the resulting steady vortex ring will become thicker than the vortex ring specified by $\alpha_{lim}$. According to the above hypothesis, the $\alpha_{lim}$ should be the lowest value for the normalized energy of the thickest vortex ring possible in a piston-cylinder apparatus. It implies that, beyond the critical point of $\alpha_{piston} = \alpha_{lim}$, not all the ejected fluids will be entrained by the final steady vortex ring because the normalized energy provided by the jet generator cannot meet the requirement of a steady translating vortex ring with $\alpha_{steady} = \alpha_{lim}$. In other words, the kinetic energy provided by the jet generator becomes too small relative to the impulse and circulation to form a steady vortex ring with $\alpha = \alpha_{lim}$. In that situation, a trailing jet with additional vorticity (and also impulse and kinetic energy) formed behind the pinched-off leading vortex ring to maintain $\alpha = \alpha_{lim}$ for the single vortex ring. The perturbation response study of O'farrell and Dabiri [42] subsequently verified this property of the Norbury-Fraenkel family of vortex rings.

By using the slug model to quantify the total normalized energy delivered by the piston-cylinder apparatus and by using the Norbury-Fraenkel family of vortex to approximate the experimentally generated vortex ring, Gharib et al. [16] successfully predicted the observed value of the formation number, in combination of the experimental data of limiting normalized energy. To avoid the use of experimental result of $\alpha_{lim}$, similar models have been developed subsequently by Mohseni and Gharib [34] to include an additional assumption on the vortex ring translational velocity (see Eqs. (9) and (10) therein), and by Linden and Turner [30] to include an additional condition on the fluid volume conservation. It is noted that Shusser et al.

[55] showed that boundary-layer growth within the hollow cylinder wall of the vortex generator was shown to lead to an increase in vorticity flux into the starting jet.

The Kelvin-Benjamin variational principle can be regarded as a before-and-after-matching of the dynamical properties between those from the total starting jet and those in the final pinched-off vortex ring. However, the dynamic process during its evolution has not been considered. To include the dynamic process of the detachment of the leading vortex ring from its trailing jet, Shusser and Gharib [54] proposed a hypothesis on the flow kinematics that "*a vortex ring completes its formation and pinches off from its generating axisymmetric jet when the translational velocity of the ring becomes equal to the jet flow near the vortex ring.*" In particular, Mohseni et al. [35] proposed an explanation for the pinch-off process by considering the very detailed translational dynamics of the leading vortex ring and its trailing shear layer. At the early stage, the vortex ring grows in size and translates downstream due to its self-induced velocity. As the vortex ring becomes thicker, it pushes the shear layer toward the axis of symmetry at which the vorticity destruction occurs due to the vorticity cancellation mechanism. Since the leading vortex ring gains its strength from the shear layer, it gradually ceases to grow if the local velocity of the shear layer is less than the velocity of the vortex ring. Alternatively, the pinch-off occurs when the shear layer is unable to enlarge the leading vortex ring, as the strength of the shear layer tends to zero at the axis of symmetry.

Several models have also been proposed according to the second theory, with some assumptions. Based on the assumption of a quasi-steady formation process, the translational velocity of the leading vortex ring is estimated by the properties of the Norbury-Fraenkel family of steady vortex rings as follows:

$$E_{ring} = \frac{1}{2}\rho R\Gamma^2\left[\ln\frac{8}{\varepsilon} - \frac{7}{4} + \frac{3}{8}\varepsilon^2\ln\frac{8}{\varepsilon}\right] \tag{12}$$

$$I_{ring} = \rho\pi\Gamma R^2\left(1 + \frac{3}{4}\varepsilon^2\right) \tag{13}$$

$$U_r = \frac{\Gamma}{4\pi R}\left[\ln\frac{8}{\varepsilon} - \frac{1}{4} + \frac{3}{8}\varepsilon^2\left(\frac{5}{4} - \ln\frac{8}{\varepsilon}\right)\right] \tag{14}$$

where $U_r$ is the translational velocity of the vortex ring, vortex ring radius $R = D_r/2$ and $\varepsilon = d_c/D_r$ is the dimensionless mean core radius of the vortex ring. On the other hand, the local trailing jet velocity is related to the piston velocity via the conservation of mass (or volume because the fluid is incompressible) and the dynamical properties of the total jet are approximated by the slug flow model (i.e., Eqs. (8)–(10)).

As a result for the system of Eqs. (8)–(10) and (12)–(14), there are six equations but seven unknowns, i.e., $E$, $I$, $\Gamma$, $L$, $U_r$, $\varepsilon$ and $R$. Thus, an extra equation is needed for the closure of the model. By comparing the vortex ring velocity $U_r$ with the trailing jet velocity near the vortex ring and by using the conservation of mass in the trailing jet, Shusser and Gharib [54] introduced an additional equation into the model as

$$U_r = \frac{D^2}{4R^2} U_0 \tag{15}$$

They found the limiting value of the dimensionless energy to be $\alpha_{lim} \approx 0.31$, which agrees well with the experimental value of $\alpha_{lim} = 0.33$.

However, there are still some concerns about the fundamental approximation in the model of Shusser and Gharib [54] that the dynamical properties of the leading vortex ring during the formation can be estimated by the slug flow model. The approximation is regarded accurate only when all the ejected fluids are entrained completely in the leading vortex ring. It is not the real flow configuration because the trailing jet appears after the leading vortex ring travels away from the nozzle exit. Part of the ejected fluid, with its kinetic energy, impulse and circulation, resides in the trailing jet during the formation stage.

To account for the actual physical process, Gao and Yu [14] proposed a revised model by determining the properties of the leading vortex from the fluxes from the trailing jet, rather than from the slug model. The initial development of the starting jet is accounted by the similar law of vortex sheet roll-up as given in Eq. (6). In the second stage, the growth of the leading vortex is determined by the flux of those integrals of motion from the trailing jet at its rear boundary, which are estimated as

$$\frac{dE_{ring}}{dt} = \frac{1}{2}\pi b^2 \rho U^2 (U - U_r) \tag{16}$$

$$\frac{dI_{ring}}{dt} = \pi b^2 \rho U (U - U_r) \tag{17}$$

$$\frac{d\Gamma_{ring}}{dt} = \frac{1}{2}(U - U_r)^2 \tag{18}$$

where $b$ is the radius of the trailing jet stem near the leading vortex ring and specifically $b = D/2$ for the piston-cylinder apparatus. Their analytical results revealed that the dynamic process of vortex ring pinch-off is signified by two time scales, i.e., the formation number, which indicates the onset of the pinch-off process, and the separation time, which corresponds to the time when the leading vortex ring becomes physically separated from the trailing jet and is therefore referred to as the end of the pinch-off process.

By considering the description of the pinch-off process presented in Sect. 2.1, it can be seen that the theory proposed by Gharib et al. [16], Mohseni and Gharib [34], and Linden and Turner [30] which is based on the Kelvin-Benjamin variation principle, is consistent with the first perspective to understand the pinch-off process. On the other hand, the theory proposed by Shusser and Gharib [54], Mohseni et al. [35] and Gao and Yu [14] is in accordance with the second perspective to understand the pinch-off process. As stated by Mohseni et al. [35], translational dynamics

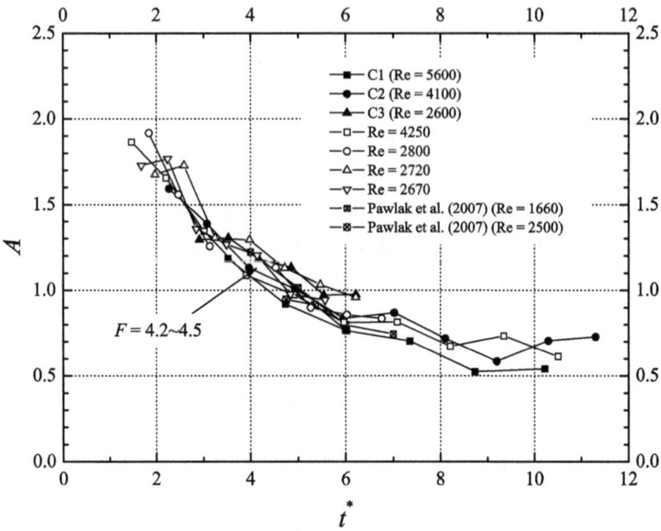


**Fig. 6** The variation of the instability parameter $A$ against the formation time $t^*$. The data of Pawlak et al. [45] are also included for comparison (graph reproduced from Gao and Yu [15])

can lead to insights about the formation process, but it is cumbersome for modeling. Relaxation approach appears to obviate the need for modeling the nonlinear dynamics of shear layer.

Finally, the pinch-off can also be explained in terms of the effects of the shear layer instability. For starting jets generated by large stroke ratio (typically greater than the formation number), it was observed both in experiments [15, 16, 41, 50], and simulations [35, 57] that secondary vortices (or trailing vortices) would be formed in the trailing shear layer as a result of the Kelvin-Helmholtz instability. Since the growth of the leading vortex ring is sustained by the influx of mass, vorticity, impulse and kinetic energy from the trailing shear layer at a specific rate, the development of the shear layer instability can certainly play an important role in the pinch-off process. By altering the characteristics of the shear layer instability by Reynolds number and initial shear layer thickness, Zhao et al. [57] found that the development of the trailing shear layer instability and the growth of secondary vortices actually accelerated the process of pinch-off and introduced approximately 20 % variation in the formation number and in the vortex ring circulation level.

After the pinch-off process, the interaction between the leading vortex ring and the trailing vortices can also greatly affect the evolution of the pinched-off vortex ring. One notable example is the merging of the trailing vortices into the leading vortex ring due to their interaction [15, 16, 41], as shown by the evolution of the first trailing vortex *V1* in Fig. 3b, c.

The effect of the trailing shear layer instability on the vortex ring pinch-off was investigated by Gao and Yu [15]. The experimental results showed that secondary vortices start to develop in the trailing jet only after the formation number is achieved. The subsequent growth of the secondary vortices reduces the vorticity flux being fed into the leading vortex ring and as a consequence, constrains the growth of leading vortex ring with larger circulation. Evolution of perturbation waves into secondary vortices is found to associate with growth and translation of the leading vortex ring during the formation process. A dimensionless parameter '*A*', defined as

$$A = \frac{\Gamma_{ring}}{x_c \Delta U} \tag{19}$$

where $\Gamma_{ring}$ is the vortex ring circulation, $x_c$ is the axial trajectory of its core and $\Delta U$ is the strength of the trailing shear layer, is therefore proposed to characterize the effect of the leading vortex ring on suppressing the nonlinear development of instability in the trailing shear layer, i.e. the initial roll-up of the secondary vortices. As shown in Fig. 6, in a starting jet, *A* follows a decreasing trend with the formation time $t^*$. A critical value $A_c = 1.1 \pm 0.1$ is identified experimentally, which physically coincides with the initiation of the first secondary vortex roll-up and this in turn indicates the onset of pinch-off process.

## 2.3 Vortex Ring Pinch-Off in Other Flow Conditions

For vortex rings generated by the classic piston-cylinder apparatus with impulsively started velocity program, the formation number was found to be approximately equal to 4. Gharib et al. [16] also discovered that the presence of the vertical wall at the nozzle plane would shift the formation number to lower values, while non-impulsive velocity program and higher Reynolds number shift the formation number to higher values. To get a deeper understanding on the dynamics of the pinch-off process, a large number of numerical and experimental studies have also been carried out so as to verify the universality of the formation number over a wide range of flow conditions. According to the numerical study by Mohseni et al. [35], it was concluded that the formation number and normalized circulation and energy of the leading vortex ring are invariant, as long as (1) the fluid ejection phase, i.e., the motion of the piston, is of sufficient long duration, (2) the Reynolds number is sufficiently high and (3) the shear layer produced by the fluid ejection is sufficiently thin. It is noted that modification in the baseline configuration of the piston-cylinder apparatus will introduce additional parameters to the starting jet flow. These parameters will be considered in the analysis of the specific flows.

First, it has been observed that vortex rings produced by non-impulsive velocity programs $U_0(t)$ can achieve higher formation numbers than those generated by impulsive velocity program. Rosenfeld et al. [47] found that the formation number increases with the slower acceleration and deceleration (the trapezoidal velocity programs therein). In the case of uniform acceleration, i.e., the linear velocity program, the formation number increases to $F = 5.22$. It is noted that all the velocity programs in their study have the same maximum stroke ratio of $L_{max}/D = 6$. For starting jets at much higher Reynolds number of 13,000, Krueger and Gharib [26] used two different velocity programs, i.e., negative sloping (NS) ramp and positive sloping (PS) ramp, in their experiments. The NS velocity program has larger initial acceleration and lower final deceleration. As expected, the vortex pinch-off in the NS cases occurs one formation time unit earlier than in the PS velocity programs. The above results indicate that the formation number can be significantly increased by using a velocity program with a steadily increasing velocity.

Another factor affecting the vortex ring formation in a starting jet is the exit velocity profile, i.e., $u(r)$ at the jet exit plane $x = 0$. In the analytical studies of the vortex ring formation in the piston-cylinder apparatus, the velocity profile is usually specified as the top-hat profile, i.e., velocity $u(r)$ is constant of $U_0$ across the nozzle exit and zero outside. In real flow cases, the velocity profile is actually time-dependent. It is close to uniform at the start of fluid ejection with a peak near the wall [12], and gradually affected by the growth of the inner boundary layer of the nozzle as the jet continues. It is noted that the velocity profile effectively determines the thickness of the separated shear layer at the nozzle edge as well as the total circulation imported into the flow field [57]. Therefore, the change in the velocity profile at the nozzle exit actually affects the characteristics of the shear layer and the magnitude of the centreline velocity. To simulate the extreme case with the thickest boundary layer, Rosenfeld et al. [47] specified a parabolic velocity profile at the nozzle exit as

$$u(r) = u_{cl}\left(1 - \frac{r^2}{(D/2)^2}\right) \tag{20}$$

where $u_{cl}$ is the jet centerline velocity. They found the formation number in this case decreases dramatically to $F = 0.91$, about one quarter of the value obtained for the uniform velocity profile. This significant discrepancy is related to the greater flux of vorticity provided by the parabolic velocity profile, which results in a rapid increase of total circulation with the formation time. The total circulation increase rate is four times larger than that of the uniform velocity case since $u_{cl} = 2U_0$, given that the top-hat and parabolic velocity profiles have the same mass flux. Rosenfeld et al. [47] also showed that this result is in good agreement with the theoretical model based on the Kelvin-Benjamin variational principle, which predicts $F = 0.87$ for the parabolic velocity profile.

Modifications to the classical piston/cylinder configuration include using a nozzle or orifice with temporally variable exit diameter, i.e., $D(t)$. According to the dynamic explanation of the pinch-off process by Mohseni et al. [35], the formation number should be delayed by temporally increasing the radial extent of the trailing shear layer. It is because increasing the nozzle diameter would delay the onset of the symmetry-axis-touching state of the shear layer, which they correlated with the vortex ring pinch-off process. This speculation was subsequently examined by the experimental studies of Dabiri and Gharib [11] (nozzle-type) and Allen and Naitoh [1] (orifice-type). Dabiri and Gharib [11] observed that temporally increasing the nozzle exit diameter as the starting flow emerges leads to a measurable increase in the radial location of peak vorticity in the leading vortex ring. However, both researches did not observe a significant delay in vortex ring pinch-off or generation of stronger vortex ring for the cases with temporally increasing exit diameter. Dabiri and Gharib [11] suggested that the discrepancy between their results and theoretical analysis might be attributed in part to the thickness of the shear layer and the relatively low Reynolds numbers in their experiments. In their experiments, the volume flux through the nozzle exit $\pi U_0(t)D(t)^2/4$ was kept as a constant. Consequently, increasing the nozzle exit diameter is accompanied by decreasing exit velocity. As given by Eqs. (3.2) and (3.3) in Dabiri and Gharib [11], temporally increasing nozzle exit diameter would dramatically decrease the vorticity flux ejected into the starting flow. In addition, nozzle with temporally increasing exit diameter would alter the vorticity distribution and result in a higher dimensionless energy of the leading vortex ring, i.e., a thinner vortex ring core. In contrast to the program of temporally increasing the exit diameter, it has been demonstrated in both researches that growth of the leading vortex ring is substantially augmented by temporally decreasing the exit diameter during fluid ejection. In these cases, vortex ring circulation can be increased by 35 %, and the normalized energy is found to be equivalent to that of Hill's spherical vortex ($\alpha_{lim} = 0.16$). In Dabiri and Gharib [11], this effect was related to the changes in the vorticity distribution of the forming leading vortex ring that are induced by the nozzle motion. Allen and Naitoh [1] explained the successful production of a vortex ring with low dimensionless energy $\alpha$ as that, by closing the orifice during jet ejection, the leading vortex ring continues to be supplied with circulation, impulse and kinetic energy via a trailing jet with increasing velocity. Therefore, it is reasonable to conclude that increasing the jet piston velocity during the formation stage is more effective to produce stronger vortex ring than expanding the radial extent of the trailing shear layer. In short, temporally increasing the exit diameter generates vortex ring with thinner core and smaller circulation, while temporally decreasing exit diameter results in thicker core and higher circulation.

The uniform bulk counter-flow and co-flow relative to the starting jet generator has significant effect on the vortex ring pinch-off. Dabiri and Gharib [9] observed a modest pinch-off delay of approximately 10 % in the limiting vortex ring formation time by immersing the piston-cylinder apparatus in a uniform bulk counter-flow. Krueger et al. [24] demonstrated the opposite effect, i.e., a reduction in the vortex ring formation number in the case of uniform bulk co-flow around a piston-cylinder apparatus. For co-flow velocity ratios $U_{co\text{-}flow}/U_0 > 0.6$, vortex ring pinch-off occurred almost immediately following the initiation of the vortex ring's formation, resulting in a much lower circulation and high normalized energy of the pinched-off vortex ring. The change to the pinch-off process by imposing a counter-flow or co-flow to the basic starting jet appears to be in consistent with the kinematic mechanism proposed by Shusser and Gharib [54] and Mohseni et al. [35].

# 3 Applications of the Starting Jet

The large scale leading vortex ring formed by boundary layer separation is the dominant feature in a starting jet. The restriction on the leading vortex ring growth imposed by the pinch-off process may suggest that there are some properties of the starting jet that can be optimized during this limiting process. In practical applications, starting jets are generally used in mixing process, mass transfer or momentum transfer (i.e., propulsion) systems. The leading vortex formation has been observed to augment entrainment of ambient fluid into the starting jet flow [10, 43]. For propulsive characteristics of a starting jet, Krueger and Gharib [26] demonstrated that the normalized average thrust of a starting jet was maximized just at the formation time. This result is intriguing because it not only provides new perspectives for analyzing the pulsatile biological flows, but also suggests design strategies for pulsed-jet technologies whose primary objective is to impart momentum to the main flow.

Besides the thrust augmentation provided by the starting jet, other possible advantages of the pulsed jet thruster could be its capability in fast tracking, effectiveness at low speed maneuvering and its non-intrusive feature for lower drag caused [23]. In addition, the traditional propeller thruster has a time delay in response to a given control force. The time delay grows inversely with the magnitude of the desired force. This causes the propeller thruster forces to be very unpredictable when creating short impulsive forces in the applications of high maneuverability. The disadvantages of the propeller thruster may be overcome by the high temporal response of the pulsed jet thruster.

In this section, the mechanism for the entrainment enhancement due to leading vortex formation will be first discussed. It is followed by the application of the optimal vortex formation principle to the unsteady-jet propulsion systems for thrust augmentation as well as for propulsive efficiency enhancement.

## *3.1 Vortex Ring Entrainment in Starting Jet*

The interest in studying the entrainment feature of the starting jet lie in the pursuit of efficient schemes for unsteady gaseous fuel, in scramjet combustors, gas turbine combustion chambers, reciprocating internal combustion engines, etc. The problem is usually addressed by considering the mass flow rate $\dot{m}$ across a section perpendicular to the jet axis, which is known to increase with the distance from the jet exit plane. Steady turbulent jets have been studied extensively in the past. For steady turbulent jets, an entrainment coefficient was defined by Ricou and Spalding [46] as

$$C_e = \frac{d\dot{m}}{dx}\left(\frac{D}{\dot{m}_0}\right) \quad (21)$$

where $\dot{m}_0$ is the initial ejected mass flux at the nozzle exit. They found that the coefficient assumes a constant value around 0.32 in the jet far field, i.e., for $x > 10D$. The existence of such a universal constant is due to the self-similarity of velocity across the jet.

For the starting jet, the study on its entrainment characteristics mainly focuses on the near-field behavior, where the unsteady vortex formation plays a crucial role in its dynamics. Early studies of unsteady injection found that mixing is greatly enhanced by the use of unsteady or fully pulsed incompressible air jets, when compared with a steady jet [5]. Auerbach [2] made the distinction between convective entrainment during shear layer roll-up phase and diffusive entrainment after the vortex ring is formed. He found that depending on the formation details as much as 40 % of the fluid carried with a steady vortex ring can be ambient fluid.

Cossali et al. [7] carried out a series of experiments on the near field development of a subsonic transient turbulent jet impulsively injected at constant rate into a large quiescent chamber. The results showed that the entrainment of the ambient air by the jet head (vortex) might take place at the beginning of the injection, but the entrainment mechanism is different from that in steady jets: "*it is likely to be due to the large wrinkling of the jet border caused by fluttering of the jet that traps surrounding air into the head.*" It is also suggested that for a transit jet there are two major mechanisms responsible for the entrainment. The first one is the unsteady mechanism closely related to the rotational motion of the leading vortex ring (a recirculation zone). Another one is the steady entrainment, which is the result of small-scale mixing (vorticity diffusion) and velocity inducement.

For the leading vortex ring in a starting jet, the rate of fluid entrainment (mainly due to the first mechanism) into the vortex ring $Q_E$ is defined as

$$Q_E = \frac{dV_E}{dt} \quad (22)$$

where $V_E$ is the volume of ambient fluid in the vortex ring spiral. In order to estimate the value of $V_E$, the Lagrangian coherent structures (LCSs) is used to identify the boundary of a vortex ring. The LCSs are obtained by the ridges in the finite time lyapunov exponent (FTLE) field, which is calculated from the velocity field data [51, 52].

Dabiri and Gharib [10] investigated vortex ring entrainment using steady bulk counter-flow to hold the rings in the field of view and streamlines obtained from DPIV to identify the bubble volume. It is found that 65 % of the vortex ring volume is from the entrained ambient fluid when its formation is completed. The results also suggested that entrainment fraction of a vortex ring can be increased by using smaller stroke ratio.

The process of entrainment during the initial ring formation and roll-up was addressed by Olcay and Krueger [43] via the analysis of attracting LCS obtained from backward time FTLE field. Their study accounted for the first mechanism of entrainment by the vortex spiral. They found that most of the entrainment occurs after the piston has stopped (for cases $L_{max}/D = 2$). It is because once the piston stops, the ejected fluid boundary is no longer held out by the jet and it may contract under the influence of the ring vorticity, making a larger area available for entrainment of ambient fluid. The quantitative results of the entrainment rate into the vortex spiral are obtained by calculating the flow rate into the entrance gap of width $\Delta r$. It was found that high entrainment rate after the jet stops should be attributed to the dramatic increase in $\Delta r$ after jet termination. By examining the entrainment characteristics among three different velocity programs, they concluded that the variation of $\Delta r$ is the key parameter responsible for the entrainment of a vortex ring. The higher entrainment rate was observed for cases with rapid initiation and termination, which led to the corresponding larger $\Delta r$. In addition, they also found that the entrainment fraction $V_E/V_B$ ($V_B$ is the volume of the vortex ring bubble) is larger for shorter stroke ratio length. Therefore, they suggested that a trapezoidal velocity program with low $L_{max}/D$ ratio might be suitable for entrainment enhancement since this program benefits from an impulsive jet initiation as well as a rapid jet termination. In addition, larger entrainment of ambient fluid is observed for rapid initiated and terminated velocity programs, as a consequence of the stronger overall Biot-Savart induction provided by the strong initial shear layer associated with the rapid jet initiation for these velocity programs [44].

## 3.2 Unsteady-Jet Propulsion

In general, the coherent vortical structures in the near-wake of a self-propelled vehicle can improve its propulsive efficiency by manipulating the local pressure

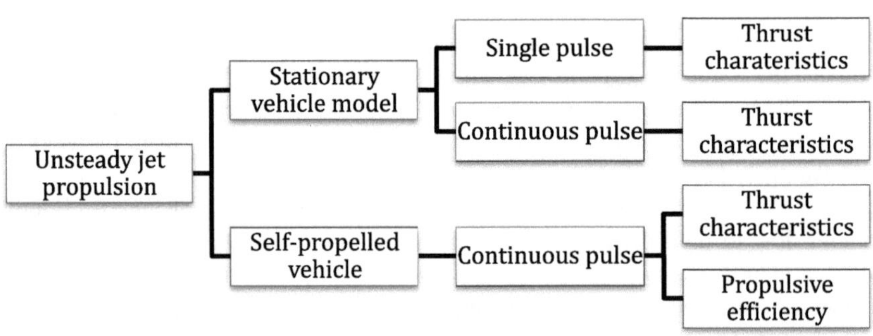


**Fig. 7** Diagram of the research areas in the unsteady jet propulsion

field and entrainment kinematics. To elucidate the main points on the study of the unsteady (or pulsed) jet propulsion, the subsection is organized as shown in Fig. 7.

The traditional propeller-type thruster is currently used extensively in various types of underwater propulsion systems. Its propulsive features are typically characterized by the mean (i.e., time-averaged) flow created by the propeller. In steady flow analysis, the thrust produced by the steady jet is assumed to be dependent only on the mean velocity profiles upstream and downstream of the vehicle. The propulsive efficiency can be expressed as the Froude efficiency as

$$\eta_f = \frac{2}{1 + \frac{U_j}{U_v}} \tag{23}$$

where $U_j$ is the jet speed relative to the vehicle and $U_v$ is the speed of the vehicle, as shown in Fig. 8. As a consequence of the steady flow assumption, schemes to enhance propulsive efficiency have focused on manipulating the mean velocity profiles upstream and downstream of the propulsor. However, for the propulsive analysis of a pulsed jet, the flow unsteadiness must be taken into account. All the parameters for the study of unsteady jet propulsion are listed in Table 1.

To determine whether thrust augmentation can be achieved by pulsing, it is necessary to compare the measurements of the time-averaged thrust from the fully pulsed jet with the thrust from an equivalent jet for which the effects of pulsing are not present [25]. For discussion on free motion jet generators, the comparison on the different concepts of the propulsive efficiency of a thruster or vehicle is very important.

Krueger and Gharib [26] provided the first demonstration of the direct relation between the vortex ring pinch-off in starting jet and the maximization principle of the jet thrust. They used a piston-cylinder mechanism attached to a force balance to investigate the impulse and thrust generated by starting jets for $L_{max}/D$ ratio in the range 2–8. Two kinds of velocity program, i.e., negative sloping (NS) ramp and positive sloping (PS) ramp, were used in the experiment. Since the desired $U_{0\_max}$

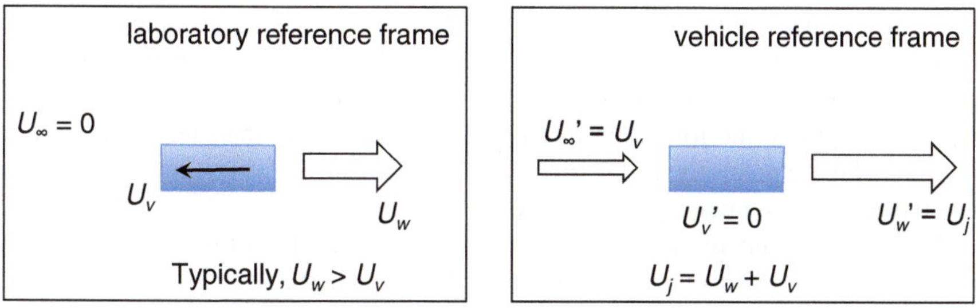


**Fig. 8** Illustration of the steady jet propulsion analysis. $U_w$ is the average wake speed. *Prime* indicates the quantities in the frame of reference on the vehicle

**Table 1** The parameter space for the study of unsteady jet propulsion

| | | |
|---|---|---|
| *Independent variables* | | |
| Jet exit diameter | $D$ | |
| Kinematic viscosity | $\upsilon$ | |
| Jet velocity program | $U_0(t)$ | |
| Pulse duration | $t_p$ | |
| **Pulse period** | $T$ | |
| *Dependent variables* | | |
| Stroke length | $L$ | $L = \int_0^{t_p} U_J dt$ |
| Maximum jet velocity | $U_{0\_max}$ | Maximum value of $U_0$(t) achieved during ejection |
| **Rest time between pulses** | $t_r$ | $t_r = T - t_p$ |
| **Pulsing frequency** | $f$ | $f = 1/T$ |
| *Dimensionless variables* | | |
| Reynolds number | Re | $\mathrm{Re} = U_{0\_max}D/\upsilon$ |
| Formation time | $t^*$ | $t^* = \int_0^{t_p} U_0 dt/D$ |
| *Maximum stroke ratio* | $L_{max}/D$ | |
| Dimensionless velocity program | $U_0^*$ | $U_0^* = U_0(t)/U_{0\_max}$ |
| **Duty cycle** | $D_c$ | $D_c = t_p/T$ |

Rows with **Bold** indicate the variables for the continuously pulsed jets

was kept constant, different stroke length of the starting jet was realized by changing the pulse duration $t_p$. As a consequence, case with smaller stroke length ratio corresponds to greater initial acceleration.

Using a force balance measuring the jet thrust $F_t$, the total impulse per pulse is obtained by

$$I_t = \int_0^{t_p} F_t(t)dt \tag{24}$$

In their experiments, a change, specifically a decrease, in the slope of the total impulse against the stroke ratio of a pulse was found to occur at the formation number, indicating the time averaged jet thrust was maximized when its stroke ratio $L_{max}/D$ is equal to the formation number. It is realized that their results should be interpreted more precisely as: the high thrust of a starting jet can be maintained until its stroke ratio becomes equal to the formation number.

In order to compare the starting jet thrust with that of a similar steady jet, the impulse due to the jet momentum flux can be calculated as

$$I_U = \int_0^{t_p} \int_0^{D/2} 2\pi r \rho u_0^2(r,t) dr\, dt \tag{25}$$

It is noted that under the assumption of steady jet propulsion, its thrust is exclusively determined by the momentum flux. As such, the normalized average thrust per pulse can be expressed as

$$F_{tn} = \frac{\bar{F}_t}{\rho A_{exit} U_{0\_max}^2} = \frac{1}{\rho A_{exit} U_{0\_max}^2} \frac{I_t}{t_p} \tag{26}$$

where $\bar{F}_t = I_t/t_p$ is the time average thrust during a pulse. The results of $F_{tn}$ for both velocity program patterns showed a peak near the corresponding formation numbers. These facts suggest a direct link between a maximum in $F_{tn}$ and vortex ring pinch-off.

It was also observed that the total impulse generated per pulse would be substantially more than that due to momentum flux from the jet alone. The difference is largely due to the overpressure effect at the nozzle exit plane (nozzle exit overpressure) during vortex ring formation. For the stroke ratio $L_{max}/D = 2$ case, the impulse due to overpressure, $I_p$, is as much as 42 % of the total impulse per pulse. In other words, the total impulse of a starting jet is equal to about 172 % of that in a steady jet. The nozzle exit overpressure is most significant at low $L_{max}/D$ where an isolated vortex ring without a trailing jet is formed by the starting jet.

They explained the overpressure effect in terms of the additional pressure required at the nozzle exit plane during the initial stage of the vortex ring formation. The pressure rise is necessary to supply impulse to the ambient fluid accelerated with the ring in the form of (1) fluid entrained into the vortex ring and (2) the added mass of the vortex ring, that is, fluid being pushed out of the way by the vortex ring during the formation process.

Although the study of starting jet with a single pulse had demonstrated the importance of vortex ring formation as the thrust augmentation mechanism, practical applications may require repeated pulsing over a long period of time to supply impulse continuously to the flow. Continuously pulsed jets with a period of no-flow between pulses are called fully pulsed jets.

It is realized that repeated pulsing is significantly different from that of isolated starting jets in that jet pulses are no longer issued into quiescent ambient fluid. This implies the possible interaction of forming vortex rings with rings formed by the previously ejected pulse. This interaction may significantly affect the impulse per pulse and the thrust performance of the pulsed jet. As listed in Table 1, this feature in a continuously pulsed jet is controlled by the periodicity of the jet pulse, i.e., the pulsing frequency $f$ (or pulse period $T = 1/f$). It is normalized as the duty cycle $D_c = t_p f$.

Thrust augmentation in a fully pulsed jet was investigated by Krueger and Gharib [25], via direct measurement of the time-averaged thrust as a function of dimensionless pulse size $L_{max}/D$ and dimensionless frequency (or duty cycle) $D_c$. The parameter ranges investigated were $2 \le L_{max}/D \le 6$ and $0.1 \le D_c \le 0.98$. The quantity they used to characterize the propulsion of a pulsed jet is the time-averaged thrust for each pulse $F_T$. This is different from the average thrust $F_t$ for the single pulsed starting jet defined in Eq. (26) because $F_T$ is averaged over the pulse period $T$ rather than the pulse duration $t_p$ ($T = t_p + t_r$ as defined in Table 1).

To highlight the effects of vortex ring formation on thrust augmentation, the time-averaged thrust was normalized by the jet thrust due to the momentum flux $I_U$ as

$$F_{Tn} = \frac{\bar{F}_T}{\bar{F}_U} = \frac{\bar{F}_T}{I_U/T} \tag{27}$$

The experimental results on $F_{Tn}$ indicated that thrust augmentation is observed for all conditions tested. The time-averaged thrust is more than 20 % higher than the thrust expected from jet momentum alone and can be as much as 90 % greater than the jet momentum flux. It is also indicated that thrust augmentation appears to be more effective for smaller $L_{max}/D$. For the pulsed jets, the general decrease in $F_{Tn}$ observed as $L_{max}/D$ increase beyond three, i.e., the value of the formation number, suggests that thrust augmentation is governed by vortex ring formation. As to the effect of the duty cycle, thrust augmentation tends to degrade with increasing $D_c$. The augmentation ratio $F_{Tn}$ has strong dependence on the duty cycle $D_c$, due to the interaction of forming vortex rings with preceding pulses, an effect distinctly different form the single-pulse jet results. Vortex formation interacts with a trailing jet and with entire pulses.

Based on the understanding of propulsive characteristics of unsteady jet produced by a fixed jet generator, the study of unsteady jet propulsion has been carried out by employing self-propelled vehicles that can move freely underwater. In these studies, another flow parameter to be considered is the jet propulsive efficiency. The parameter is important for power consumption and endurance of small underwater vehicles. As shown above, the additional thrust can be obtained from the unsteady pulsed jets. As over-pressure increases the thrust provided by pulsed jets, additional energy is supplied to the flow via over-pressure. The balance between the increased thrust and the increased excess kinetic energy will determine whether the enhanced jet thrust also comes with improved propulsive efficiency.

Propulsive efficiency is generally defined as the ratio of the useful propulsive power output to the total power input (i.e., the total power exerted propelling the vehicle). For the pulsed jets, it is given by

$$\eta_p = \frac{W_u}{W_u + E_{ex}} \tag{28}$$

where $W_u$ is the rate of useful work done by thrust during propulsion, $E_{ex}$ is the rate at which excess kinetic energy is shed into the wake [3, 4]. Given the periodic feature of the self-propelled pulsed jet underwater vehicle, the rate of useful work can be averaged over one jet pulse as

$$W_u \approx \bar{F}_T \bar{U}_v \tag{29}$$

where $\bar{F}_T$ is the time-average thrust and $\bar{U}_v$ is the time average vehicle velocity. Using these results to Eq. (28) gives

$$\eta_p \approx \frac{\bar{F}_T \bar{U}_v}{\bar{F}_T \bar{U}_v + \bar{F}_{ex}} \tag{30}$$

Both $F_T$ and $E_{ex}$ can be obtained from the DPIV measurement. The average thrust per cycle is determined from $F_T = I/T$, where $I$ is the hydrodynamic impulse required to generate each pulse and is determined from

$$I = \rho\pi \int \omega_\theta r^2 drdx \tag{31}$$

Assuming axisymmetric of the measured flow field about the jet centerline, excess kinetic energy per jet pulse can be computed from DPIV measurements as

$$E_{ex} = \rho\pi \int V^2 rdrdx \tag{32}$$

where $V$ is the velocity magnitude in the laboratory frame of reference.

The propulsive efficiency of pulsed-jet thruster on a self-propelled underwater vehicle (dubbed as Robosquid) was studied by Moslemi and Krueger [37]. Their study focused on the comparison of the isosceles triangular and trapezoidal velocity programs. Similar to the stationary pulsed-jet generators, three key parameters were the stroke ratio, the velocity program and the pulsing duty cycle. In their experiments, the stroke ratio $L_{max}/D$ was changed by changing the pulse duration $t_p$, and duty cycle $D_c$ was changed independent of $L_{max}/D$ by adjusting the rest time $t_r$ while holding $t_p$ and $U_{0\_max}$ constant. They found that the Robosquid's velocity was approximately constant for triangular velocity program at each $D_c$.

The propulsive efficiency ratio $\eta_p/\eta_{p,ss}$ ($\eta_{p,ss}$ is the propulsive efficiency given in Eq. (30) for the steady jet) increases as $D_c$ increases for a fixed $L_{max}/D$. The increase in propulsive efficiency as $D_c$ increases is related to an increase in $U_v$ due to an increase in average thrust. A transition from decreasing propulsive efficiency ratio $\eta_p/\eta_{p,ss}$ to approximately constant $\eta_p/\eta_{p,ss}$ exists as $L_{max}/D$ increases above about 4, which is the approximate value of the formation number for these jets. Unlike the condition for the optimal thrust, the optimal propulsive efficiency occurs for $L_{max}/D$ less than the formation number where isolated vortex rings are presented in the jet flow. The increase in propulsive efficiency as $D_c$ increases is related to an increase in the average vehicle velocity .$\bar{U}_v$ due to an increase in average thrust. Finally, the ratio of pulsed-jet to steady jet propulsive efficiencies was found to be higher than one for $D_c > 0.4$ and $L/D > 4$ in both isosceles triangular and trapezoidal velocity programs, indicating better performance of a pulsed jet over equivalent steady jet propulsion under these conditions.

Recently, Ruiz et al. [48] have carried out detailed flow and power consumption measurements with a self-propelled vehicle in water. The vehicle has a propulsion mechanism that allows a comparison of the efficiency of a pulsed and steady jet to be made. In terms of the Froude propulsive efficiency given in Eq. (23), the unsteady propulsor consistently outperformed the baseline steady propulsor. Increase in efficiency exceeding 40 % was achieved, accompanying by modest decrease in the propulsion enhancement at higher motor speed. For comparison, the propulsive performance was also presented in terms of the hydrodynamic efficiency defined as

$$\eta_h = \frac{F_D U_v}{F_D U_v + (1/2)\rho A U_w (U_w - U_\infty)^2} \tag{33}$$

where $F_D$ is the hydrodynamic drag measured by using a load cell. The hydrodynamic efficiency based on the vehicle drag is consistent with the general definition of propulsive efficiency given by Eq. (28). In this case, the magnitude of vortex enhancement in propulsion was even greater, exceeding 70 % over the baseline steady case. The results suggested a potential mechanism to further enhance the performance of the existing propulsion systems.

## 4 Summary

The basic features of the starting jet generated by a piston-cylinder apparatus and other flow configurations are presented in this chapter. Unlike the steady jet, the studies on the starting jet focus on the near-field development of the flow structure, which is dominated by the formation and evolution of a large scale leading vortex ring. The formation of the leading vortex ring in a starting jet is restricted by the pinch-off process by which a forming vortex ring gradually detaches from the trailing jet and is no longer able to entrain additional vorticity from the jet generator. The vortex ring pinch-off is characterized by a dimensionless time scale, i.e., the formation number $F$, which indicates the onset of the pinch-off process. The specific value of the formation number was observed to be dependent on several flow conditions, i.e., velocity program, velocity profile at nozzle exit plane, Reynolds number, background flow conditions and so on.

The existence of a limiting dimensionless time scale for vortex ring formation can be understood in terms of the Kelvin-Benjamin variational principle for a steady, axis-touching vortex ring, or the translational dynamics of the starting jet flow. From the first perspective, the pinch-off process occurs when the apparatus is no longer able to deliver energy at a rate compatible with the requirement that a steadily translating vortex ring have maximum energy with respect to impulsive-preserving iso-vortical perturbations. In other words, the formation number corresponds to the time when normalized energy of the total jet $\alpha_{\text{piston}}$ drops to the limiting value $\alpha_{\text{lim}}$. From the second perspective, a vortex ring completes its formation and pinches off from the starting jet when the translational velocity of the vortex ring becomes equal to that of the trailing jet. Both theories lead to several analytical models for the pinch-off process and the prediction of the formation number.

The restricted vortex ring formation also represents a maximization principle for the properties of the vortex ring. In biological flows characterized by the form of starting jet, such as the blood flow in the left ventricle and unsteady jet propulsion system of several aquatic creatures, the vortex rings are found to be optimized for maximum thrust or efficiency by being generated at the critical formation number. Due to its enhanced propulsive performance over the steady jet, several prototypes of the unsteady jet thruster have been developed for the propulsion and maneuverability of underwater vehicles. Further researches on the unsteady jet propulsion system are required to examine the effect of several addition parameters introduced by the flow unsteadiness on the thrust and propulsive efficiency of the devices.

## References

1. Allen, J. J. & Naitoh, T. (2005). Experimental study of the production of vortex rings using a variable diameter orifice. *Physics of Fluids, 17*(6), 061701.
2. Auerbach, D. (1991). Stirring properties of vortex rings. *Phys Fluids A-Fluid, 3*, 1351–1355.
3. Bartol, I. K., Krueger, P. S., Stewart, W. J., & Thompson, J. T. (2009). Hydrodynamics of pulsed jetting in juvenile and adult brief squid Lolliguncula brevis: Evidence of multiple jet 'modes' and their implications for propulsive efficiency. *Journal of Experimental Biology, 212*, 1889–1903.
4. Bartol, I. K., Krueger, P. S., Thompson, J. T., & Stewart, W. J. (2008). Swimming dynamics and propulsive efficiency of squids throughout ontogeny. *Integrative and Comparative Biology, 48*, 720–733.
5. Bremhorst, K., & Watson, R. D. (1981). Velocity-field and entrainment of a pulsed core jet. *Journal of Fluids Engineering T ASME, 103*, 605–608.
6. Cantwell, B. J. (1986). Viscous starting jets. *Journal of Fluid Mechanics, 173*, 159–189.
7. Cossali, G. E., Coghe, A., & Araneo, L. (2001). Near-field entrainment in an impulsively started turbulent gas jet. *AIAA Journal, 39*, 1113–1122.
8. Dabiri, J. O. (2009). Optimal vortex formation as a unifying principle in biological propulsion. *Annual Review of Fluid Mechanics, 41*, 17–33.
9. Dabiri, J. O., & Gharib, M. (2004). Delay of vortex ring pinchoff by an imposed bulk counterflow. *Physics of Fluids, 16*, L28–L30.
10. Dabiri, J. O., & Gharib, M. (2004). Fluid entrainment by isolated vortex rings. *Journal of Fluid Mechanics, 511*, 311–331.
11. Dabiri, J. O., & Gharib, M. (2005). Starting flow through nozzles with temporally variable exit diameter. *Journal of Fluid Mechanics, 538*, 111–136.
12. Didden, N. (1979). Formation of vortex rings—Rolling-up and production of circulation. *Zeitschrift fur Angewandte Mathematik und Physik, 30*, 101–116.
13. Fraenkel, L. E. (1972). Examples of steady vortex rings of small cross-section in an ideal fluid. *Journal of Fluid Mechanics, 51*, 119–135.
14. Gao, L., & Yu, S. C. M. (2010). A model for the pinch-off process of the leading vortex ring in a starting jet. *Journal of Fluid Mechanics, 656*, 205–222.
15. Gao, L., & Yu, S. C. M. (2012). Development of the trailing shear layer in a starting jet during pinch-off. *Journal of Fluid Mechanics, 700*, 382–405.

16. Gharib, M., Rambod, E., & Shariff, K. (1998). A universal time scale for vortex ring formation. *Journal of Fluid Mechanics, 360*, 121–140.
17. Glezer, A. (1988). The formation of vortex rings. *Physics of Fluids, 31*, 3532–3542.
18. Heeg, R. S., & Riley, N. (1997). Simulations of the formation of an axisymmetric vortex ring. *Journal of Fluid Mechanics, 339*, 199–211.
19. Hettel, M., Wetzel, F., Habisreuther, P., & Bockhorn, H. (2007). Numerical verification of the similarity laws for the formation of laminar vortex rings. *Journal of Fluid Mechanics, 590*, 35–60.
20. Hill, M. J. M. (1894). On a spherical vortex. *Proceedings of the Royal Society of London, A185*, 213–245.
21. James, S., & Madnia, C. K. (1996). Direct numerical simulation of a laminar vortex ring. *Physics of Fluids, 8*, 2400–2414.
22. Kaden, H. (1931). Aufwicklung einer unstabilen Unstetigkeitsfläche. *Ingenieur Archiv, 2*, 140–168.
23. Krieg, M., & Mohseni, K. (2008). Thrust characterization of a bioinspired vortex ring thruster for locomotion of underwater robots. *IEEE Journal of Oceanic Engineering, 33*, 123–132.
24. Krueger, P. S., Dabiri, J. O., & Gharib, M. (2006). The formation number of vortex rings formed in uniform background co-flow. *Journal of Fluid Mechanics, 556*, 147–166.
25. Krueger, P. S., & Gharib, A. (2005). Thrust augmentation and vortex ring evolution in a fully pulsed jet. *AIAA Journal, 43*, 792–801.
26. Krueger, P. S., & Gharib, M. (2003). The significance of vortex ring formation to the impulse and thrust of a starting jet. *Physics of Fluids, 15*, 1271–1281.
27. Lawson, J. M., & Dawson, J. R. (2013). The formation of turbulent vortex rings by synthetic jets. *Physics of Fluids, 25*(10), 105113.
28. Leweke, T., & Williamson, C. H. K. (1998). Cooperative elliptic instability of a vortex pair. *Journal of Fluid Mechanics, 360*, 85–119.
29. Lim, T. T., Nickels, T. B. (1995). Vortex rings. In S. Green (Ed.), *Fluid Vortices* (pp. 95–153). Netherlands: Springer.
30. Linden, P. F., & Turner, J. S. (2001). The formation of 'optimal' vortex rings, and the efficiency of propulsion devices. *Journal of Fluid Mechanics, 427*, 61–72.
31. List, E. J. (1982). Turbulent jets and plumes. *Annual Review of Fluid Mechanics, 14*, 189–212.
32. Maxworthy, T. (1972). Structure and stability of vortex rings. *Journal of Fluid Mechanics, 51*, 15–32.
33. Maxworthy, T. (1977). Some experimental studies of vortex rings. *Journal of Fluid Mechanics, 81*, 465–495.
34. Mohseni, K., & Gharib, M. (1998). A model for universal time scale of vortex ring formation. *Physics of Fluids, 10*, 2436–2438.
35. Mohseni, K., Ran, H. Y., & Colonius, T. (2001). Numerical experiments on vortex ring formation. *Journal of Fluid Mechanics, 430*, 267–282.
36. Moore, D. W., & Saffman, P. G. (1973). Axial-flow in laminar trailing vortices. *Proceedings of the Royal Society of London. A. Mathematical and Physical Sciences, 333*, 491–508.
37. Moslemi, A. A., Krueger, P. S. (2010). Propulsive efficiency of a biomorphic pulsed-jet underwater vehicle. *Bioinspiration and Biomimetics*, *5*(3), 036003.
38. Nitsche, M. (1993) Numerical simulation of axisymmetric vortex sheet roll-up. *Vortex flows and related numerical methods*. In: Proceedings *NATO Advanced Res. Workshop*. (eds. Beale, J. T., Collet, G. H. and Hubersou, S.) Kluwer Acad. Publishers.
39. Nitsche, M., & Krasny, R. (1994). A numerical study of vortex ring formation at the edge of a circular tube. *Journal of Fluid Mechanics, 276*, 139–161.
40. Norbury, J. (1973). Family of steady vortex rings. *Journal of Fluid Mechanics, 57*, 417–431.
41. O'farrell, C., & Dabiri, J. O. (2010). A Lagrangian approach to identifying vortex pinch-off. *Chaos, 20*, 017513.
42. O'farrell, C., & Dabiri, J. O. (2012). Perturbation response and pinch-off of vortex rings and dipoles. *Journal of Fluid Mechanics, 704*, 280–300.

43. Olcay, A. B., & Krueger, P. S. (2008). Measurement of ambient fluid entrainment during laminar vortex ring formation. *Experiments in Fluids, 44*, 235–247.
44. Olcay, A. B., & Krueger, P. S. (2010). Momentum evolution of ejected and entrained fluid during laminar vortex ring formation. *Theoretical and Computational Fluid Dynamics, 24*, 465–482.
45. Pawlak, G., Marugan-Cruz, C., Martinez-Bazan, C., & Garcia-Hrdy, P. (2007). Experimental characterization of starting jet dynamics. *Fluid Dynamics Research, 39*, 711–730.
46. Ricou, F. P., & Spalding, D. B. (1961). Measurements of entrainment by axisymmetrical turbulent jets. *Journal of Fluid Mechanics, 11*, 21–32.
47. Rosenfeld, M., Rambod, E., & Gharib, M. (1998). Circulation and formation number of laminar vortex rings. *Journal of Fluid Mechanics, 376*, 297–318.
48. Ruiz, L. A., Whittlesey, R. W. & Dabiri, J. O. (2011). Vortex-enhanced propulsion. *Journal of Fluid Mechanics, 668*, 5–32.
49. Saffman, P. G. (1978). Number of waves on unstable vortex rings. *Journal of Fluid Mechanics, 84*, 625–639.
50. Schram, C., & Riethmuller, M. L. (2001). Vortex ring evolution in an impulsively started jet using digital particle image velocimetry and continuous wavelet analysis. *Measurement Science and Technology, 12*, 1413–1421.
51. Shadden, S. C., Dabiri, J. O., & Marsden, J. E. (2006). Lagrangian analysis of fluid transport in empirical flows. *Physics of Fluids, 18.*
52. Shadden, S. C., Lekien, F., & Marsden, J. E. (2005). Definition and properties of Lagrangian coherent structures from finite-time Lyapunov exponents in two-dimensional aperiodic flows. *Physica D: Nonlinear Phenomena, 212*, 271–304.
53. Shariff, K., & Leonard, A. (1992). Vortex rings. *Annual Review of Fluid Mechanics, 24*, 235–279.
54. Shusser, M., & Gharib, M. (2000). Energy and velocity of a forming vortex ring. *Physics of Fluids, 12*, 618–621.
55. Shusser, M., Gharib, M., Rosenfeld, M., & Mohseni, K. (2002). On the effect of pipe boundary layer growth on the formation of a laminar vortex ring generated by a piston/cylinder arrangement. *Theoretical and computational fluid dynamics, 15*, 303–316.
56. Weigand, A., & Gharib, M. (1997). On the evolution of laminar vortex rings. *Experiments in Fluids, 22*, 447–457.
57. Zhao, W., Frankel, S. H., & Mongeau, L. G. (2000). Effects of trailing jet instability on vortex ring formation. *Physics of Fluids, 12*, 589–596.

# The Impact of a Vortex Ring on Porous Surfaces—A Review

**T.T. Lim and D. Adhikari**

**Abstract** A vortex ring interacting with a solid surface have been studied extensively, both experimentally and numerically, for many years, however, there are relatively few studies made on the interaction with porous surfaces. Most of these studies were carried out recently and limited to laboratory investigations. This chapter provides an overview of the current state of knowledge on the interaction of a vortex ring on a porous surface and future research direction.

**Keywords** Vortex ring · Vortex ring/wall interaction · Porous surface

## 1 Introduction

A vortex ring or smoke ring refers to a bounded region of vorticity in a fluid where vortex lines form close loops. It comes in various shapes and sizes, and occurs naturally in situations such as rising of warm air or falling of cold air in atmosphere (for example, microburst), volcanic eruption, in the left ventricle of human heart during cardiac relaxation, nuclear explosion and insect flight to name just a few. Smokers sometimes produce them for amusement by blowing smoke through their mouth in a certain manner [7]. Also, cetaceans such as dolphins and beluga whales produce vortex ring bubbles occasionally for entertainment by puffing a burst of air through their blow hole in water [35, 37]. Vortex rings have fascinated scientists and engineers for more than 100 years (see Thomson [49]). The fascination is due partly to numerous engineering and biological applications, and partly to their compact

T.T. Lim (✉)
Department of Mechanical Engineering, National University of Singapore, Singapore, Singapore
e-mail: mpelimtt@nus.edu.sg

D. Adhikari
School of Civil and Environmental Engineering, Georgia Institute of Technology, 790 Atlantic Drive, Atlanta, GA 30332, USA



D.T.H. New and S.C.M. Yu (eds.), *Vortex Rings and Jets*,
Fluid Mechanics and Its Applications 111, DOI 10.1007/978-981-287-396-5_2

nature and profoundly interesting flow properties. Their compact nature make them ideal as a simpler building block in modelling more complex flows, such as turbulent free jets, hazardous downburst and propulsion of jelly fish and squids [16, 17]. On a more technological level, cavitating vortex rings, produced by exciting cavitating jets, have been used for underwater cleaning and rock cutting [10], and Akhmetov et al. [3] suggested using vortex rings for fighting oil well fires. Turner [51] and Fohl [22] proposed that vortex rings might be used to transport smoke or other effluents to higher atmosphere, thus reducing the necessity for tall chimneys. Numerous experimental, theoretical and numerical studies have been conducted on the generation, motion and interaction of vortex rings, and comprehensive reviews can be found in Shariff and Leonard [48], Lim and Nickels [33], and in a book by Saffman [47] and more recently by Akhmetov [2]. In this chapter, we focus primarily on the interaction of a vortex ring with a porous surface. The interaction has attracted considerable attention in recent years due partly to scientific interests and partly to its relevance in various complex flows such as static mixers and filters, flow transition to turbulence, etc.

There are various techniques of producing vortex ring in a laboratory, but the two most often used methods are by impulsively ejecting a slug of fluid through an orifice opening or a nozzle in quiescent fluid. The vortex generator can be a piston-cylinder arrangement [33] driven by a stepper motor (see Fig. 1) or by a pressurized

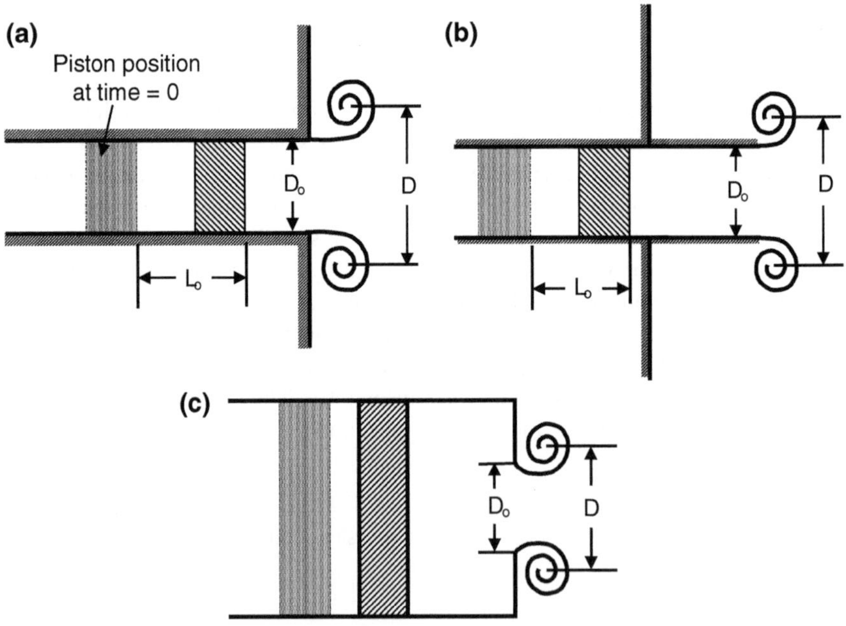


**Fig. 1** Schematics of some typical vortex ring generators using piston-cylinder arrangement. **a** Nozzle exit flushed with the side wall of a tank. **b** Nozzle exit protrudes into a tank. **c** Orifice opening

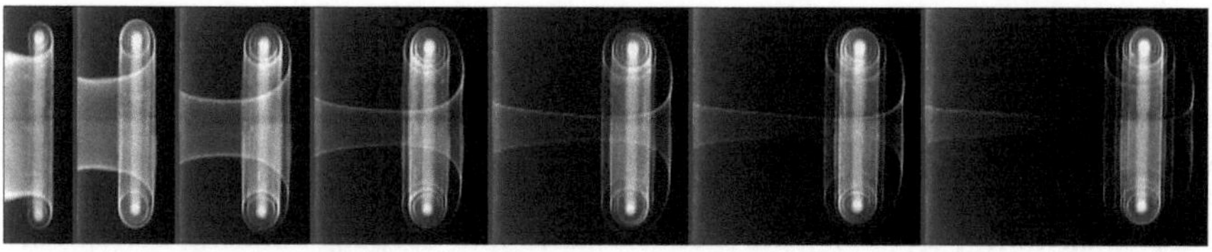

**Fig. 2** A vortex ring generated by nozzle type vortex ring generator. Here the roll up of a cylindrical shear layer into toroidal vortex can be clearly seen

air/water regulated vessel [41] or direct ejection of fluid using a vertical column of fluid [28]. The exit geometry of the orifice or nozzle (i.e. circular, elliptic or triangle, etc.) determines the shape, motion and stability of the resulting vortex ring [for example see 18, 29]. For a circular nozzle or orifice, the motion of the piston will cause a cylindrical vortex sheet to form at the edge of the nozzle or orifice before rolling up into a scroll as shown in Fig. 2. The fully developed vortex ring then propagates downstream through self-induced velocity.

In general, the diameter of a fully developed vortex ring ($D$) is dependent on the ratio of stroke length ($L_o$) and the nozzle diameter ($D_o$). For a circular nozzle using the geometry shown in Fig. 1b, Auerbach [5] obtained the following empirical relations.

$$\frac{D}{D_o} = 1.18\left(\frac{L_o}{D_o}\right)^{1/3} \text{ for } 0.3 \leq \frac{L_o}{D_o} \leq 1.0$$

$$\frac{D}{D_o} = 1.18\left(\frac{L_o}{D_o}\right)^{1/5} \text{ for } 0.3 \leq \frac{L_o}{D_o} \leq 3.3$$

These expressions are restricted to $1500 \leq Re_n \leq 4500$, where $Re_n$ is defined based on ejection velocity and nozzle diameter. In a subsequent experimental study by Gharib et al. [25] using a similar nozzle arrangement depicted in Fig. 1b, it is shown that a vortex ring cannot keep increasing in size with increasing $L_o/D_o$. They showed that the formation of an axisymmetric vortex ring is characterized by a time-scale that the vortex ring pinches off from its generating axisymmetric jet to form a well-defined vortex ring. Beyond this time-scale, larger vortex rings are not possible and the leading vortex ring is followed by a trailing jet. The transition between these two states occurs in a narrow range of stroke ratio ($L_o/D_o$) between 3.6 and 4.5. They refer to this stroke ratio as “formation number”. At this formation number, the circulation that a vortex ring can acquire is at its maximum.

Once a vortex ring is fully developed, usually 1.5 nozzle diameter downstream [19], it propagates with a translation velocity that depends on the circulation ($\Gamma$), vortex core radius ($a$) and radius ($R_o$) of the ring. For a thin core axisymmetric vortex ring, Saffman [44] derived an expression for the translation velocity ($U$) given by

$$U \approx \frac{\Gamma}{4\pi R_o}\left\{\log\left(\frac{4R_o}{\sqrt{\nu t}}\right) - 0.5880\right\}$$

where ν is the kinematic viscosity of fluid. The above expression is restricted to a very small time $\left(\frac{\nu t}{R_o^2} \ll 1\right)$ only. It also takes into account diffusing vortex core, which is assumed to have a Gaussian distribution of vorticity ($\omega_\phi$)

$$\omega_\phi = \frac{\Gamma}{4\pi\nu t}\exp^{-\frac{r^2}{4\nu t}}$$

where *r* is the core radius and *t* is time. In a more recent study, Fukumoto and Moffatt [23] extended the expression derived by Saffman [44] to a higher order using the variation principle and obtained the translation velocity as

$$U \approx \frac{\Gamma}{4\pi R_o}\left\{\log\left(\frac{4R_o}{\sqrt{\nu t}}\right) - 0.5880 - 3.6716\frac{\nu t}{R_o^2}\right\}$$

Past experimental studies have shown that if the Reynolds number (based on translation velocity and maximum ring diameter) is less than 600, a laminar vortex ring is formed. Once fully developed, the vortex ring propagates downstream with decreasing velocity while its volume increases with time. Reynolds [43] attributed the increased in volume to the entrainment of the surrounding fluid, which has the effect of causing the vortex ring to slow down in order to conserve momentum. However, Maxworthy [38] argued that the decrease in the propagation velocity may also be due to the shedding of vorticity in the wake. Under certain circumstances, a laminar vortex ring may undergo flow instability and develop azimuthal waves around its circumference. These instability waves grow with time and eventually transform a laminar vortex ring into a turbulent vortex ring, shedding vorticity in its wake as it propagates downstream. Detailed analysis of the azimuthal instability waves can be found in Widnall et al. [53], Tsai and Widnall [50], Widnall and Tsai [55] and Saffman [45] and the investigations on turbulent vortex rings can be found in Maxworthy [39], Glezer and Coles [26], Lim [32], and Gan and Nickels [24]. See also review articles by Shariff and Leonard [48] and Lim and Nickels [33].

In this chapter, we focus our attention only on the impact of a laminar vortex ring on porous surfaces. At the time of writing, the authors are not aware of similar studies where turbulent vortex rings are used. For completeness, we also include the case of a vortex ring interacting with a solid surface as this can be considered a special case of a porous surface with zero porosity. Besides, many of the flow features encountered during vortex ring/solid surface interaction are also observed during vortex ring/porous surface interaction.

# 2 Interaction of a Vortex Ring Normally with a Solid Surface

This particular interaction has been studied extensively both experimentally and numerically (for example, see [8, 9]; Walker et al. [14, 52]; Lim et al. [34, 42]; Doligalski et al. [20]; Chu et al. [15]; Fabris et al. [21]; Cheng et al. [13]). The interest is motivated partly by the fundamental physics involved and partly by its relevance in the study of turbulence wall bounded flow and in applications, such as flight of helicopters, turbulent flows in turbo-machines, geophysical flow, just to name a few.

An early experimental study of vortex ring/solid surface interaction was conducted by Magarvey and MacLatchy [36]. Using smoke visualization technique, they observed the formation of a secondary vortex ring during the impact of a primary vortex ring on a plane surface. However, they erroneously attributed the formation of the secondary vortex ring to the uneven distortion of the vortex core of the primary vortex ring during the impact. In a similar experimental study using a vortex ring produced by impacting a drop of dyed water with the surface of the water, Boldes and Ferreri [8] observed the vortex ring rebounded from the solid surface, which they correctly attributed to the generation and separation of wall boundary layer. This finding is consistent with a related and earlier experimental study by Harvey and Perry [27], which show that boundary layer separation is responsible for the rebound of a pair of trailing vortices approaching the wall. Yamada et al. [56] arrived at the same conclusion from their experimental investigation using smoke visualization technique.

However, the most conclusive evidence showing that boundary layer separation and the subsequent generation of secondary vortex is responsible for the rebound of a primary vortex ring was provided by Lim et al. [34]. In their attempt to resolve the contentious issue of whether vortex rebound is caused by the generation and separation of wall boundary layer [8, 27] or the distortion of vortex core [6, 36], they performed two sets of experiments; one involved head-on collision of a vortex ring with a solid boundary and the other one involved head-on collision of two identical vortex rings. The latter experiment is equivalent to the interaction of a vortex ring with a "surface" in the absence boundary layer (i.e. free-slip condition). They argued that if vortex rebound is indeed caused by the distortion of vortex cores, then the phenomenon should be observed in both sets of experiments. However, their result did not show any evidence of vortex ring rebound during the head-on collision of two vortex rings, and this led them to conclude that boundary layer separation was indeed responsible for the vortex ring rebound.

The most comprehensive experimental investigation of the impact of a vortex ring on a solid surface was performed by Walker et al. [52] (see also [9]) for a range of Reynolds number (based on translation velocity and diameter of a fully developed vortex ring) from less than 250 to about 3000. Using dye visualization technique, they observed detailed flow features induced by a variety of laminar vortex. For a vortex ring with a low initial Reynolds number of less than approximately 250, they found that the ring slowed down and expanded in diameter as it approached a solid surface. Due to low circulation of the primary vortex ring, the boundary layer generated underneath the vortex core was not sufficiently strong to roll up to form a secondary vortex ring. A schematic drawing of this scenario is shown in Fig. 3a. This behavior is in contrast to inviscid analysis that takes no account of the boundary layer process and predicts ever expanding vortex ring diameter. In the presence of no slip condition at the surface, the radial expansion of the primary vortex ring was eventually arrested by cancellation with vorticity of opposite sign at the surface. However, when the initial Reynolds number was increased to greater than 250, Walker et al. [52] found that the induced velocity of the primary vortex was strong enough to cause boundary layer to separate and roll up to form a secondary vortex ring of opposite circulation. The presence of the secondary vortex ring influences the trajectory of the primary vortex ring and vice versa. In fact, it can be deduced using the Biot-Savart law that the primary vortex ring, with its higher circulation, caused the secondary vortex ring to orbit around it and shrunk in diameter. At the same time, the induced velocity of the secondary vortex caused the primary vortex ring to rebound from the wall as can be seen in Figs. 3b and 4a. For Reynolds number less than 600, the rebound of the primary vortex ring is accompanied by a reduction in the radial expansion of the vortex core. But as the Reynolds number of the primary vortex ring is increased, the enhanced

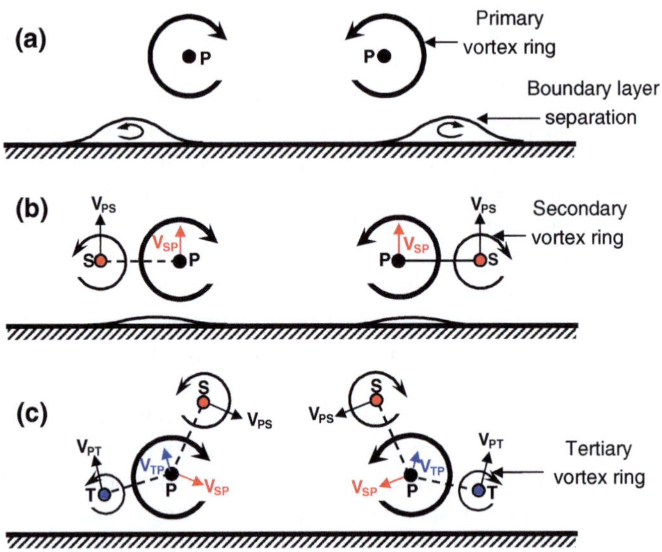


**Fig. 3** Schematic presentation of the interaction between primary vortex ring (P), secondary vortex ring (S) and tertiary vortex ring (T). $V_{PS}$ and $V_{PT}$ are the velocities induced by the primary vortex ring on the secondary and tertiary vortex rings, respectively. $V_{SP}$ is the induced by the secondary vortex ring on the primary vortex ring, and $V_{TP}$ is the induced by the tertiary vortex ring on the primary vortex ring. Note that the mutual induced velocities of the secondary and tertiary vortex rings are not shown to avoid complication

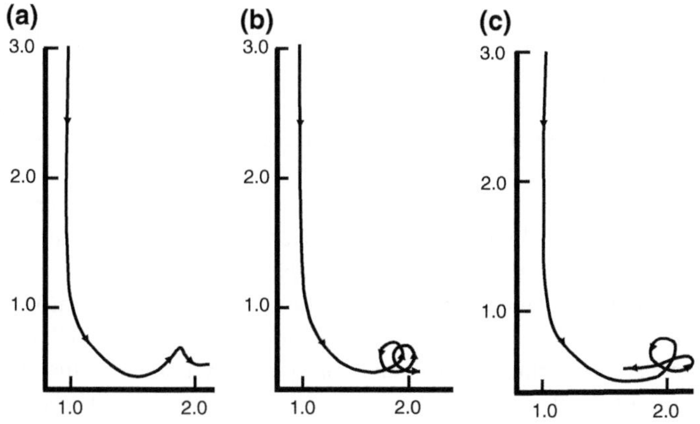


**Fig. 4** Trajectories of the right cores of the primary vortex rings based on the results of Walker et al. [52]. **a** $Re$ = 564. **b** $Re$ = 1680. **c** $Re$ = 2550

strength of the secondary vortex ring may be sufficiently strong to cause a momentary reversal in the radial velocity of the primary ring as can be seen in Fig. 4b. Similarly, a sufficient strong tertiary vortex ring can also cause the primary vortex ring to rebound. This is also reflected in Figs. 3c and 4b, where the first loop in the trajectory is due to the secondary vortex ring and the second loop is due to the tertiary vortex ring. For this particular case, despite the reversal in the core trajectory, the primary vortex ring, on the whole, continue to expand. However, at Reynolds number higher than $Re$ = 2550, the trajectories can end up in permanent reversal as can be seen in Fig. 4c.

At low Reynolds number, the secondary vortex ring does not develop wavy instabilities during its interaction with the primary vortex ring, but at higher Reynolds number, secondary vortex ring can develop wavy instabilities that can affect the overall flow features. Walker et al. [52] attributed the instabilities of the secondary vortex ring to the compression of the secondary vortex filament as it orbits into the interior of the primary vortex ring, and not due to the azimuthal wavy instabilities often observed during the transition from a laminar vortex ring to a turbulent vortex ring as was discussed by Widnall and Sullivan [54]. The present wavy instabilities are found to induce an apparent waviness in the primary vortex ring, and the whole process eventually leads to degeneration into three dimensional chaotic 'turbulent' flow. These wavy instabilities occur in the Reynolds number range of 470–1600 only. At much higher Reynolds numbers, Walker et al. [52] observed another secondary vortex ring instabilities, which they termed as kink instabilities. Prior to the formation of the kink stability, the formation of the secondary vortex ring follows the same manner as that observed at lower Reynolds number. This kink instability does not appear to grow (in amplitude) once it orbits into the interior of the primary vortex ring and pauses for some time until the tertiary instability is formed and orbited around the primary vortex ring. This event triggers the ejection of the secondary vortex ring, and through its induced velocity taking the tertiary vortex ring along with it. Due to the fact that the secondary and tertiary vortex rings possess circulation of the same sign, they eventually merge and propagate away from the surface.

Besides the experimental investigations, numerical simulations of the vortex ring/solid surface interaction have also been conducted by numerous researchers (for example, see Walker et al. [42, 52]; Chu et al. [15]; Fabris et al. [21]). The most recent study was performed by Cheng et al. [13], who used lattice Boltzmann method to compute the hydrodynamic behavior of a vortex ring during its interaction with a surface for a range of incidence angle (θ) from 0° to 40° and Reynolds number (based on translational velocity and initial diameter of the ring) from 100 to 1000. In this chapter, we restrict our discussion to the case of $\theta = 0°$, in which the ring approaches the surface with its axis of symmetry normal to the surface. The salient feature of the flow, such as secondary and tertiary vortex rings are captured by the simulation and compare favorably with the experimental observation. In addition to the vortex structures, Cheng et al. [13] computed the dynamics of vortex stretching and surface pressure distribution during the interaction, which are difficult to obtain experimentally. Their results show that a vortex ring approaching a surface undergoes a slow rate of vortex stretching initially and a monotonic decrease in the rate of change of centre vorticity strength. As the vortex ring moves closer to the surface, the rate of radial expansion increases rapidly, and this is accompanied by a reduction in core size and vorticity intensification. This process enhances viscous diffusion and vorticity cancellation with the vorticity of opposite sign from boundary layer. As expected, higher Reynolds number produces higher rate of radial expansion, higher adverse surface pressure gradient and higher maximum stagnation pressure. These computed results provide valuable insight into the evolution of pressure and vorticity field during the impact.

## 3 Interaction of a Vortex Ring Normally with a Porous Surface

As far as the authors are aware, one of the earliest observations of vortex ring/porous surface interaction was recorded by Morton in his unpublished work (see Adhikari and Lim [1]). Morton commented that when "*the plane wall is replaced by a fine gauze (such as a fly screen), the motion of the ring is little affected at low Reynolds numbers but at higher Reynolds numbers the ring passes through the screen and continues as a modified vortex ring in its lee*". Compared to vortex ring interacting with a solid surface, there are fewer studies on the interaction with porous surfaces. Most of them are conducted relatively recently via experimental investigation (see [1, 28, 41]). These studies show that although vortex ring/porous surface interaction shares many of the salient flow features found in vortex ring/solid surface interaction, the final outcomes of the interaction is very much dependent on screen porosity, wire diameter and Reynolds number of the approaching primary vortex ring. The dependency on the Reynolds number and screen porosity for a fixed wire diameter was investigated by Adhikari and Lim [1] for screen porosity of 62 and 81 % and Reynolds number of a primary vortex ring from 384 to 2369. Their results of the vortex core trajectories versus non-dimensional time are presented in Fig. 5. Adhikari and Lim [1] categorized the interaction into four broad scenarios based on the transmitted flow through the screen. We now discuss each of these scenarios with reference to 62 % porosity screen.

Scenario 1 occurs when a low Reynolds number primary vortex ring impacts normally on a low porosity screen. One such example is presented in Figs. 5a and 6 for the case of $Re$ = 384 and $\beta$ = 62 %, where the core trajectory and snapshots during the interaction are depicted. It can be seen from Fig. 5a that the trajectory of the vortex core interacting with the porous screen follows closely that of the corresponding interaction with a solid surface except near the end of the interaction where the two trajectories deviate. The deviation arises from the fact that the porous screen allows the self-induced velocity of the primary vortex ring to pass through it

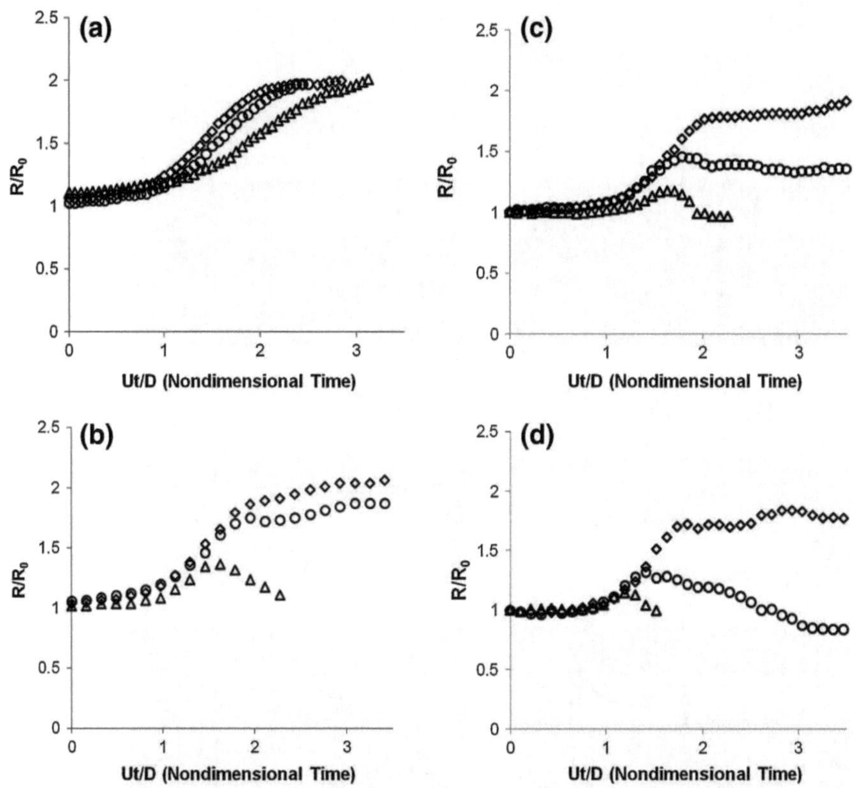


**Fig. 5** Radial expansion of the primary vortex ring versus non-dimensional time ($t^*$ = Ut/D) on a solid wall and porous screens. *Diamond* Solid Wall; *circle* 62 % Porosity; *triangle* 81 % Porosity. **a** $Re_\Gamma$ = 384. **b** $Re_\Gamma$ = 779. **c** $Re_\Gamma$ = 1597. **d** $Re_\Gamma$ = 2369. Reproduced from Adhikari and Lim [1]

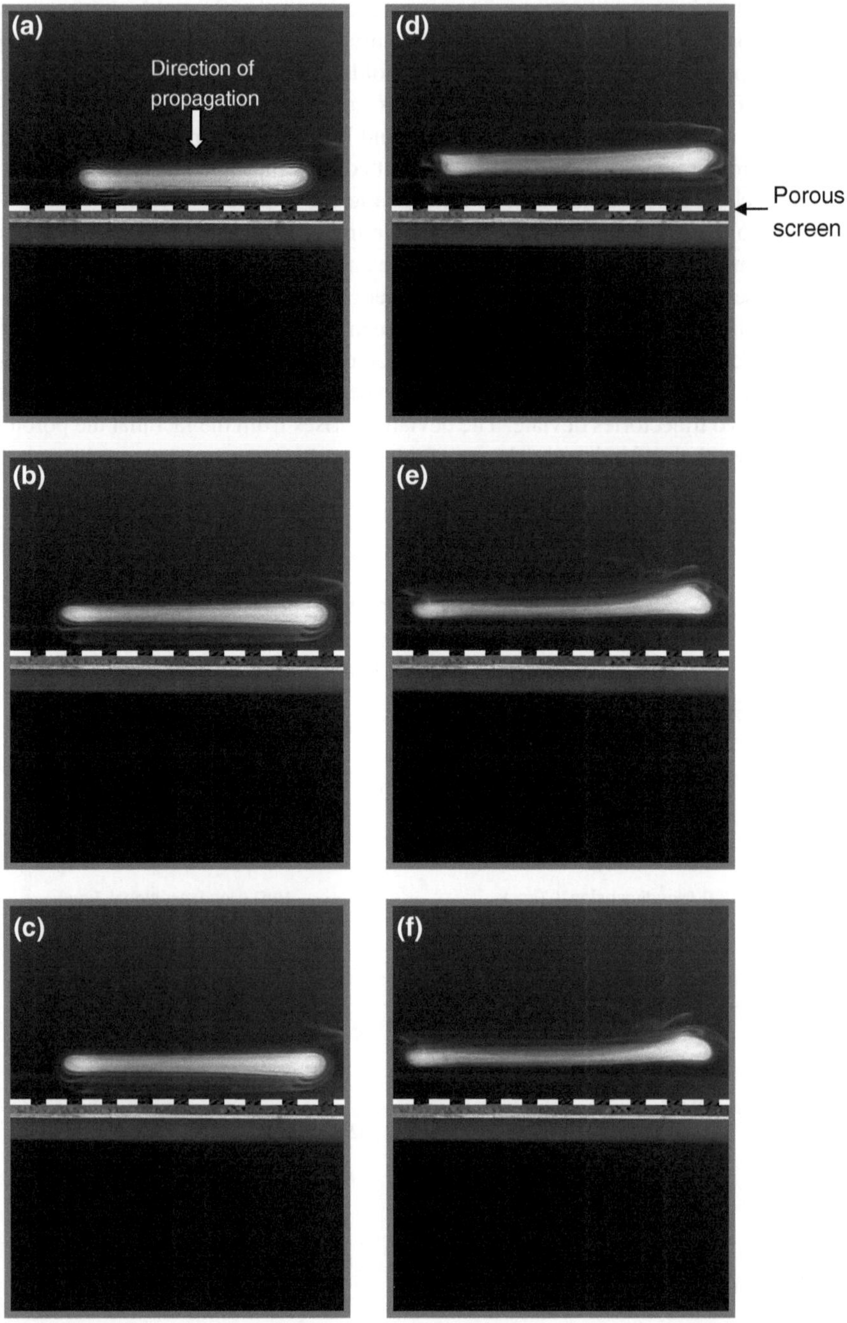


**Fig. 6** Impact of a vortex ring on a 62 % porosity screen for $Re_{\Gamma} = 384$. Time increases from (**a**) to (**f**). $t^* = 0.0$ corresponds to the instance when the vortex ring is one ring diameter from the screen. Reproduced from Adhikari and Lim [1]. **a** $t^* = 2.21$. **b** $t^* = 2.72$. **c** $t^* = 2.83$. **d** $t^* = 3.86$. **e** $t^* = 6.07$. **f** $t^* = 6.75$

to form a jet-like flow. Adhikari and Lim [1] referred to this kind of interaction, which produces the jet-like flow on the leeside of the screen as Scenario 1. Their interpretation of the interaction is illustrated in Fig. 7. This transmitted jet carried with it some of the original vorticity leaving behind a much weaker primary vortex

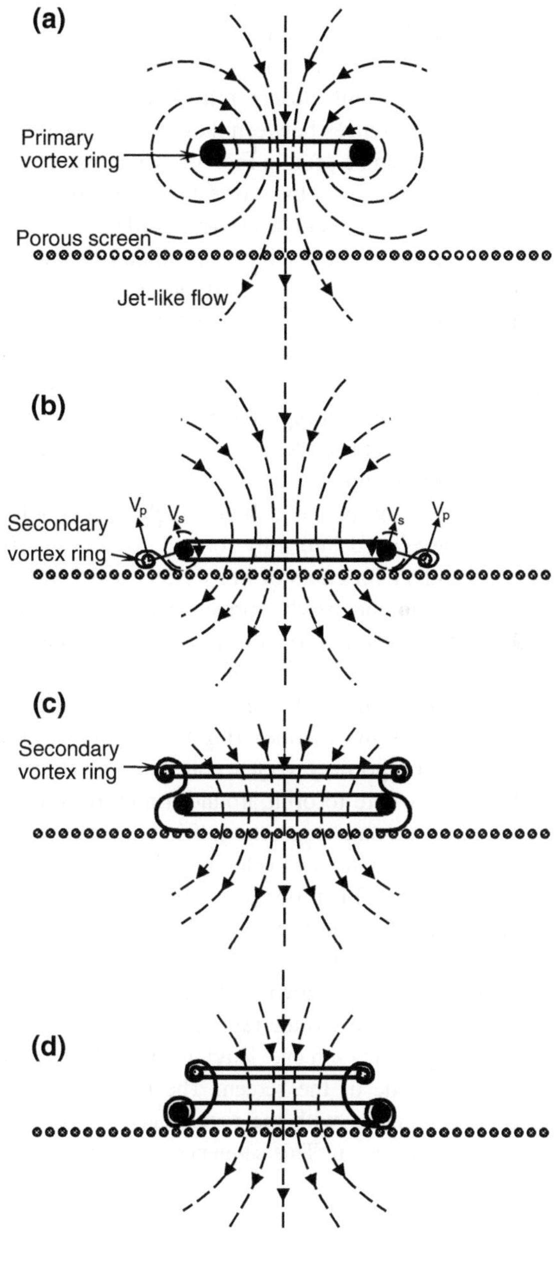


**Fig. 7** Adhikari and Lim [1]'s interpretation of the vortex ring/porous screen interaction at $Re_{\Gamma} = 384$ (Scenario 1). Time increases from (**a**) to (**d**). Note that broken lines merely represent conceptual streamlines depicting the transmission of the induced flow of the toroidal ring through the screen. The actual streamline pattern is more complex. *Single arrows* represent direction of flow field. Reproduced from Adhikari and Lim [1]

ring to interact with the screen. As a consequent, a much weaker secondary vortex ring is generated resulting in a much weaker vortex rebound compared to a solid surface. The fact that the fluid passes through the screen suggests that the "symmetry plane" is broken and the method of images can no longer be applied generally throughout the whole flow field to explain the observed behavior. However, it can still be used as an approximation locally to explain the initial radial expansion of the primary vortex ring as it interacts with the screen.

When the Reynolds number of the primary vortex ring is increased to 779, the evolution of the primary vortex ring during the initial stages of the interaction with a porous screen still closely follow that of the interaction with a solid surface. Significant differences between these two surfaces emerged only near the end of the interaction as can be seen in Fig. 5b, which depicts the vortex core trajectory and in Fig. 8, which shows the corresponding flow visualization images during the interaction. Although not as clearly visible in Fig. 8, Adhikari and Lim [1] observed during the experiment that the transmitted jet for this case is now sufficiently strong, in terms of circulation, to roll up into a ring-like structure. This ring-like structure has circulation of the same sign as the primary vortex ring. Due to the presence of the porous screen, its influence on the trajectory of the primary vortex ring is limited compared to the influence of the secondary vortex ring. This interaction is schematically illustrated in Fig. 9 and Adhikari and Lim [1] referred to it as Scenario 2. The difference between Scenario 1 and 2 lies in the transformation of jet-like flow to ring-like structure.

Scenario 3 occurs at a much higher Reynolds number, where the ring-like structure developed into fully formed vortex ring. This can be clearly observed in Figs. 10 and 11 for cases of $Re_{\Gamma} = 1597$ and 2369, respectively. Adhikari and Lim [1] referred to the newly formed ring as "regenerated vortex ring". As in the previous cases, the porous screen limits the influence of the regenerated vortex ring on the trajectory of the primary vortex ring, but the close proximity of the secondary vortex ring to the primary vortex ring leads to their mutual interaction, which causes primary vortex ring to rebound from the screen and contract. At the same time, the primary vortex core, through its induced velocity, causes the secondary vortex core to orbit around it and towards the inner region of the primary vortex ring. At some stage, the self-induced velocity of the secondary vortex ring is able to escape the induced velocity field of the primary vortex ring and travels away from the screen. This process is similar to vortex ring/solid surface interaction and is clearly illustrated in Fig. 12.

Scenario 4 occurs at even higher Reynolds number and screen porosity. An example of which is shown in Fig. 13. Here, the primary vortex ring, after losing some of the vorticity to the regenerated vortex ring, is still sufficiently energetic to pass through the screen, leaving behind the secondary vortex ring on the approaching side of the screen. The modified "transmitted" primary vortex ring eventually pairs up with the regenerated vortex ring, and propagates as a whole away from the screen. This sequence of event is schematically presented in Fig. 14.

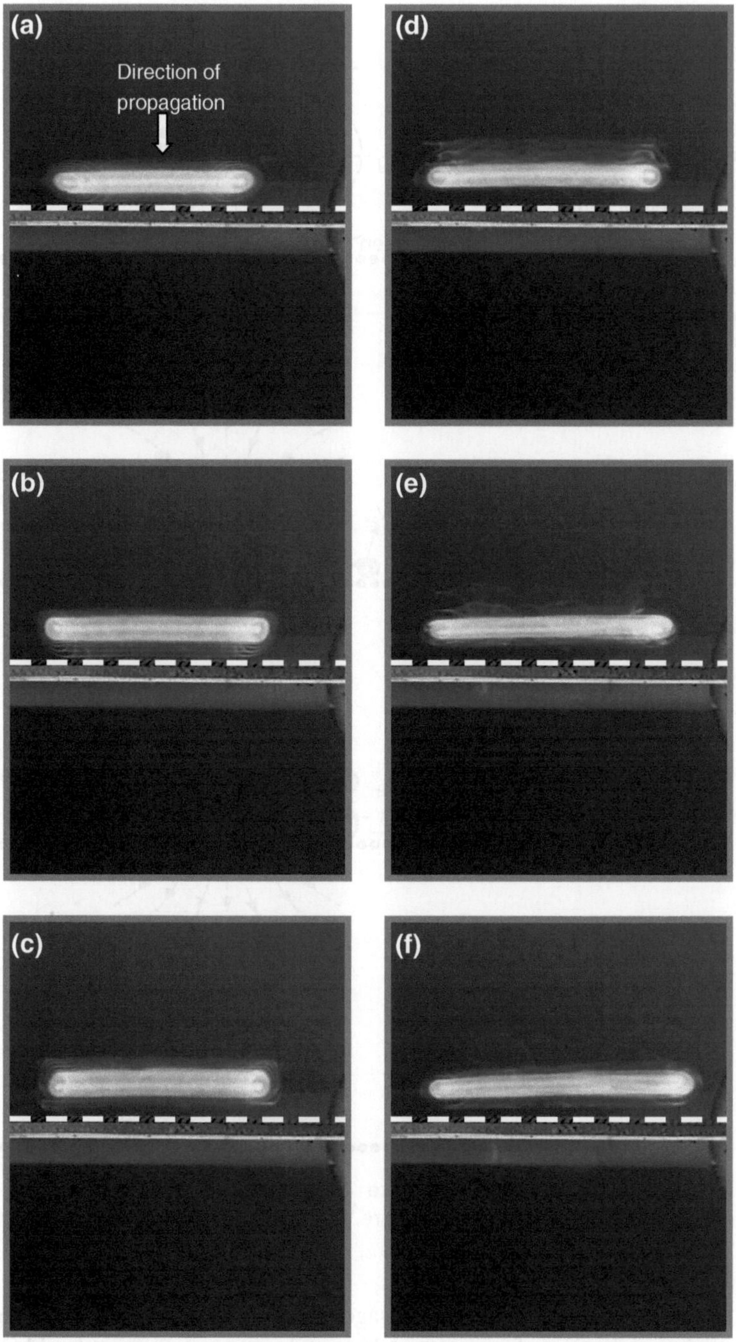


**Fig. 8** Impact of a vortex ring on a 62 % porosity screen for $Re_{\Gamma} = 779$. Time increases from (**a**) to (**f**). $t^* = 0.0$ corresponds to the instance when the vortex ring is one ring diameter from the screen. Reproduced from Adhikari and Lim [1]. **a** $t^* = 1.55$. **b** $t^* = 1.97$. **c** $t^* = 2.44$. **d** $t^* = 2.93$. **e** $t^* = 4.07$. **f** $t^* = 6.59$

**Fig. 9** Adhikari and Lim [1]'s interpretation of the vortex ring/porous screen interaction at $Re_{\Gamma} = 779$ (Scenario 2). Time increases from (**a**) to (**d**). $V_{ps}$ = induced velocity of the primary vortex ring on the secondary vortex ring, $V_{sp}$ = induced velocity of the secondary vortex ring on the primary vortex ring. *Single arrows* represent direction of flow field. Reproduced from Adhikari and Lim [1]

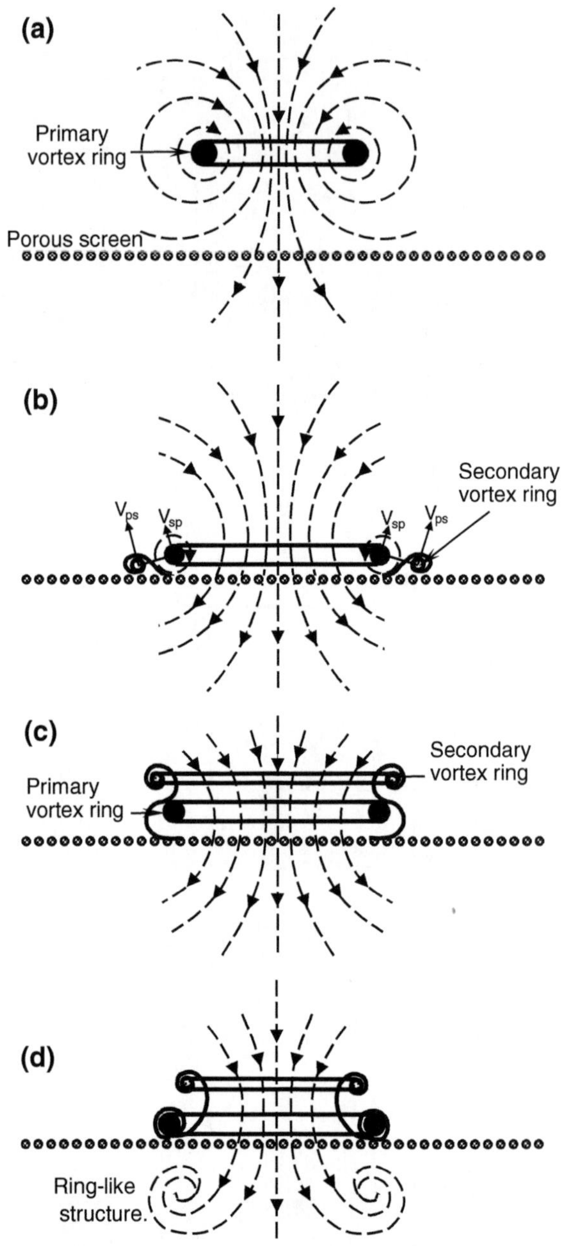


Adhikari and Lim [1] attributed the passage of the primary vortex ring through the screen to the mechanism of vortex stretching and reconnection (see also [4, 11, 30, 31, 40, 46]), and proposed a simplified model to explain it. When the Reynolds number is increased further, the interaction follows the same sequence of event as

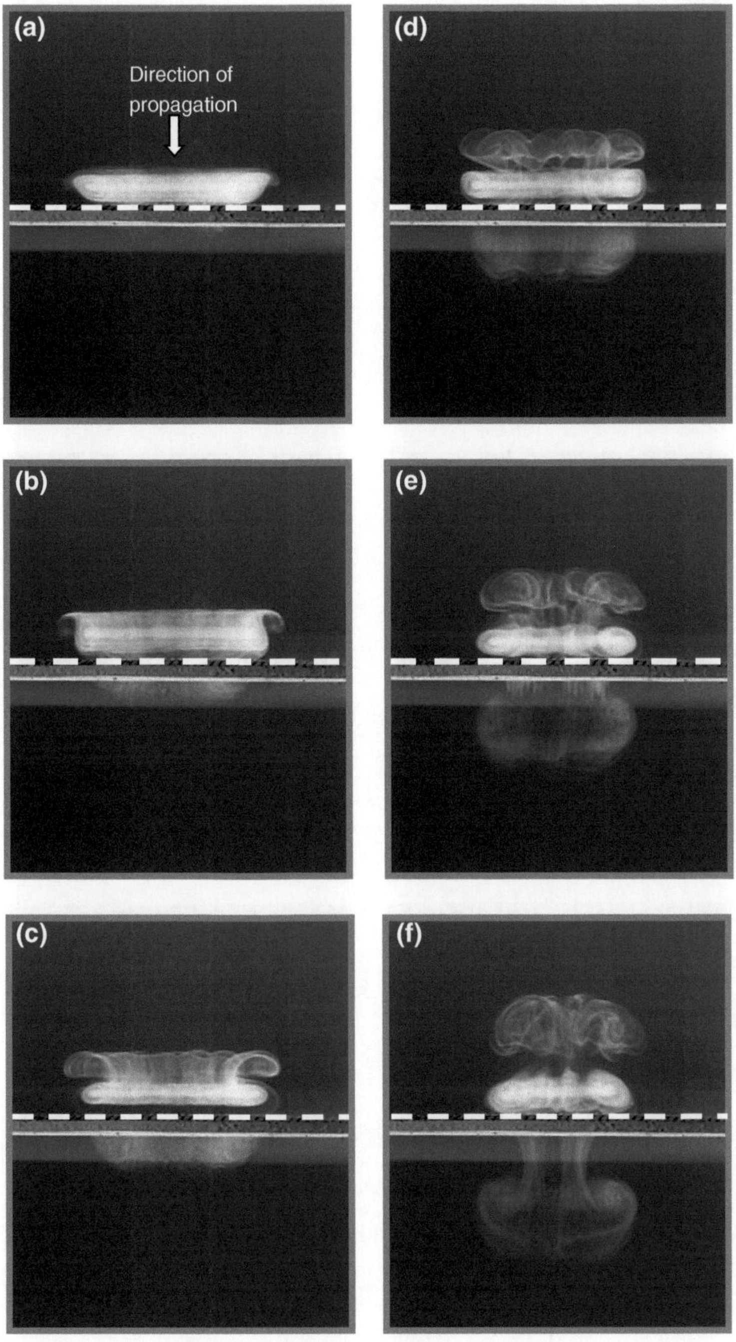


**Fig. 10** Impact of a vortex ring on a 62 % porosity screen for $Re_{\Gamma} = 1597$. Time increases from (**a**) to (**f**). $t^* = 0.0$ corresponds to the instance when the vortex ring is one ring diameter away from the wall. Reproduced from Adhikari and Lim [1]. **a** $t^* = 1.47$. **b** $t^* = 1.84$. **c** $t^* = 2.45$. **d** $t^* = 3.65$. **e** $t^* = 5.73$. **f** $t^* = 9.11$

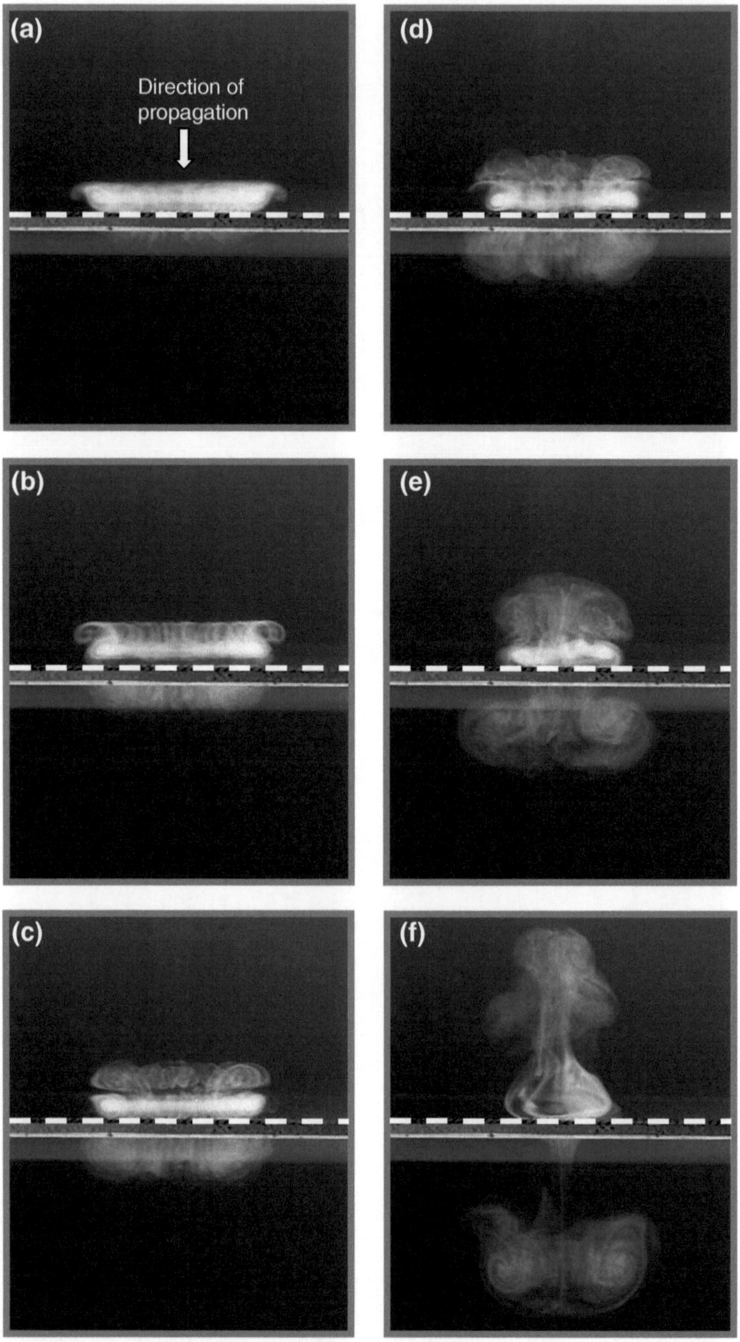


**Fig. 11** Impact of a vortex ring on a 62 % porosity screen for $Re_{\Gamma} = 2369$. Time increases from (**a**) to (**f**). $t^* = 0.0$ corresponds to the instance when the vortex ring is one ring diameter away from the wall. Reproduced from Adhikari and Lim [1]. **a** $t^* = 1.19$. **b** $t^* = 1.57$. **c** $t^* = 2.14$. **d** $t^* = 2.56$. **e** $t^* = 4.16$. **f** $t^* = 9.49$

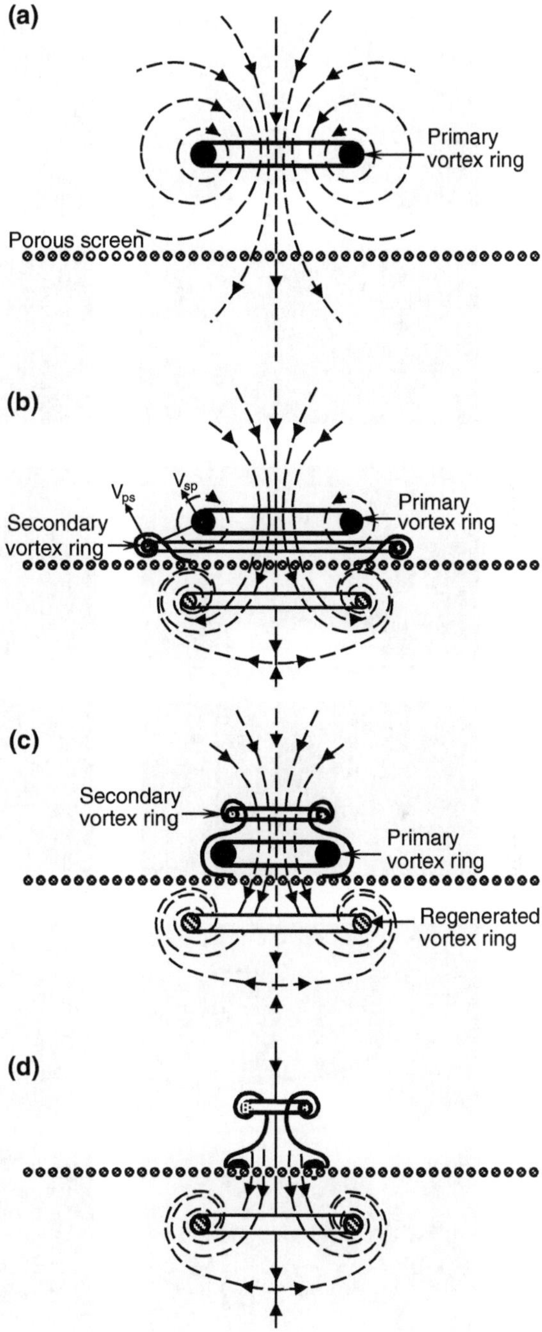


**Fig. 12** Adhikari and Lim [1] 's interpretation of Figs. 12 and 13 (Scenario 3). Time increases from (**a**) to (**d**). $V_{ps}$ = induced velocity of the primary vortex ring on the secondary vortex ring, $V_{sp}$ = induced velocity of the secondary vortex ring on the primary vortex ring. Only the induced velocity vectors on the left hand side of the vortices are shown in (**b**). *Single arrows* represent direction of flow field. Reproduced from Adhikari and Lim [1]

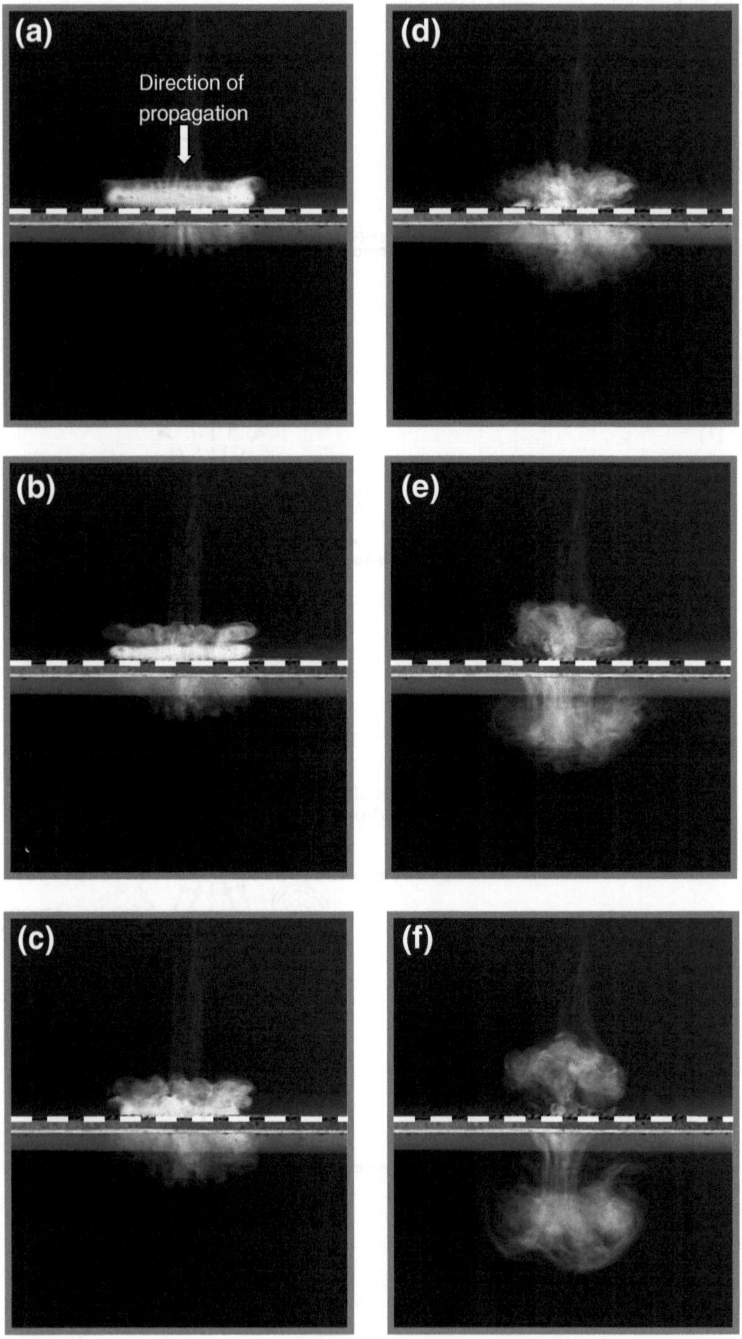


**Fig. 13** Impact of a vortex ring on a 62 % porosity screen for $Re_{\Gamma} = 4026$. Time increases from (**a**) to (**f**). $t^* = 0.0$ corresponds to the instance when the vortex ring is one ring diameter away from the wall. Reproduced from Adhikari and Lim [1]. **a** $t^* = 1.13$. **b** $t^* = 1.28$. **c** $t^* = 1.45$. **d** $t^* = 1.77$. **e** $t^* = 2.44$. **f** $t^* = 3.67$

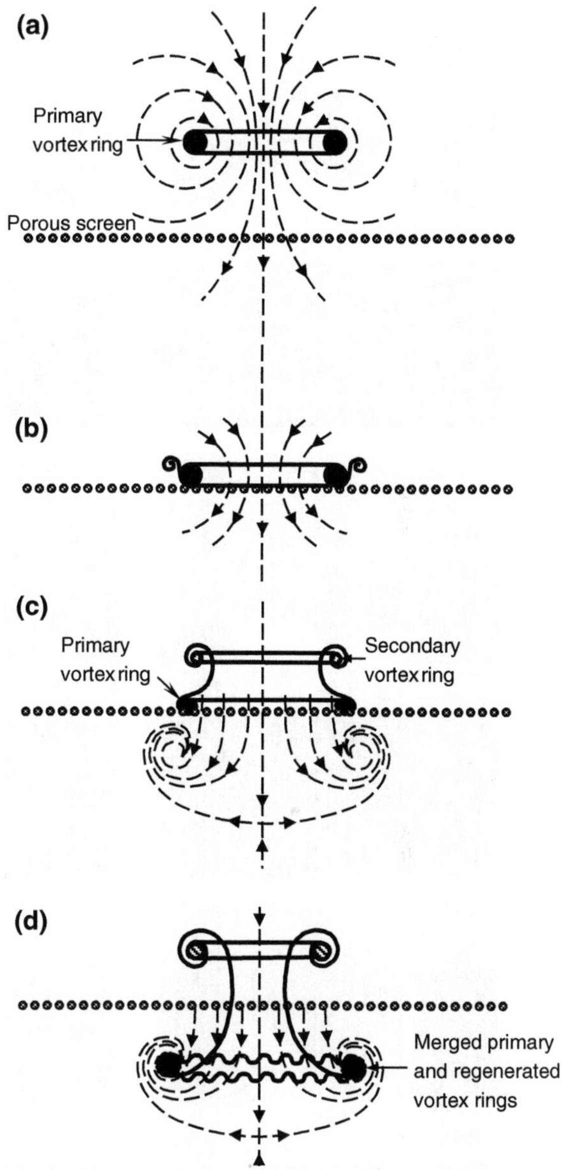


**Fig. 14** Adhikari and Lim [1] 's interpretation of Fig. 15 (Scenario 4). Time increases from (**a**) to (**d**). The passage of the primary vortex ring through the screen is via a possible mechanism of vortex stretching and reconnection as detailed in Fig. 17. Note that the wavy ring in (**d**) is only a conceptual representation and the bridges as indicated in Fig. 17 are not shown here to avoid complication to the figure. *Single arrows* represent direction of flow field. Reproduced from Adhikari and Lim [1]

that in scenario 4, except that higher Reynolds number vortex ring is able to pass through the screen quicker and emerges on the leeside of the screen as a modified ring.

For higher porosity screen, Adhikari and Lim [1] found that the same four scenarios exist except that each scenario occurs at lower Reynolds number (see for example Fig. 15).

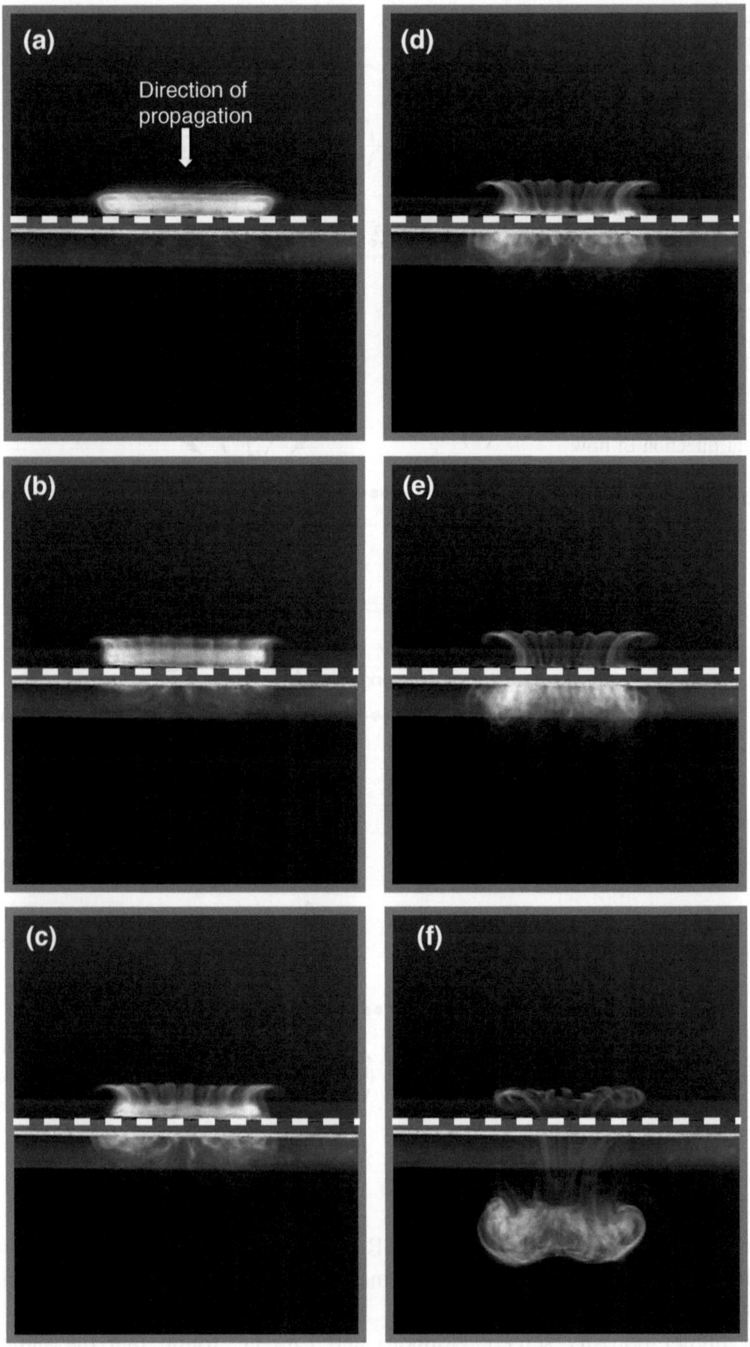


**Fig. 15** Impact of a vortex ring on a 81 % porosity screen for $Re_{\Gamma} = 1597$. Time increases from (**a**) to (**f**). $t^* = 0.0$ corresponds to the instance when the vortex ring is one diameter away from the wall. Reproduced from Adhikari and Lim [1]. **a** $t^* = 1.49$. **b** $t^* = 1.67$. **c** $t^* = 1.89$. **d** $t^* = 2.07$. **e** $t^* = 2.28$. **f** $t^* = 4.23$

## 4 Effect of Porosity

Along the same line of investigation as Adhikari and Lim [1], Naaktgeboren et al. [41] conducted flow visualization studies and DPIV measurements to investigate the effect of screen porosity on a vortex ring as it interacts with a porous screen. Their flow visualization is limited to only one screen with an open–area ratio of about 58 %, wire diameter of 0.267 mm and pore aspect ratio of 1.57, whereas their DPIV measurements covered a wider range of open-area ratios from 0.44 to 0.79 and same nominal wire diameter of 0.7 mm and pore aspect ratio of one (square mesh). The test conditions were stroke-to-diameter ratio ($L/D$) of 1.0 and 3.0 and nominal Reynolds numbers ($Re_o = U_oD/\nu$), based on jet exit velocity ($U_o$) and nozzle diameter ($D$), of 3000 and 6000. Define in terms of initial circulation of vortex ring, the Reynolds number ($Re_\Gamma = \Gamma/\nu$) for the case of $Re_o = 3000$ is $Re_\Gamma = 2500$ for $L/D = 1$ and $Re_\Gamma = 4800$ for $L/D = 3$. For the higher Reynolds number of $Re_o = 6000$, they are $Re_\Gamma = 5700$ and 10000 for $L/D = 1$ and 3, respectively. Unlike the porous screens used by Adhikari and Lim [1], which are specially fabricated so that the joints of the wires are fused together, the ones used by Naaktgeboren et al. [41] were purchased commercially with the wires woven similar that woven strings on a tennis racquet. It is not entirely clear whether the uneven surface affects the details of the interaction, but their flow visualization results appear to be broadly similar to that reported by Adhikari and Lim [1]. Quantitatively, their DPIV measurements show that the kinetic energy dissipation ($\Delta E$) and reduction in impulse ($\Delta I$) of a vortex ring after the interaction with a porous screen are dependent primarily on $\phi$ with a slight decrease as $L/D$ and/or $Re$ is increased. They further found that the kinetic energy dissipation ($\Delta E$) is strongly non-linear dependent in $\phi$ whereas reduction in impulse ($\Delta I$) is approximately linearly dependent in $\phi$.

## 5 Effect of Wire Diameter

Hrynuk et al. [28] extended the studies of Adhikari and Lim [1] and Naaktgeboren et al. [41] to include the effect of wire diameters on a vortex ring during the interaction with porous surfaces. The study was conducted using flow visualization technique at Reynolds number ($Re_m$) of the vortex ring (based on maximum ejection velocity ($V_m$) and nozzle diameter ($D_o$)) of 2300–4200. The construct of the porous surfaces is similar to that used by Naaktgeboren et al. [41] and consists of woven stainless steel wires with square meshes. The open area of the porous surface or porosity was kept relatively constant while the wire diameter was varied from 0.018 to 0.267 cm. The detailed experimental parameters are depicted in Table 1. The visual results obtained by Hrynuk et al. [28] are broadly similar to those reported by

**Table 1** Experimental parameters used by Hrynuk et al. [28]

| Mesh (wire per cm) | Porosity (%) | Wire diameter (cm) |
|---|---|---|
| 10.2 | 65 | 0.018 (fine) |
| 4.3 | 64.5 | 0.046 |
| 2.4 | 62.7 | 0.089 |
| 2.0 | 63.2 | 0.104 (medium) |
| 1.6 | 64 | 0.122 |
| 1.2 | 63.8 | 0.160 |
| 0.8 | 62.4 | 0.267 (Coarse) |

Adhikari and Lim [1] and Naaktgeboren et al. [41] with a few exceptions which are wire diameter dependent. Based on the latter, they classified the vortex ring behavior into several regimes. For small wire diameter surface (i.e. $d_w = 0.018$ cm), they found that the vortex ring that passed through the porous surface immediately reformed downstream. The transmitted ring was coherent and had a similar mean diameter as the primary vortex ring but lower circulation as is reflected in the slower propagation velocity (see Fig. 16). Their results also revealed the existence of secondary and tertiary vortex rings which orbited around the primary vortex ring before they merged to move away from the porous surface towards the vortex ring generator (see Fig. 16). With the wire diameter increased to $d_w = 0.046$ cm, the vortex ring reformed further downstream of the surface (see Fig. 17), and as the wire diameter was increased further (i.e. $d_w = 0.104$ cm), the transmitted vortex ring shows significant distortion downstream. It is not as coherent as in the case of small wire size and the ring also possess more fine-scale structures (see Fig. 17). For the coarse wire ($d_w = 0.160$ cm), the transmitted ring is so highly distorted that it lost its coherent and accompanied by significantly more fine-scale structures. Generally, the larger mesh size or wire diameter requires a longer distance for the transmitted vortex ring to reform, and there are also more fine-scale scale structures on the transmitted ring. In all cases, secondary vortex ring was generated upstream of the porous surface but the extent of vortex ring rebound and speed of convection away from the porous surface decreased with larger wire size.

Hrynuk et al. [28] proposed a criterion to determine whether a primary vortex ring would immediately reform and/or breakup after passing through a porous surface. The criteria is based on an interaction Reynolds number ($Re_i$), which they defined as

$$\mathrm{Re_i} = \frac{\mathrm{U_c d_w}}{\phi \nu}$$

where $\phi$ is the screen porosity, $U_c$ is pre-interaction propagation velocity of the vortex ring and $d_w$ is the wire diameter. To test their hypothesis, two sets of experiments were conducted using vortex rings with initial Reynolds numbers of

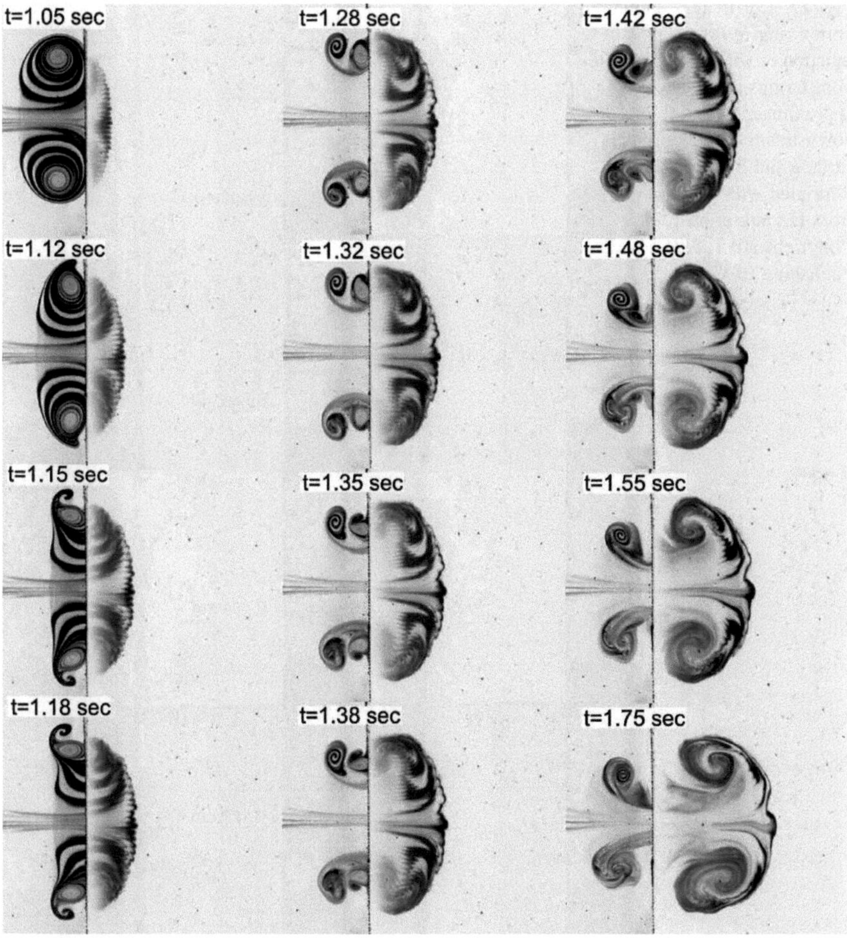


**Fig. 16** Detailed images of vortex ring interaction with fine screen. Reprinted with permission from Hrynuk et al. [28]. Copyright 2012, AIP Publishing LLC

($Re_m$) of 2300 and 4200, and the results are presented in Fig. 18 for different interaction Reynolds number ($Re_i$). Based on their findings, it appears that when $Re_i < 63$, the primary vortex ring, after passing through the porous surface, immediately reforms into coherent vortex ring. For $63 < Re_i < 79$, the reformation of the primary vortex ring is delayed downstream, which they attributed to the destabilization of the primary vortex ring caused by the generation of vortical sub-structures by the wires. For $90 < Re_i < 164$, the larger wire size causes the length-scale of vortical sub-structures to increase, resulting in partial reformation of the primary vortex ring. When $Re_i > 203$, the length-scale of the sub-structures is sufficiently large that they completely disrupt the vortex ring and the flow

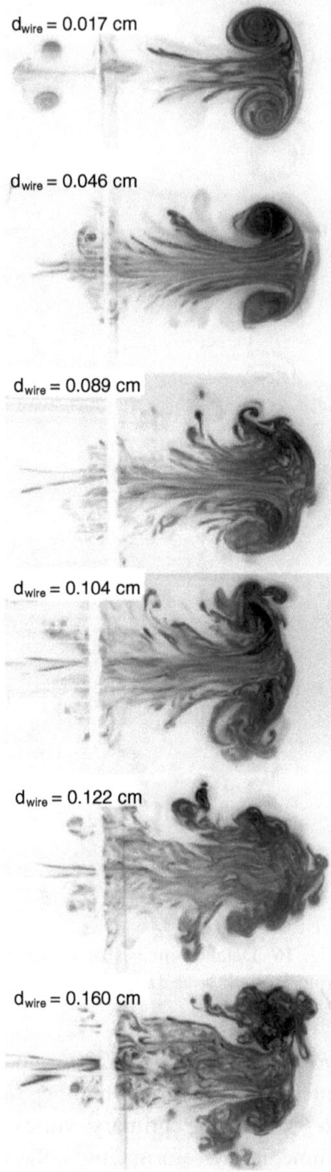


**Fig. 17** LIF images showing vortex ring reformation as a function of wire diameter. The vortex rings are located at approximately $3D_o$ downstream of the porous surface and $Re_m = 4200$. Reprinted with permission from Hrynuk et al. [28]. Copyright 2012, AIP Publishing LLC

eventually becomes turbulence. While the results of Hrynuk et al. [28] seems to suggest, at least qualitatively, the dependency of the state of the transmitted primary vortex ring on the interaction Reynolds number ($Re_i$), further quantitative investigation is certainly needed to establish the exact relationship.

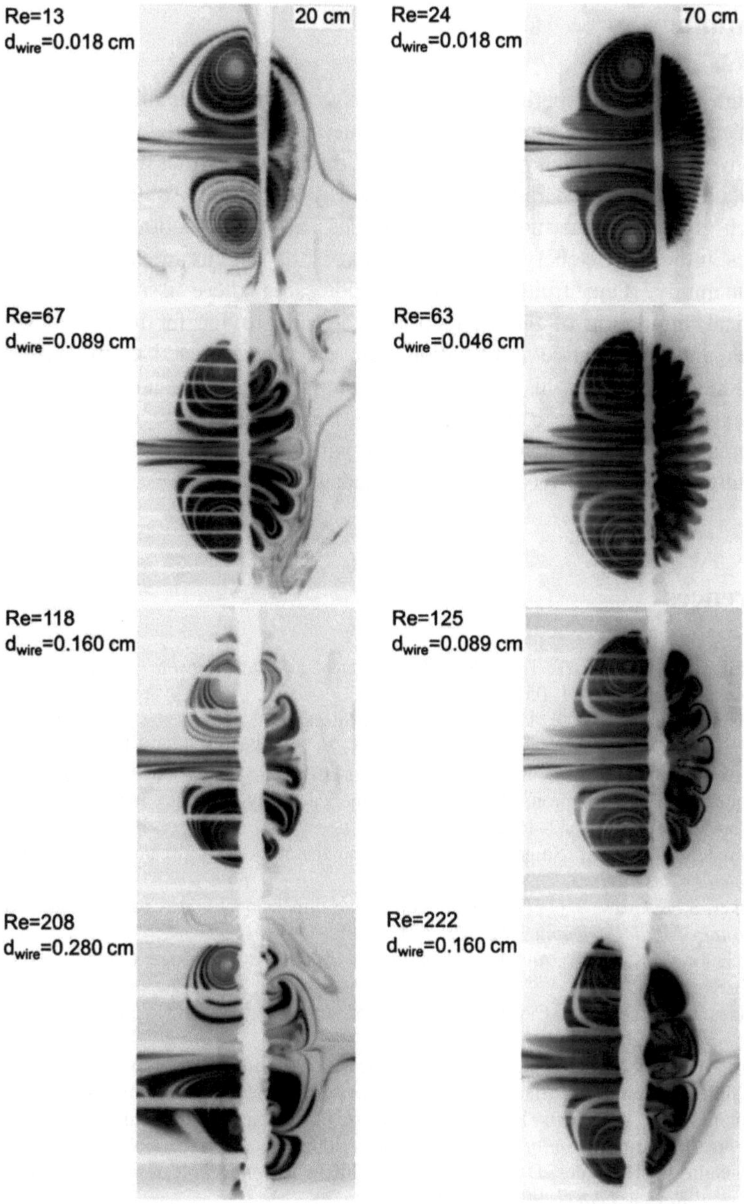


**Fig. 18** LIF images showing vortex ring reformation as a function of wire diameter. The vortex rings are located at approximately $3D_o$ downstream of the porous surface and $Re_m$ = 4200. Reprinted with permission from Hrynuk et al. [28]. Copyright 2012, AIP Publishing LLC

# 6 Summary

Although the above experimental studies have provided valuable insights into the dynamics of vortex ring/porous surface interaction, a shortcoming arising from these investigations is that porous screens of different construct were used. Hence, it is difficult to make meaningful comparison of the results obtained by different research groups. Furthermore, flow quantities such as evolution of core vorticity, changes in the kinetic energy and entropy across the screens are not easily measured experimentally. One fruitful mean of addressing these issues is to conduct numerical simulation of the interaction as it can provide far more detailed flow features and quantitative information than can be extracted from experiments. During the final editing this chapter, Cheng et al. [12] reported a numerical study on a vortex ring impacting a permeable wall over a wide range of parameters (i.e. open-area ratio, wire structure dimensions, wall thickness and Reynolds numbers) and addressed some of the above-mentioned issues.

# References

1. Adhikari, D., & Lim, T. T. (2009). The impact of a vortex ring on a porous screen. *Fluid Dynamics Research, 41*, 051404.
2. Akhmetov, D. G. (2009). *Vortex ring*. Berlin: Springer.
3. Akhmetov, D. G., Lugovtsov, B. A., & Tarasov, V. F. (1980). Extinguishing gas and oil well fires by means of vortex rings. *Combustion Explosion and Shock Waves, 16*(5), 490–494.
4. Ashurst, W. T., & Meiron, D. I. (1987). Numerical study of vortex reconnection. *Physical Review Letters, 58*, 1632–1635.
5. Auerbach, D. (1988). Some open questions on the flow of a circular vortex rings. *Fluid Dynamics Research, 3*, 209–213.
6. Barker, S., & Crow, S. (1977). The motion of two-dimensional vortex pairs in a ground effect. *Journal of Fluid Mechanics, 82*, 659–671.
7. Batchelor, G. K. (1974). *An introduction to fluid dynamics*. Cambridge: Cambridge University Press.
8. Boldes, U., & Ferreri, J. C. (1973). Behavior of vortex rings in the vicinity of a wall. *Physics of Fluids, 16*, 2005–2006.
9. Cerra, A. W., & Smith, C. R. (1983). Experimental observations of vortex ring interaction with the fluid adjacent to a surface. *Technical Report FM-4*. Bethlehem: Lehigh University.
10. Chahine, G. L., & Genoux, P. F. (1983). Collapse of a cavitating vortex ring. *Journal of Fluids Engineering Transaction of the ASME, 105*(4), 400–405.
11. Chatelain, P., Kivotides, D., & Leonard, A. (2003). Reconnection of colliding vortex rings. *Physical Review Letters, 90*(5), 054501.
12. Cheng, M., Lou, J., & Lim, T. T. (2014). A numerical study of a vortex ring impacting a permeable wall. *Physics of Fluids, 26*, 103602.
13. Cheng, M., Lou, J., & Luo, L. S. (2010). Numerical study of a vortex ring impacting a flat wall. *Journal of Fluid Mechanics, 660*, 430–455.
14. Chu, C. C., & Falco, R. E. (1998). Vortex ring and viscous wall layer interaction model and turbulence production process near wall. *Experiments in Fluids, 6*, 305–325.
15. Chu, C. C., Wang, C. T., & Chang, C. C. (1995). A vortex ring impinging on a solid plane surface-vortex structure and surface force. *Physics of Fluids A, 7*(60), 1391–1401.

16. Dabiri, J. O. (2009). Optimal vortex formation as a unifying principle in biological propulsion. *Annual Review of Fluid Mechanics, 42*, 17–33.
17. Dabiri, J. O., Colin, S. P., & Costello, J. H. (2006). Fast-swimming hydomedusae exploit velar kinematics to form an optimal vortex wake. *Journal of Experimental Biology, 209*, 2025–2033.
18. Dhanak, M. R., & Bernardinis, D. E. B. (1981). The evolution of an elliptic vortex ring. *Journal of Fluid Mechanics, 109*, 189–216.
19. Didden, N. (1979). On the formation of vortex rings: Rolling-up and production of circulation. *Journal of Applied Mechanics Physics (ZAMP), 30*, 101–116.
20. Doligalski, T. L., Smith, C. R., & Walker, J. D. A. (1994). Vortex interactions with walls. *Annual Review of Fluid Mechanics, 26*, 573–616.
21. Fabris, D., Liepmann, D., & Marcus, D. (1996). Quantitative experimental and numerical investigation of a vortex ring impinging on a wall. *Physics of Fluids, 8*, 2640–2649.
22. Fohl, T. (1967). Optimization of flow for forcing stack wastes to high altitudes. *Journal of Air Pollution Control Association, 17*, 730–733.
23. Fukumoto, Y., & Moffatt, H. K. (2008). Kinematic variational principle for motion of vortex ring. *Physics D, 237*, 2210–2217.
24. Gan, L., & Nickels, T. B. (2010). An experimental study of turbulent vortex rings during their early development. *Journal of Fluid Mechanics, 64*, 467–496.
25. Gharib, M., Rambod, E., & Shariff, K. (1998). A universal time scale for vortex ring formation. *Journal of Fluid Mechanics, 360*, 121–140.
26. Glezer, A., & Coles, D. (1990). An experimental study of a turbulent vortex ring. *Journal of Fluid Mechanics, 211*, 243–283.
27. Harvey, J., & Perry, F. (1971). Flowfield Produced by Trailing Vortices in the Vicinity of the Ground. *AIAA Journal, 9*(8), 1659–1660.
28. Hrynuk, J. T., Luipen, J. V., & Bohl, D. (2012). Flow visualization of a vortex ring interaction with porous surfaces. *Physics of Fluids, 24*, 037103.
29. Kambe, T., & Takao, T. (1971). Motion of distorted vortex rings. *Journal of the Physical Society of Japan, 31*, 591–599.
30. Kida, S., & Takaoka, M. (1994). Vortex reconnection. *Annual Review of Fluid Mechanics, 26*, 169–189.
31. Kida, S., Takaoka, M., & Hussain, F. (1991). Collision of two vortex rings. *Journal of Fluid Mechanics, 230*, 583–646.
32. Lim, T. T. (1997). On the role of Kelvin-Helmholtz-like instability in the formation of turbulent vortex rings. *Fluid Dynamics Research, 21*(1), 47–56.
33. Lim, T. T., & Nickels, T. B. (1995). Vortex rings. In S. I. Green (Ed.), *Fluid Vortices* (pp. 95–153). Netherlands: Springer.
34. Lim, T. T., Nickels, T. B., & Chong, M. S. (1991). A note on the cause of rebound in the head-on collision of a vortex ring with a wall. *Experiments in Fluids, 12*, 41–48.
35. Lundgren, T. S., & Mansour, N. N. (1991). Vortex ring bubbles. *Journal of Fluid Mechanics, 224*, 177–196.
36. Magarvey, R. H., & MacLatchy, C. S. (1964). The disintegration of vortex rings. *Canadian Journal of Physics, 42*, 678–689.
37. Marten, K., Shariff, K., Psarakos, S., & White, D. J. (1996). Ring bubbles of dolphins. *Scientific American, 275*, 83.
38. Maxworthy, T. (1972). The structure and stability of vortex rings. *Journal of Fluid Mechanics, 51*, 15–32.
39. Maxworthy, T. (1974). Turbulent vortex rings. *Journal of Fluid Mechanics, 64*, 227–240.
40. Melander, M., & Hussain, F. (1989). Cross-linking of two antiparallel vortex tubes. *Physics of Fluids A, 1*, 633–636.
41. Naaktgeboren, C., Krueger, P. S., & Lage, J. L. (2012). Interaction of a laminar vortex ring with a thin permeable screen. *Journal of Fluid Mechanics, 707*, 260–286.
42. Orlandi, P., & Verzicco, R. (1993). Vortex ring impinging on a walls: Axisymmetric and three-dimensional simulations. *Journal of Fluid Mechanics, 256*, 615–646.

43. Reynolds, O. (1876). On the resistance encountered by vortex rings and relation between the vortex rings and the streamlines of a disk. *Nature, 14*, 477–479.
44. Saffman, P. G. (1970). The velocity of viscous vortex rings. *Studies in Applied Mathematics, 49*, 71–380.
45. Saffman, P. G. (1978). The number of waves on unstable vortex rings. *Journal of Fluid Mechanics, 84*, 625–639.
46. Saffman, P. G. (1990). A model of vortex reconnection. *Journal of Fluid Mechanics, 212*, 395–402.
47. Saffman, P. G. (1992). *Vortex dynamics*. Cambridge: Cambridge University Press.
48. Shariff, K., & Leonard, A. (1992). Vortex Rings. *Annual Review of Fluid Mechanics, 24*, 235–279.
49. Thomson, W., & Kelvin, L. (1867). On vortex atoms. *Philosophical Magazine, 34*, 15–24.
50. Tsai, C. Y., & Widnall, S. E. (1976). The stability of short waves on a straight vortex filament in a weak externally imposed strain field. *Journal of Fluid Mechanics, 73*, 721–733.
51. Turner, J. S. (1960). Intermittent release of smoke from chimneys. *Journal of Mechanical Engineering Science, 2*, 97–100.
52. Walker, J. D. A., Smith, C. R., Cerra, A. W., & Doligalski, T. L. (1987). The impact of a vortex ring on a wall. *Journal of Fluid Mechanics, 181*, 99–140.
53. Widnall, S. E., Bliss, D. B., & Tsai, C. Y. (1974). The instability of short waves on a vortex ring. *Journal of Fluid Mechanics, 66*, 35–47.
54. Widnall, S. E., & Sullivan, J. P. (1973). On the stability of vortex rings". *Proceedings of the Royal Society of London A, 332*, 335–353.
55. Widnall, S. E., & Tsai, C. Y. (1977). The instability of a thin vortex ring of constant vorticity. *Philosophical Transactions of the Royal Society of London A, 287*, 273–305.
56. Yamada, H., Kohsaka, T., & Yamabe, H. (1982). Flowfield produced by a vortex ring near a plane wall. *Journal of the Physical Society of Japan, 51*, 1663–1670.

# Experimental Studies of the Near-Field Spatio-Temporal Evolution of Zero-Net-Mass-Flux (ZNMF) Jets

**J. Soria**

**Abstract** This chapter describes the results of an investigation of the near-field spatio-temporal evolution of a zero-net-mass-flux (ZNMF) jet at Re = 13,329 and a St = 0.03. This ZNMF jet has been found to be a turbulent jet in the far field (da C Andrade in On the circulations caused by the vibration of air in a tube, pp. 445–470, 1931 [6]). The experimental investigation used state-of-the art time resolved high spatial resolution 2-component 2-dimensional (2C-2D) particle image velocimetry (PIV). Advanced multigrid cross-correlation digital PIV (MCCDPIV) analysis has been employed to yield high spatial resolution with improved velocity uncertainty PIV measurements (Soria in Multigrid approach to cross-correlation digital PIV and HPIV analysis. World Scientific Publication, Hackensack, pp. 309–347, 2006 [25]). The measured time series of the (x, r) in-plane velocity fields and the azimuthal out-of-plane vorticity fields have been used to investigate the temporal spectra in the jet shear layer of this ZNMF jet. Triple decomposition has been employed to investigate the axial and radial distributions of the mean, phase correlated and turbulence intensities and Reynolds stress. The topology of the near field of this ZNMF jet and its phase evolution have been investigated via the streamline pattern of the phase-average velocity fields and the phase correlated azimuthal vorticity.

**Keywords** Zero-net-mass flux jets · Time-resolved PIV · Topology · Self-similarity

J. Soria (✉)
Laboratory for Turbulence Research in Aerospace and Combustion,
Department of Mechanical and Aerospace Engineering, Monash University,
Melbourne, VIC 3800, Australia
e-mail: julio.soria@monash.edu

J. Soria
Department of Aeronautical Engineering, King Abdulaziz University,
Jeddah, Kingdom of Saudi Arabia



D.T.H. New and S.C.M. Yu (eds.), *Vortex Rings and Jets*,
Fluid Mechanics and Its Applications 111, DOI 10.1007/978-981-287-396-5_3

## 1 Introduction

The zero-net-mass-flux (ZNMF) jet is typically generated by a sinusoidal oscillating membrane or piston in a cylinder to alternatively force fluid from the piston-cylinder cavity through an orifice into the external quiescent fluid and back into the cavity. During the ejecting stroke (or phase), the ejected fluid separates at the sharp edges of the orifice and rolls up to form a vortex pair or ring. When the membrane or piston begin its suction stroke (or phase), the vortex pair or ring can find itself far enough downstream from the orifice that it keeps propagating downstream due to its self-induced velocity, without being sucked back through the orifice into the ZNMF jet cavity. This is followed by subsequent propagating vortex pairs or rings coalescing into a jet with net axial momentum transfer to the surrounding fluid [4].

This type of streaming flow, which is a consequence of an oscillating generation mechanism, has been known and studied for many years, in particular, due to acoustic sources as has been reviewed by [16]. One of the earliest studies related to ZNMF jets due to acoustic forcing is due to [14], a study, which was preceding many years before by the investigation of [3], who showed that steady fluid streaming motion around solid bodies can be established by sound waves in a tube. However, it has only been in recent years that the formation and evolution of free single ZNMF jets have been investigated experimentally using detailed measurements in the seminal works of [20] for plane jets and [4] for the round jet. This latter study is interesting because it is the only study using detailed particle image velocimetry (PIV) measurements where a direct comparison between the turbulent ZNMF jet and the equivalent turbulent continuous jet at the same Reynolds number formed an integral part of the characterisation of the turbulent ZNMF jet flow.

ZNMF jet in cross-flow have also received a fair attention due to their relevance in a range of flow control applications by [2, 9, 17, 18, 21, 28]. The detailed spatiotemporal structure of the round ZNMF jet in cross-flow was measured using phase-average PIV by [11], while the mean structure of elliptic ZNMF jet in cross-flow was investigated using Planar-laser-induced fluorescence (PLIF) by [1]. PLIF measurements were also employed to investigate the mean passive scalar field in round zero-net-mass-flux jets in cross-flow by [10].

In the present study the next level of PIV technology is employed to investigate the spatio-temporal structure of the near field of the free round ZNMF jet at a momentum Reynolds number of $Re$ = 13,329 and a $St$ = 0.03, which yields a turbulent ZNMF jet in the far field [6]. Specifically, time-resolved high-spatial resolution 2-component 2-dimensional (2C-2D) PIV is used to measure the instantaneous axial and radial velocity components and the azimuthal vorticity component with a bandwidth of 312 Hz. This has allowed the spectral analysis and the development of the temporal axial velocity spectra along the jet shear layer using the measured time-series of the axial velocity fields.

Triple decomposition of the measured instantaneous velocity and vorticity field was carried out to investigate the spatial structure of the mean, phase correlated and turbulent components, their intensities and the Reynolds stress. Their radial and axial distribution were also investigated, as was the evolution of the phase-correlated components as a function of the phase angle of the periodic motion of the piston. The topology of the phase-average velocity field was investigated through an analysis of the streamline pattern and the phase correlated vorticity as a function of the phase angle of the periodic motion of the piston.

Section 2 presents the geometry, the governing equations and defines pertinent parameters. This section also proves via an integral momentum analysis of the mean axial momentum equation that the mean axial momentum flow crossing the ZNMF jet orifice plane is equivalent to the momentum flow imparted by the oscillating piston on the fluid. It is worth noting that the mean axial momentum flow is independent of the frequency of the ZNMF jet generation mechanism. A natural definition of the appropriate Reynolds number (*Re*) and Strouhal number (*St*) for the ZNMF jet via the momentum velocity ($U_o$) as a characteristic velocity and the orifice diameter ($D_o$) as the characteristic length scale follows immediately from this analysis. Note that in addition to lengths being scaled by $D_o$ in the results presented in this chapter, $r_{1/2}(0) = D_o/2$ is also used for this purpose. Section 2 concludes with the presentation of the concepts underlying triple decomposition and its mathematical description.

Section 3 describes all aspects of the experimental apparatus and methods, starting with a detailed description of the ZNMF jet facility. This is followed by a description of the PIV experimental equipment, the optical and imaging arrangement and procedure. Information on the image pre-processing and the MCCDPIV analysis, the vorticity calculation method and uncertainty estimates of the velocity and vorticity measurements concludes this section.

The results of the measurements and discussion of them are presented in Sect. 4, beginning with the spectral analysis of the axial velocity, followed by the triple decomposition of the velocity and vorticity fields and concluding with an analysis of the topology of the phase correlated flow of the ZNMF jet. The chapter comes to a close with the conclusions given in Sect. 5.

## 2 Theoretical Considerations and Parameterisation

### *2.1 Geometry and Governing Equations*

The geometry and the pertinent parameters of the round ZNMF jet are shown in Fig. 1. A cylindrical coordinate system $(x, r, \theta)$ is the appropriate coordinate system to describe the round ZNMF jet flow, with the origin of the coordinate system located at the centre of the orifice of the ZNMF jet as shown in Fig. 1. Since the measurements described in this chapter are made in the plane that includes the semi-infinite half-planes defined by $\theta = 0$ and $\theta = \pi$, the following simplified nomenclature will be adopted: the coordinate $(x, r)$ corresponding to the semi-infinite half plane $(x, r)$ is implied when $r \geq 0$ while the semi-infinite half plane $(x, r, \pi)$ is

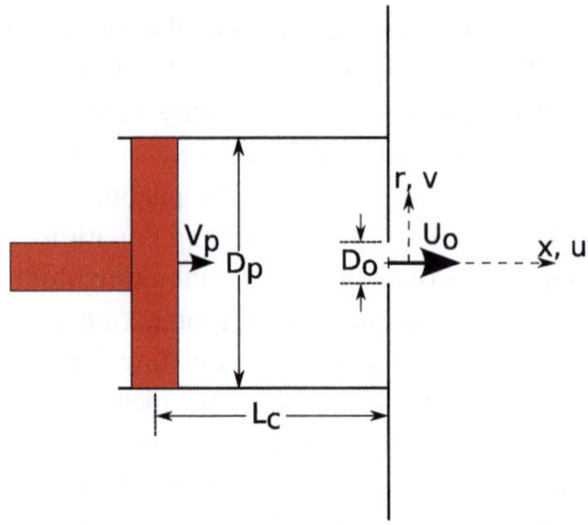


**Fig. 1** Round ZNMF jet geometry and the pertinent parameters governing this flow: $V_p$—piston velocity, $D_p$—piston diameter, $L_c$—cavity length, $D_o$—orifice diameter, $U_o$—momentum velocity at the orifice, $x$—axial direction, $r$—radial direction, $u$—axial velocity component and $v$—radial velocity component

understood when $r < 0$. The current experimental investigation employed 2C-2D PIV which provides a measurement of the axial and radial velocity components $(u, v)$ and the azimuthal component of vorticity $\omega_\theta$. The azimuthal component of velocity $w$ and axial and radial components of vorticity, $\omega_x$ and $\omega_r$ respectively are not available from these measurements. Given the axisymmetric nature of the round ZNMF jet geometry, the mean flow is expected to be nominally axisymmetric and swirl free and therefore the 2C-2D PIV should in principle provide sufficient information to investigate the statistics of the velocity field and the phase-correlated spatio-temporal structure of the near-field of the ZNMF jet.

The governing equations for the ZNMF jet are the incompressible Navier-Stokes equations which in cylindrical coordinates are given by

$$\begin{aligned}
&\frac{\partial u}{\partial x}+\frac{1}{r}\frac{\partial (rv)}{\partial r}+\frac{1}{r}\frac{\partial w}{\partial \theta}=0\\
&\frac{\partial u}{\partial t}+u\frac{\partial u}{\partial x}+v\frac{\partial u}{\partial r}+\frac{w}{r}\frac{\partial u}{\partial \theta}=-\frac{1}{\rho}\frac{\partial p}{\partial x}+\nu\left[\frac{\partial^2 u}{\partial x^2}+\frac{1}{r}\frac{\partial}{\partial r}\left(r\frac{\partial u}{\partial r}\right)+\frac{1}{r^2}\frac{\partial^2 u}{\partial \theta^2}\right]\\
&\frac{\partial v}{\partial t}+u\frac{\partial v}{\partial x}+v\frac{\partial v}{\partial r}+\frac{w}{r}\frac{\partial v}{\partial \theta}-\frac{w^2}{r}=-\frac{1}{\rho}\frac{\partial p}{\partial r}+\nu\left[\frac{\partial^2 v}{\partial x^2}+\frac{1}{r}\frac{\partial}{\partial r}\left(r\frac{\partial v}{\partial r}\right)-\frac{v}{r^2}+\frac{1}{r^2}\frac{\partial^2 v}{\partial \theta^2}-\frac{2}{r^2}\frac{\partial w}{\partial \theta}\right]\\
&\frac{\partial w}{\partial t}+u\frac{\partial w}{\partial x}+v\frac{\partial w}{\partial r}+\frac{w}{r}\frac{\partial w}{\partial \theta}-\frac{vw}{r}=-\frac{1}{r\rho}\frac{\partial p}{\partial \theta}+\nu\left[\frac{\partial^2 w}{\partial x^2}+\frac{1}{r}\frac{\partial}{\partial r}\left(r\frac{\partial w}{\partial r}\right)-\frac{w}{r^2}+\frac{1}{r^2}\frac{\partial^2 w}{\partial \theta^2}-\frac{2}{r^2}\frac{\partial v}{\partial \theta}\right]
\end{aligned} \tag{1}$$

and represent the conservation of mass and the conservation of linear momentum principle in the coordinate directions $x$, $r$ and $\theta$ respectively.

Integrating the axial x-direction linear momentum equation of Eq. 1 in the piston cylinder cavity which is bounded by the piston and the orifice, using Gauss's divergence theorem and integrating over one period of the piston oscillation, $T$, it can be shown that for a rigid body oscillatory piston velocity with a velocity amplitude $\hat{V}_p$ given by

$$V_p(t) = \hat{V}_p \sin\left(\frac{2\pi}{T}t\right), \tag{2}$$

the average axial momentum flow across the orifice due the piston oscillation is given by

$$A_o U_o^2 = \frac{A_p \hat{V}_p^2}{2}. \tag{3}$$

In this equation $U_o$ is defined as the momentum velocity, which is given by

$$U_o = \sqrt{\frac{1}{A_o T}\int_0^T \left[\int_0^{D_o/2} \left(\int_0^{2\pi} 2\pi r u(0, r, \theta) d\theta\right) dr\right] dt} \tag{4}$$

The piston area is given by $A_p = \pi D_p^2/4$ while the orifice area is given by $A_o = \pi D_o^2/4$. Using these relationships it can be shown using Eq. 2 that the momentum velocity, $U_o$, given formally by Eq. 4 can also be calculated from the peak velocity of the piston and the diameters of the piston and orifice using the relationship

$$U_o = \frac{D_p}{D_o}\frac{\hat{V}_p}{\sqrt{2}}. \tag{5}$$

Alternatively using the root-mean-square piston velocity, $V_{p,rms}$, $U_o$ is also given by

$$U_o = \frac{D_p}{D_o} V_{p,rms}. \tag{6}$$

The momentum velocity, $U_o$, is the characteristic velocity of the ZNMF jet as it is related to the average momentum flow from the jet generation mechanism, i.e. it is directly related to the linear momentum source, which is also the definition of a jet source—a compact linear momentum source. This definition of the characteristic velocity of a ZNMF jet is consistent with a continuous jet as it provides the correct velocity scale also for a continuous jet [4]. We are now in a position to define the Reynolds number, $Re$, of a ZNMF jet as follows,

$$Re = \frac{U_o D_o}{\nu}, \tag{7}$$

and the Strouhal number, *St*, as,

$$St = \frac{D_o f}{U_o}, \tag{8}$$

where $f$ is the temporal frequency of the piston oscillation given by $f = 1/T$.

Furthermore, noting that

$$U_o = \pi\sqrt{2}\frac{D_p f a}{D_o} \tag{9}$$

and substituting for $U_o$ in Eqs. 7 and 8 and using $V_p = 2\pi fa$, where $a$ is the piston oscillation displacement amplitude, yields

$$Re = \pi\sqrt{2}\left(\frac{D_p f a}{\nu}\right) \tag{10}$$

$$St = \frac{1}{\pi\sqrt{2}}\left(\frac{D_o}{D_p}\frac{D_o}{a}\right) \tag{11}$$

Equations 10–11 show interestingly that the *St* is independent of the oscillating ZNMF driving frequency *f*, and only depends on the geometric ratio of the piston diameter to the office diameter and the ratio of the piston oscillating amplitude to the orifice diameter. In contrast *Re* does not depend on the initial jet length scale represented by the orifice diameter $D_o$, but only depends on the piston diameter, its oscillating amplitude and frequency. This permits the independent variation of these non-dimensional numbers.

The product of the *Re* and *St*, which is independent of the piston diameter and oscillating amplitude, can also be interpreted as a Reynolds number, $Re_f$, by defining the velocity scale:

$$U_f \equiv D_o f \tag{12}$$

which can be thought of as an oscillating velocity scale and thus allows us to write,

$$Re_f = \frac{U_f D_o}{\nu}. \tag{13}$$

Using the *Re* definition the governing equations can now be rewritten as:

$$\begin{aligned}
&\frac{\partial u}{\partial x} + \frac{1}{r}\frac{\partial (rv)}{\partial r} + \frac{1}{r}\frac{\partial w}{\partial \theta} = 0 \\
&\frac{\partial u}{\partial t} + u\frac{\partial u}{\partial x} + v\frac{\partial u}{\partial r} + \frac{w}{r}\frac{\partial u}{\partial \theta} = -\frac{\partial p}{\partial x} + \frac{1}{Re}\left[\frac{\partial^2 u}{\partial x^2} + \frac{1}{r}\frac{\partial}{\partial r}\left(r\frac{\partial u}{\partial r}\right) + \frac{1}{r^2}\frac{\partial^2 u}{\partial \theta^2}\right] \\
&\frac{\partial v}{\partial t} + u\frac{\partial v}{\partial x} + v\frac{\partial v}{\partial r} + \frac{w}{r}\frac{\partial v}{\partial \theta} - \frac{w^2}{r} = -\frac{\partial p}{\partial r} + \frac{1}{Re}\left[\frac{\partial^2 v}{\partial x^2} + \frac{1}{r}\frac{\partial}{\partial r}\left(r\frac{\partial v}{\partial r}\right) - \frac{v}{r^2} + \frac{1}{r^2}\frac{\partial^2 v}{\partial \theta^2} - \frac{2}{r^2}\frac{\partial w}{\partial \theta}\right] \\
&\frac{\partial w}{\partial t} + u\frac{\partial w}{\partial x} + v\frac{\partial w}{\partial r} + \frac{w}{r}\frac{\partial w}{\partial \theta} - \frac{vw}{r} = -\frac{1}{r}\frac{\partial p}{\partial \theta} + \frac{1}{Re}\left[\frac{\partial^2 w}{\partial x^2} + \frac{1}{r}\frac{\partial}{\partial r}\left(r\frac{\partial w}{\partial r}\right) - \frac{w}{r^2} + \frac{1}{r^2}\frac{\partial^2 w}{\partial \theta^2} - \frac{2}{r^2}\frac{\partial v}{\partial \theta}\right]
\end{aligned} \tag{14}$$

where it is now understood that the independent spatial variables are non-dimensionalized by the orifice diameter, $D_o$, the dependent velocities by the momentum velocity, $U_o$, and the pressure by $\rho U_o^2$.

## 2.2 Triple Decomposition

In the context of the ZNMF jet it is appropriate to decompose the flow field using the triple decomposition [13, 15],

$$u_i(\mathbf{x},t) = U_i(\mathbf{x}) + \tilde{u}_i(\mathbf{x},t) + u_i'(\mathbf{x},t) \tag{15}$$

where $u_1 = u$, $u_2 = v$, $u_3 = w$ and $\mathbf{x} = (x, r, \theta)$. $U_i$ is the time average, $\tilde{u}_i$ is the phase-correlated organized contribution to the velocity and $u_i'$ is the turbulence velocity fluctuation. The time average is defined as

$$U_i(\mathbf{x}) = \lim_{T_a \to \infty} \frac{1}{T_a} \int_{t_o}^{t_o+T_a} u_i(\mathbf{x},t)dt, \tag{16}$$

while the phase-average is defined by

$$\langle u_i(\mathbf{x},t) \rangle = \lim_{N \to \infty} \frac{1}{N} \sum_{n=0}^{N-1} u_i(\mathbf{x},t+nT) \tag{17}$$

where $T$ is the period of the organized contribution, which in this study corresponds to the ZNMF jet oscillation period. The phase-average velocity is thus the average at any point in space of the velocity that correspond to a particular phase $\phi$ in the cycle of the oscillatory ZNMF jet generation. The organized oscillatory component of the velocity is then given by

$$\tilde{u}_i(\mathbf{x},t) = \langle u_i(\mathbf{x},t) \rangle - U_i(\mathbf{x}) \tag{18}$$

and the turbulence velocity oscillation is given by

$$u_i'(\mathbf{x},t) = u_i(\mathbf{x},t) - \langle u_i(\mathbf{x},t) \rangle. \tag{19}$$

It can be shown that on the average the organized velocity, $\tilde{u}_i(\mathbf{x},t)$ and the turbulence, $u_i'(\mathbf{x},t)$ are uncorrelated [13].

# 3 Experimental Details

This section of the chapter provides details of the experimental facility, the PIV experimental set-up and the PIV analysis.

## *3.1 The LTRAC ZNMF Jet Facility*

The ZNMF jet experiments were carried out in the LTRAC ZNMF jet facility at Monash University which was developed over a number of years and has been previously described in [4, 5]. Here, the facility is only briefly described for completeness and in as far as the driving mechanism differs from previous studies. The LTRAC ZNMF jet facility consists of an acrylic tank 1000 mm long, 500 mm wide and 500 mm deep as shown in plan view in Fig. 2. It is filled with filtered water to the top of the tank. In order to avoid air within the facility, the tank has a riser tube with an inner diameter of 56.5 mm located on the Perspex roof at the far end wall of the tank opposite to where the jet orifice is located. The riser tube also serves the purpose of alleviating the water mass injected and extraction during the ZNMF jets experiments.

For each experiment, the ZNMF jet flow was generated by operating a piston within a circular tube with an inner diameter $D_p$ = 50 mm which discharges water through an orifice plate with a diameter $D_o$ = 10 mm and a thickness of 2 mm as

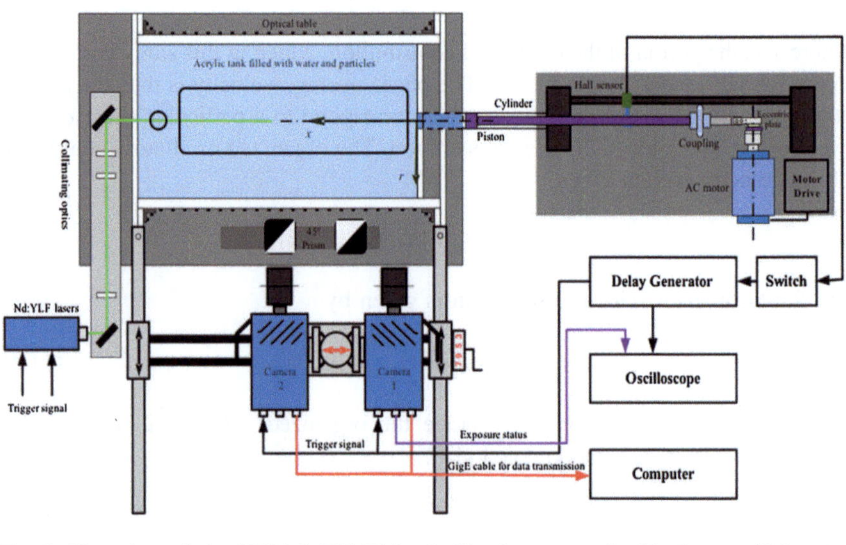


**Fig. 2** Plan view of the LTRAC ZNMF jet facility instrumented with the two high-speed 2016 × 2016 $px^2$ CMOS arrays and the high repetition high energy Nd:YLF twin cavity laser in their optical arrangement

shown in Fig. 2. For the present experiments the piston was located at a nominally mean distance of 3 piston diameters from the orifice, i.e. $L_c = 3D_p$. The orifice and circular tube are positioned in the centre of one of the end wall of the tank. The piston is connected via a linear rod which is supported by two linear bearings through a flexible coupling and an eccentric plate to an AC motor. The rotational speed of the AC motor is controlled through an AC motor drive. The desired Reynolds and Strouhal numbers characterizing the ZNMF jet are set by adjusting the eccentricity radius on the eccentric plate and the AC motor drive frequency for the fixed orifice diameter used in the present investigation. For the present investigation the forcing frequency was set at 4 Hz, which produced a ZNMF jet with a $Re$ = 13,329 and a $St$ = 0.03. This jet is turbulent in the far-field according to the classifications by [4] and is in the neighbourhood of the *Re*–*St* combination used in that study.

## 3.2 PIV Experimental Set-up and Procedures

Prior to each PIV experiment, the water in the tank was seeded with hollow glass spheres with nominal diameter of 11 μm and a density of 1100 $kg/m^3$ (Potter spherical). These tracer particles have an estimated relaxation time of 7.39 μs and are therefore expected to faithfully follow the local water velocity. The particles were illuminated using a Quantronix Nd:YLF twin cavity laser system producing 527 nm wavelength laser pulses at up to 100 W. The laser beam was shaped into a 1 mm thick laser sheet using appropriate spherical and cylindrical lenses and was aligned vertically along the jet axis as shown in Fig. 2. The exposure of the single-exposed PIV images was controlled by the duration of the laser pulse which is typically less than 210 ns with the time between the PIV image pair exposures set to 20 μs for these experiments and the time between image pairs set to 1.603 ms.

The scattered laser light from the particles was recorded using two synchronized PCO-DIMAX high speed CMOS cameras with an array size 2016 × 2016 $px^2$. Each camera was fitted with a 200 mm Micro Nikkor lens set at a f-stop number of 8. The optical arrangement of these two cameras was such that there was a small overlap in the axial imaging domain of each camera with the camera imaging closest to the ZNMF jet orifice using a reproduction ratio of 1.35 and imaging a domain that extended nominally up to $3.05D_o$ from the orifice, while the second camera employed a reproduction ratio of 2.75 which represents an imaging domain of $6.15D_o$ in the axial direction. The two cameras were operated in the "double shutter" mode which allowed for a maximum image pair acquisition rate of 624 per second. Five sets of 3121 image pairs were acquired. The PIV single-exposed image pair acquisition of each set was triggered using a Hall based device connected to the drive shaft of the piston. The TTL trigger pulse was sent to the cameras 41.667 ms before the piston reached its maximum suction stroke location. Each set corresponds to 20 ZNMF jet periods with each period represented by 156 2C-2D instantaneous velocity fields deduced from the 156 single-exposed PIV image pairs.

In this paper only the results from the camera that acquired the 2C-2D velocity field in the near field of the ZNMF jet are presented. The spatial resolution of this camera is estimated to be $14.99 \pm 0.09$ μm/px based on a calibration grid. These five sets of image data for this one camera corresponds to 0.25 TB of data storage.

## 3.3 Image Pre-processing and PIV Analysis

The raw PIV image data was pre-processed prior to cross-correlation PIV analysis. The average image of the corresponding image of the image pair was subtracted and the images were cropped to remove solid wall region from the image. The images were rotated by $-0.37°$ to correct for a small misalignment of the camera with respect to the jet axis. The single exposed image pairs were analyzed using the multi-grid cross-correlation digital PIV (MCCDPIV) algorithm described in [27], which has its origin in an iterative and adaptive cross-correlation algorithm introduced by [22–24]. Details of the performance, precision and experimental uncertainty of the MCCDPIV algorithm with applications to the analysis of single exposed PIV and holographic PIV (HPIV) images have been reported in [7, 25], respectively. The present single-exposed image acquisition experiments were designed for a two-pass MCCDPIV analysis. The first pass used an IW = 48 px ($0.072D_o$), while the second pass used an IW = 16 px ($0.024D_o$) with discrete IW offset to minimize the measurement uncertainty [29]. The sampling spacing between the centres of the interrogation windows was set to 8 px ($0.012D_o$).

The MCCDPIV algorithm incorporates the local cross-correlation function multiplication method introduced by [12] to improve the search for the location of the maximum value of the cross-correlation function. A two dimensional Gaussian function model was used to find, in a least square sense, the two-dimensional cross-correlation function peak to sub-pixel accuracy [22]. Each MCCDPIV velocity field was subsequently validated using the threshold cross-correlation peak value criterion ($\rho > 0.6$) and the dynamic mean value operator test described in [19]. The tests were applied in the specified order. Following data validation, the in-plane velocity components $(u, v)$ in the $(x, r)$ coordinate directions respectively were computed by taking the optical magnification into account and by dividing the measured MCCDPIV displacement in each interrogation window by the time between the exposures of the image pair. The uncertainty relative to the maximum velocity in the velocity components at the 95 % confidence level for these measurements is 0.03 % [24]. The version of the current MCCDPIV algorithm uses MPI to implement parallelization of the MCCDPIV analysis. The MCCDPIV analysis of the results presented in this paper was carried out using 36 CPU cores on the NCI Massive HPC.

The out-of-plane vorticity, $\omega_\theta$, was calculated from the MCCDPIV velocity field measurements using a local least-squares fit procedure to the velocity field, followed by analytic differentiation using the relationship

$$\omega_\theta = \partial v/\partial x - \partial u/\partial r. \tag{20}$$

A thirteen point, two-dimensional, local fit to the velocity field data was used [24, 26]. This calculation is an approximation that introduces a bias and a random error into the computed vorticity. These errors have been investigated and discussed in [8, 26]. For a vorticity distribution with a characteristic length scale of $0.191D_o$, the bias error is estimated as −0.3 % and the random error is estimated as ±2.4 % at the 95 % confidence level, while for a vorticity distribution with a characteristic length scale of $0.048D_o$, the bias error is estimated as −4.9 % and the random error is estimated as ±0.6 % at the 95 % confidence level.

## 4 Results and Discussion

In this section the experimental results are presented and discussed starting with the temporal spectra of the axial velocity. This is followed by a description of the statistics of the velocity and azimuthal vorticity from the triple decomposition of the 2C-2D velocity fields. Finally the evolution of the phase-average structure of the near field of this ZNMF jet is investigated by considering the streamline pattern and the phase-correlated azimuthal vorticity distribution.

### *4.1 Spectra of the ZNMF Jet*

The time-resolved PIV measurements have been used to extract the spatial evolution of the auto-spectrum of the different kinematic quantities. Figure 3 presents the average auto-spectra of the axial velocity fluctuations at a number of selected locations corresponding to: (a) approximately the centreline of the ZNMF jet orifice, (b) the edge of the office but approximately half an orifice diameter downstream, (c) following the edge of the orifice but approximately $D_o$ downstream and (d) a downstream location that is $2.5D_o$ downstream from the orifice but still at a distance from the jet axis corresponding to the office edge.

The spectra in Fig. 3b–d can be considered as moving along the mean shear layer of the ZNMF jet. At the orifice axis the dominant peak in the spectrum is at the forcing frequency of 4 Hz, followed by a spectral peak at the 3rd harmonic of the driving frequency located at 12 Hz and smaller amplitude harmonics. At the location along the shear layer nearest to the ZNMF jet orifice a similar spectra is found, but at this location a more noticeable 2nd harmonic located at 8 Hz is also observed, although the 3rd harmonic is still larger. There is also a dominant peak at 32 and 36 Hz which has the same energy as the 2nd and 3rd harmonics respectively. Further downstream at $x = D_o$ harmonics are found to increase in energy while at $x = 2.5D_o$ sum and differences of the harmonics start developing. At $x = D_o$, the 3rd harmonic

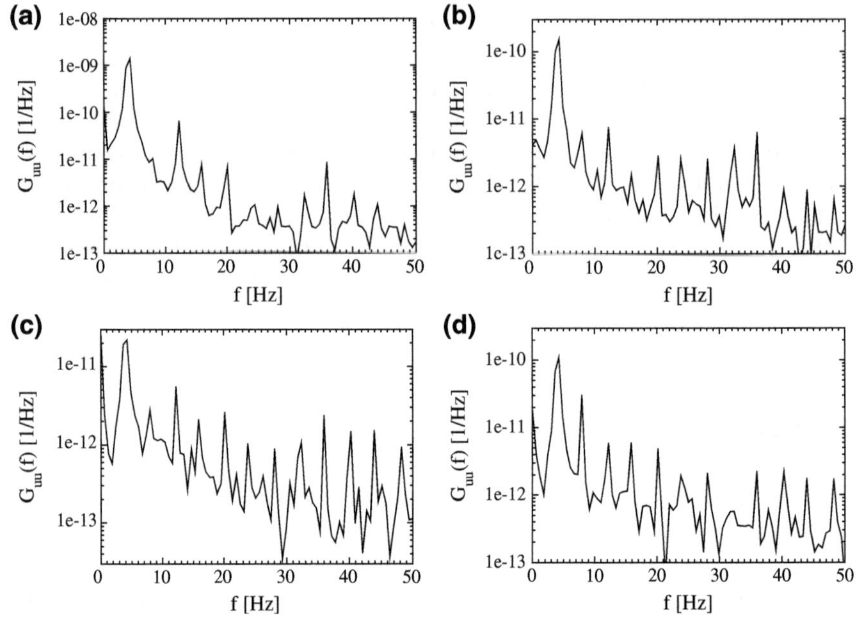


**Fig. 3** Average axial velocity spectra at a number of locations in the near field of the ZNMF jet. **a** $\frac{x}{D_o} \approx 0$, $\frac{r}{D_o} \approx 0$. **b** $\frac{x}{D_o} \approx 0.4$, $\frac{r}{D_o} \approx 0.5$. **c** $\frac{x}{D_o} \approx 1$, $\frac{r}{D_o} \approx 0.6$. **d** $\frac{x}{D_o} \approx 2.5$, $\frac{r}{D_o} \approx 0.6$

component has more energy than the second but by $x = 2.5D_o$ this has switched with the 2nd harmonic containing more energy that the 3rd. Furthermore, at this location it is found that the energy content of the harmonics decays on the average from the fundamental driving frequency, which is still the dominant frequency.

## *4.2 Triple Decomposition of the Velocity and Vorticity Fields*

The 2C-2D velocity field, $(u, v)$, is triple decomposed following the definition given in Eq. (15). The vorticity $\omega_z$ has an analogous triple decomposition. Figure 4 shows the mean velocity field and mean streamline pattern for the domain of the ZNMF jet of interest. These measurements have been used to extract the variation with axial direction, $x$, of the mean axial centreline velocity, $U_c(x)$ and the half width of the mean axial velocity profile of the ZNMF jet, $r_{1/2}(x)$, which is defined as the radius where the mean axial velocity has decayed from its centreline value by half, as shown in Fig. 5. $r_{1/2}(x)$ is found to decrease slightly from its value of $D_o$ at the orifice reaching a minimum at $x/r_{1/2}(0) \approx 1$. From this point onwards, $r_{1/2}(x)$ increases slowly reaching a value slightly larger than the orifice radius at $x \approx 6r_{1/2}(0)$.

The mean axial centreline velocity is also plotted as its inverse relative to the momentum velocity $U_o$ in Fig. 5. This inverse is observed to decrease very quickly

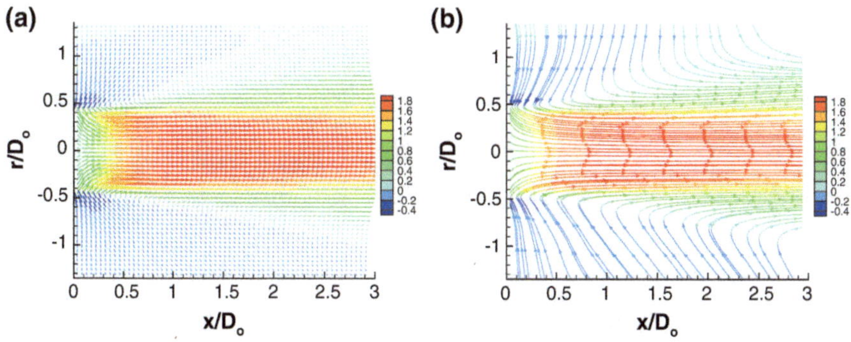


**Fig. 4** Mean velocity vector field (only every 4th vector in direction is shown) (**a**) and mean streamlines (**b**) coloured by the mean axial velocity $U/U_o$

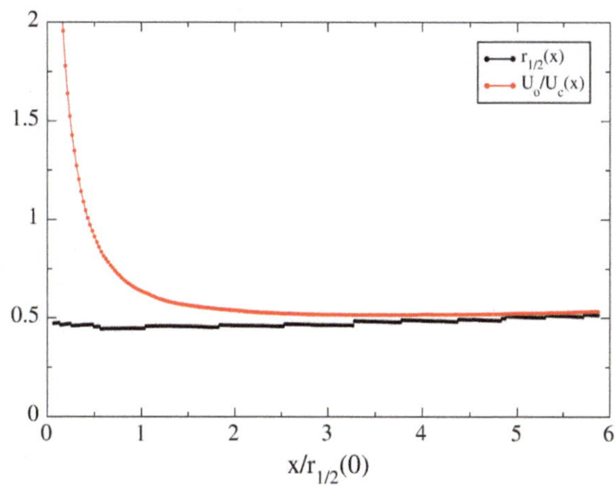


**Fig. 5** Half-width, $r_{1/2}(x)$, and axial mean centre-line velocity, $U_c(x)$, variation with axial direction $x$

from the orifice exit with $x$ reaching a flat plateau at approximately $x \approx 3r_{1/2}(0)$ with little change until x $\approx 6r_{1/2}(0) = 3D_o$. The statistics presented in the subsequent graphs are normalised by $U_c(x)$ and $r_{1/2}(x)$.

Figure 6 shows the mean velocity vector field coloured by the level of the second order statistics of the time-dependent triple decomposition components of the axial and radial velocity and the azimuthal vorticity. Figure 6a, e show that the axial phase correlated velocity fluctuation level is significantly larger than any axial turbulence velocity fluctuation level. The axial phase correlated velocity oscillation level is found to be highly concentrated along a cylindrical core of the jet of diameter reaching to approximately $D_o$ from the orifice with little spread in this near field domain of the ZNMF jet. In contrast the axial turbulence oscillation level in the core of the jet up to $x \approx 2.5D_o$ is very low. The peak axial turbulence oscillation level is located on the edge of the orifice and is found to spread from there in a

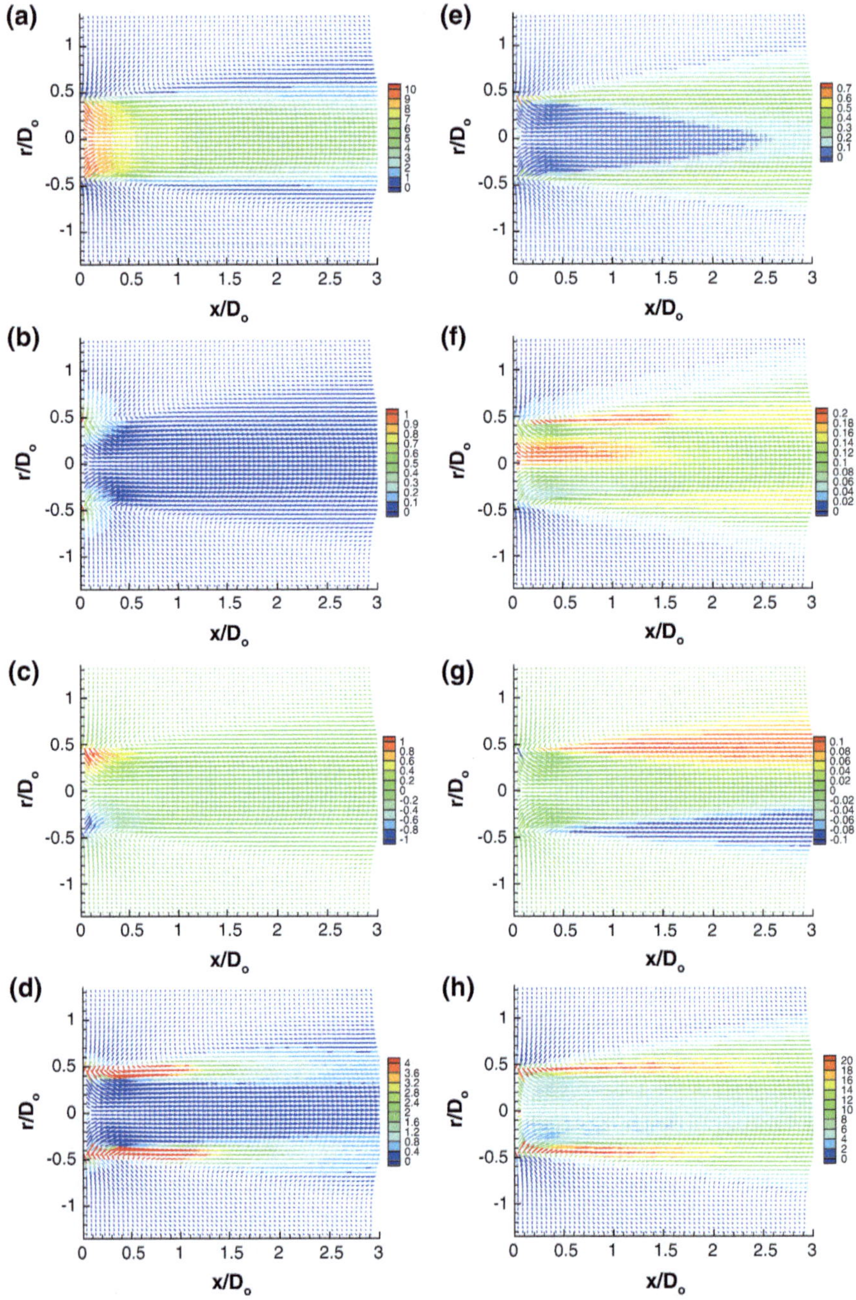
(a)
(b)
(c)
(d)
(e)
(f)
(g)
(h)
r/D$_o$
x/D$_o$
0 0.5 1 1.5 2 2.5 3
1 0.5 0 -0.5 -1
10 9 8 7 6 5 4 3 2 1 0
1 0.9 0.8 0.7 0.6 0.5 0.4 0.3 0.2 0.1 0
1 0.8 0.6 0.4 0.2 0 -0.2 -0.4 -0.6 -0.8 -1
4 3.6 3.2 2.8 2.4 2 1.6 1.2 0.8 0.4 0
0.7 0.6 0.5 0.4 0.3 0.2 0.1 0
0.2 0.18 0.16 0.14 0.12 0.1 0.08 0.06 0.04 0.02 0
0.1 0.08 0.06 0.04 0.02 0 -0.02 -0.04 -0.06 -0.08 -0.1
20 18 16 14 12 10 8 6 4 2 0

◀**Fig. 6** Mean velocity vector field (only every 4th vector in each direction is shown). The colour range represents in: **a** phase correlated axial mean square oscillation field $\overline{\tilde{u}^2}/U_o^2$, **b** phase correlated radial mean square oscillation field $\overline{\tilde{v}^2}/U_o^2$, **c** phase correlated Reynolds stress $\overline{\tilde{u}\tilde{v}}/U_o^2$, **d** phase correlated mean square vorticity oscillations $\overline{\tilde{\omega}_z^2}/\left(U_o/r_{1/2}\right)$, **e** turbulence axial mean square oscillation field $\overline{u'^2}/U_o^2$, **f** turbulence radial mean square oscillation field $\overline{v'^2}/U_o^2$, **g** turbulence Reynolds stress $\overline{u'v'}/U_o^2$ and **h** turbulence mean square vorticity oscillations $\overline{\omega_z'^2}/\left(U_o/r_{1/2}\right)$

conical fashion in the downstream direction until the inner edge of the cone region of the axial turbulence oscillations merge at $x \approx 2.5D_o$.

The energy in the radial velocity oscillation of the phase correlated component is found to be approximately one order of magnitude higher than those corresponding to the radial turbulence velocity fluctuations, but are found to be concentrated around the edge of the ZNMF jet orifice with no phase correlated radial velocity oscillations found outside this region. By comparison the radial turbulence velocity oscillations are found in the core of the jet with larger radial turbulence levels found along the shear layer emanating from the orifice edge of the ZNMF jet. An asymmetry is clearly observable in the radial turbulence velocity oscillation whose cause is more likely attributable to imperfections of the ZNMF jet generation mechanism rather than any specific fluid mechanics mechanism. The phase correlated and turbulence Reynolds stress fields shown in Fig. 6c, g respectively have a similar magnitude level relationship and spatial distribution to the radial velocity oscillation shown in Fig. 6b, f.

The mean square azimuthal vorticity of the phase correlated component and the turbulence component fields are shown in Fig. 6d, h respectively. High levels of azimuthal vorticity fluctuation for both components are found to be concentrated in an annular shear layer emanating from the edge of the ZNMF jet. The level of the turbulence vorticity oscillation is higher than that of the phase correlated vorticity oscillation, with the latter found not to spread radially very much with $x$ compared to the former which has a larger radial spread and a much thinner high level annular sheet-like structure which extends up to $x \approx 2D_o$.

The radial dependence of the mean velocity and vorticity profiles is shown in Fig. 7. Note that the profiles shown in these graphs closest to the orifice are located at $x = 0.07r_{1/2}(0)$. All profiles for all statistics presented for this $x$ location have been scaled to allow for proper analyses of the $x$-dependence of the normalized profiles. The axial mean velocity when scaled with the local axial mean centreline velocity and the local half-width, Fig. 7a, shows that as the mean flow evolves downstream it approaches a self-similar profile. At $x = 0.07r_{1/2}(0)$ there is a very thin annular region with large positive velocity just inside the orifice, which is found to decay towards the axis of the jet reaching approximately 20 % of the annular peak at the centerline. There is a negative axial velocity that ranges from just inside the orifice edge to one orifice diameter from the jet axis. These large positive and negative mean axial velocity deviations are smoothed out as the flow evolves in the axial direction, disappearing by $x \approx r_{1/2}(0)$, where a negative mean axial velocity is still

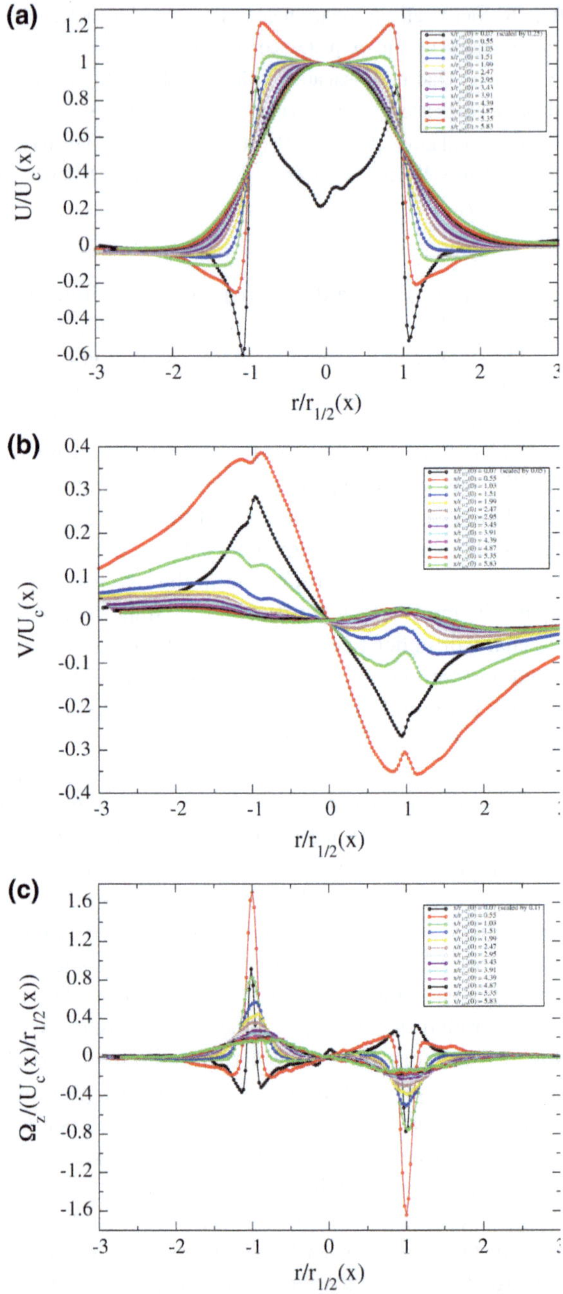


**Fig. 7** Mean flow: **a** axial velocity, **b** radial velocity and **c** azimuthal vorticity

present at the edge of the jet, with a positive mean axial velocity top-hat profile that extends over most of the core of the jet. This profile evolves towards a Gaussian-like mean axial velocity profile with no negative velocity present from $x \geq 2r_{1/2}(0)$.

The mean radial velocity profiles shown in Fig. 7b are always negative, i.e. indicating that the velocity is towards the axis for $r \geq 1.5r_{1/2}(x)$ (when $\theta = 0$) and $r \geq r_{1/2}(x)$ (when $\theta = \pi$) for all $x$. For $x \leq 2r_{1/2}(0)$, the radial velocity is negative everywhere, while for $x > 2r_{1/2}(0)$, the inner core of the jet has positive radial velocity extending from the jet axis to $r \approx 1.5r_{1/2}(x)$ (when $\theta = 0$) and $r \approx r_{1/2}(x)$ (when $\theta = \pi$). This implies that in the near field mean entrainment of environmental fluid, i.e. fluid which comes from the region far from the jet, is found to occur only very close to the jet orifice within one $D_o$. These results also show that there is a small mean asymmetry in this ZNMF jet, which as has already been suggested is more likely a consequence of the jet apparatus operational imperfections. Figure 7c shows the mean azimuthal vorticity profiles. Within half an orifice diameter from the orifice there is a thin negative mean vorticity sheet at the edge of the jet (i.e. orifice edge) which is surrounded by a positive mean vorticity sheet of much smaller magnitude. By $x \approx r_{1/2}(x)$, there is only a negative mean vorticity sheet which continues spreading in the axial direction and approaches a self-similar shape by the downstream edge of the measurement domain.

Figure 8 shows the mean square velocity fluctuation level, Reynolds stress and mean square vorticity level associated with the phase-correlated component in (a–c) and the turbulence component in (d–f) respectively. The phase-correlated mean square axial velocity oscillation profile starts as a top-hat radial profiles near the orifice and develops into a Gaussian-like self-similar profile in the $x$ direction as shown in Fig. 8a. In contrast the turbulence mean square axial velocity oscillation begin with a double peak structure at the orifice as shown in Fig. 8e. One peak is located on the axis and a much larger peak is located at $r = r_{1/2}(x)$. The magnitude of the turbulence peaks is more than one magnitude smaller than those of the phase correlated axial fluctuation levels. The axial velocity peak decays quickly with $x$, while the off-axis peak decays slowly and spreads radially as it approaches a self-similar profile with the location of the maximum moving towards the axis asymptotically approaching $r \approx 0.8r_{1/2}(x)$.

The radial velocity fluctuation levels are shown in Fig. 8b, f. The phase correlated radial velocity fluctuation also have a maximum that starts at $r \approx D_o/2$ near the ZNMF jet orifice and moves radially outwards as the profiles develop with $x$, as the value of the maximum in this scaled plot decreases. A clear asymmetry is observed in the position of the peak, but not in the magnitude of the peak of the radial phase-correlated fluctuation levels. The radial turbulence fluctuation levels have a maximum located on the axis and a maximum located at $r = D_o/2$. The level of the axis maximum is larger than the off-axis maximum but decays much quicker with $x$, while the off-axis maximum decay slower until it remains as the only maximum and the entire profile approaches a self-similar profile in the scaled variables. In contrast to the axial velocity fluctuation levels, the radial velocity fluctuation levels asymptote for the phase-correlated and turbulence components to similar magnitude levels with $x$.

The same general patterns of observation are found in the scaled profiles of the Reynolds stress shown in Fig. 8c, g and the azimuthal vorticity fluctuation profiles shown in Fig. 8d, h, with all profiles approaching self-similar profiles with maxima near $r = r_{1/2}(x)$.

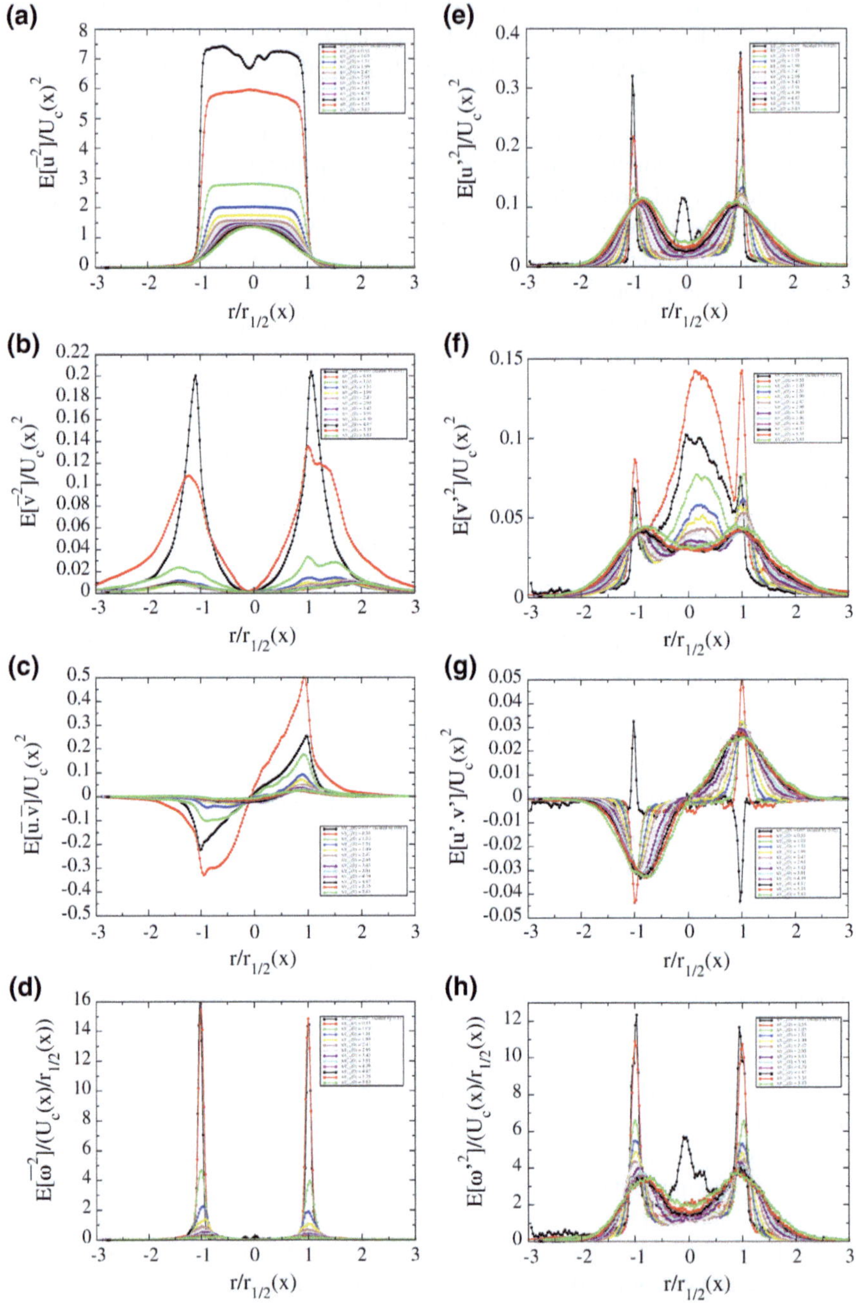


**Fig. 8** Mean square velocity and vorticity fluctuations and Reynolds stress profiles and their variation with axial distance from the orifice. **a–d** correspond to the phase correlated oscillation component of the velocity and vorticity, while **e–h** correspond to the turbulence component of the velocity and vorticity

In the case of the Reynolds stress the magnitude level of the asymptotic phase correlated profiles is approximately one order larger than the turbulence Reynolds stress, while the opposite is true for the azimuthal vorticity fluctuation magnitude of the asymptotic profile in the axial direction.

The evolution of the phase-correlated radial profiles with $x$ as a function of the phase relative to the periodic cycle of the piston motion that generates the ZNMF jet flow is shown for the axial velocity in Fig. 9. In these graphs $\phi = 0$ and $2\pi$ corresponds to the location when the piston is at its maximum negative displacement, i.e. the end of the suction stroke. The results in Fig. 9a–h are shown at eight equally spaced phase locations separated by $\Delta\phi = \pi/4$. At $\phi = 0$, which corresponds to zero velocity of the piston, $\tilde{u}/U_c$ has a negative core with a positive boundary layer connecting to the negative core via a thin shear layer located around $r = r_{1/2}(x)$ near the orifice. With axial development the axial profile at $\phi = 0$ acquires a Gaussian-like self-similar profile with negative magnitude. The positive axial velocity at the edge of the jet disappearing by $x \approx D_o$. The magnitude of the value of the peaks of the profiles are found to have a magnitude equal to that of the mean centreline velocity $U_c(x)$.

All $\tilde{u}/U_c$ profiles become positive by $\phi = \pi/4$, starting as top-hat profiles near the ZNMF jet orifice and developing into a self-similar Gaussian-like profile with a peak that is $1.5U_c(x)$ at this phase of the ZNMF jet temporal evolution. The radial dependence of $\tilde{u}$ for $\pi/4 \leq \phi < \pi$ is similar with the only difference being that the profiles decay in magnitude with $\phi$ during the phase shown for the ejection phase of the ZNMF jet. At $\phi = \pi$ the suction phase of the ZNMF jet begins and the profiles become negative in value with positive $\tilde{u}/U_c$ around the edge of the jet as is the case for $\phi = 0$. The profiles develop into a self-similar negative Gaussian-like profile with axial direction, with no positive velocity observed for $x \geq D_o$. For $\phi \geq 5\pi/4$ or possibly earlier until the suction stroke of the piston completes, the phase-correlated axial velocity profiles have negative values which approach self-similar Gaussian-like profiles with $x$. The magnitude of the peak of the asymptotic profiles at $\phi = \pi$ is $0.5U_c(x)$, which increases to approximately $U_c(x)$ during the remainder of the suction stroke.

The phase correlated evolution of the radial velocity and azimuthal vorticity have an analogous development and approach self-similar profiles in the axial direction.

## 4.3 The Topology of the Phase Correlated Flow of the ZNMF Jet

The structure and phase evolution of the topology of the phase-correlated component of the flow field has been investigated by considering the streamline pattern of the phase-average velocity field and the phase-correlated azimuthal vorticity field. Since the kinematic quantities are approximately axisymmetric the following description applies for $r > 0$. Figure 10 shows the evolution of the instantaneous streamline pattern of the phase-average velocity field $(U(x) + \tilde{u}(x, r, t)$,

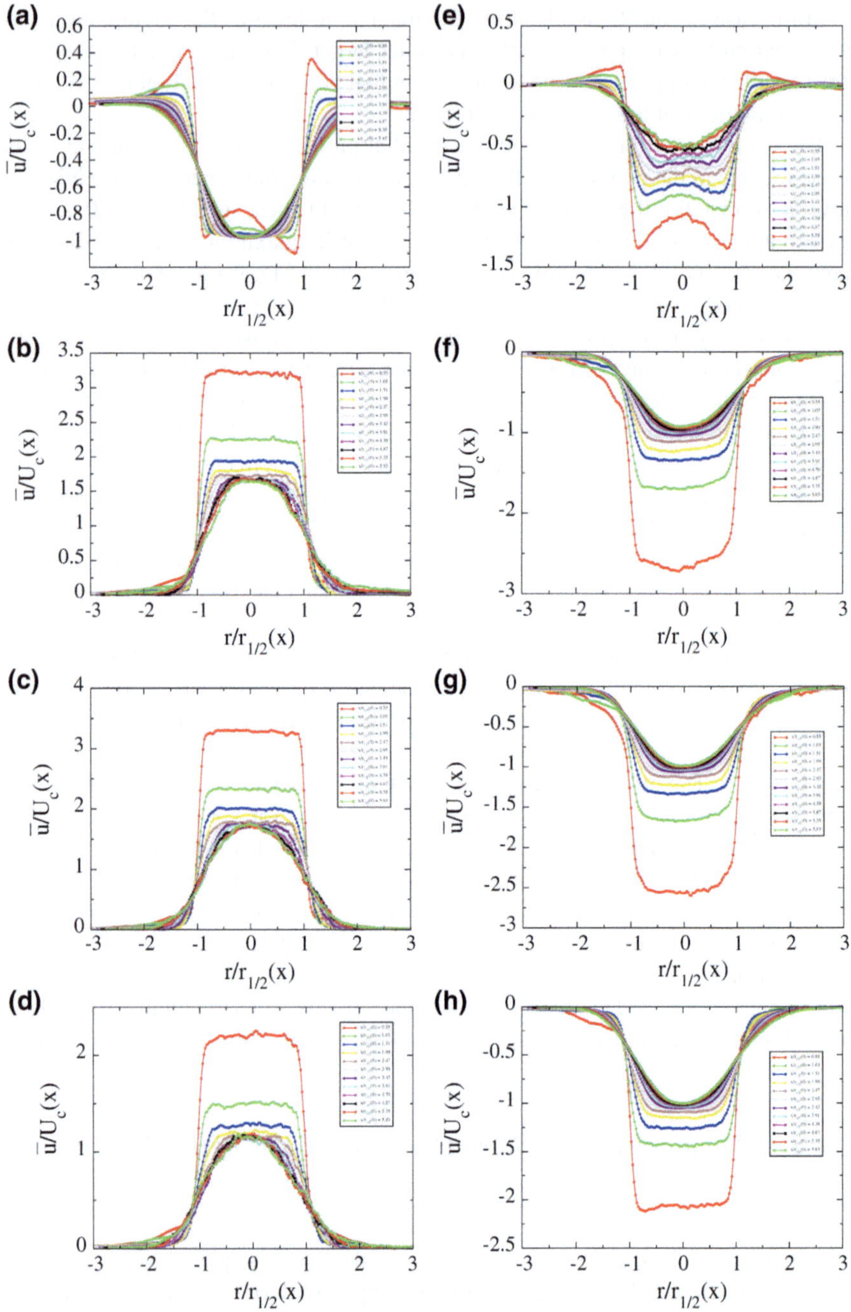
(a)
(b)
(c)
(d)
(e)
(f)
(g)
(h)
ū/U_c(x)
r/r_1/2(x)

◀**Fig. 9** Phase correlated axial velocity $\tilde{u}(x, r, t)$ profiles for the $x$-domain ranging from the orifice up to $x \approx 3D_o$. The radial direction is normalized by the local half width, $r_{1/2}(x)$ and $\tilde{u}(x, r, t)$ is normalized by the local mean axial centre-line velocity, $U_o(x)$. The time evolution of $\tilde{u}(x, r, t)$ is shown for the 8 phases $\phi_n = \frac{n\pi}{4}$. **a** $\phi = 0, 2\pi$, **b** $\phi = \frac{\pi}{4}$, **c** $\phi = \frac{\pi}{2}$, **d** $\phi = \frac{3\pi}{4}$, **e** $\phi = \pi$, **f** $\phi = \frac{5\pi}{4}$, **g** $\phi = \frac{3\pi}{2}$, **h** $\phi = \frac{7\pi}{4}$

$V(x) + \tilde{v}(x, r, t))$. The colour grading in Fig. 10 represents the phase-average axial velocity distribution along the streamline. The phase samples shown correspond to the same eight equally spaced sample phase points during one period of the ZNMF jet used in Sect. 4.2. These sample points clearly show that during the ejection phase of the ZNMF jet and after the initial transient, i.e. $\pi/4 \le \phi < \pi$, the streamline pattern is that of a typical laminar jet flow with radial entrainment from the far field into the jet. The magnitude of the phase-average axial velocity during this period ranges up to $5U_o$, reducing to virtual zero in the quiescent radial far-field of the jet. These sample point do not show the initial transient phase of the ejection phase of the ZNMF jet which has completed by $\phi = 15\pi/78$ as will be shown later in detail.

Once the suction phase starts a saddle point develops in the neighbourhood of the ZNMF jet orifice located on the jet axis as shown in Fig. 10e. Upstream of this saddle point the fluid is sucked into the ZNMF jet orifice, while downstream of the saddle point the flow has positive axial velocity feeding the ZNMF jet in the far field with radial entrainment. Figure 10f shows that as the ZNMF jet evolves in time, the saddle point moves downstream and the region from which environmental fluid is entrained into the ZNMF jet cavity through the orifice increases resulting in a nearly spherically symmetric radially inward flowing streamline pattern in the neighbourhood of the ZNMF jet orifice.

Figure 11 depicts the detailed evolution of the instantaneous phase-average streamline pattern during the initial transient of the ZNMF jet ejection phase which occurs during $0 \le \phi < 15\pi/78$ (i.e. $0 \le t/T < 0.1$). The topology of the ejection starting phase of the ZNMF jet resembles an axisymmetric starting vortex ring analogous to the starting phase observed in the impulsively started continuous jet [27]. The vortex ring is observed to grow as it evolves with $x$ and is found to have traversed the near field which extends to $2.5D_o$ by the time the piston ejection phase has reached $\phi = 0.9$ as shown in Fig. 11h.

The detailed streamline pattern evolution from the beginning of the suction phase of the ZNMF jet is shown in Fig. 12. At $\phi = 77\pi/78$ the generation of a saddle point very close to the ZNMF jet orifice on the jet axis is observed in Fig. 12a. The edge of the positive axial velocity region downstream of the saddle point has a parabolic shape. The radial separatrices of the saddle mark the demarcation plane between fluid that can be entrained into the ZNMF jet and fluid that is entrained into the downstream flowing jet during the suction phase of the ZNMF jet. As can be observed in Fig. 12a–h the saddle point moves quickly along the jet axis in the downstream direction reaching $x \approx 1.5D_o$ by the time $\phi \approx 87\pi/78$. From then on, the downstream movement of the saddle point location is at a considerably slower axial velocity commensurate with the deceleration of the piston movement. The positive

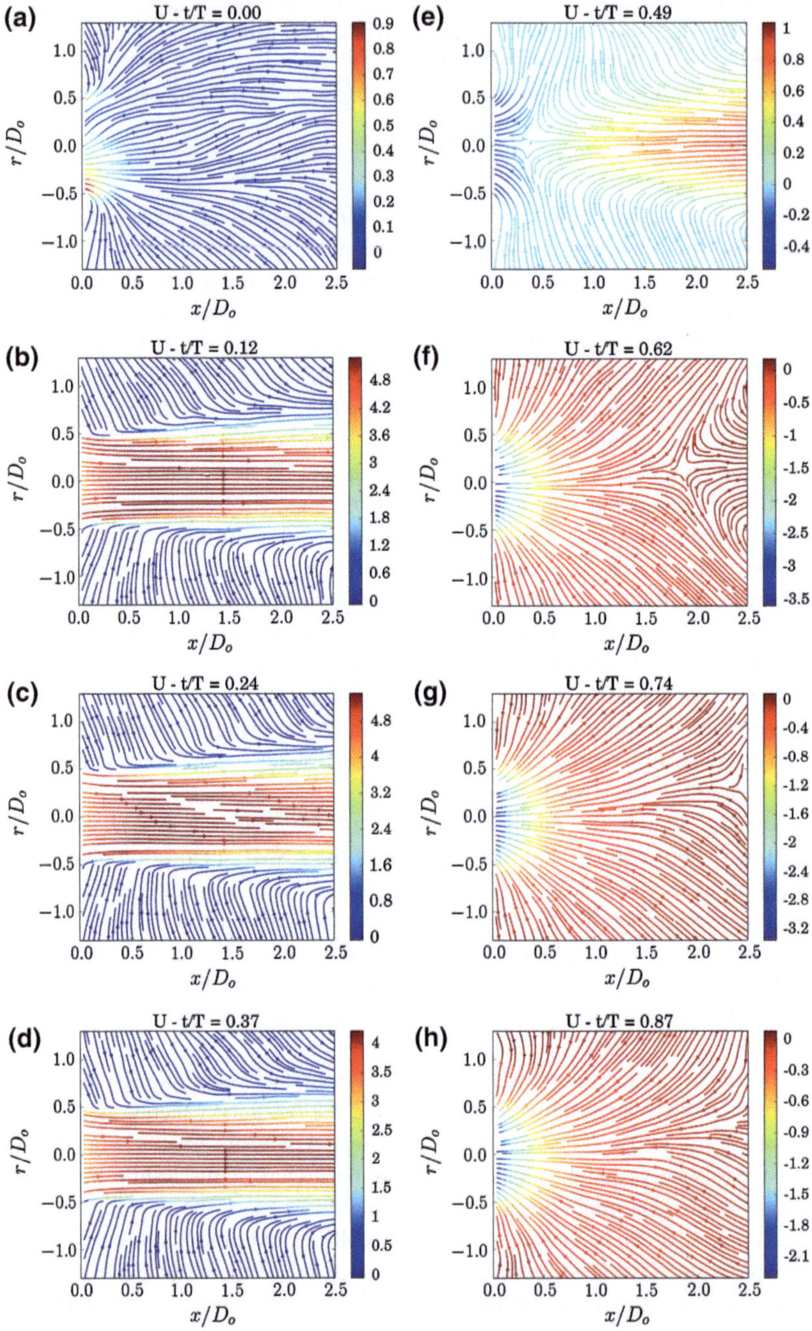


**Fig. 10** Streamline plot of the phase-average velocity field coloured by the phase-average axial velocity, $U(x) + \tilde{u}(x, r, t)$ normalized by $U_o$. The time evolution is shown for the 8 phases $\phi_n = \frac{n\pi}{4}$. **a** $\phi = 0, 2\pi$, **b** $\phi = \frac{\pi}{4}$, **c** $\phi = \frac{\pi}{2}$, **d** $\phi = \frac{3\pi}{4}$, **e** $\phi = \pi$, **f** $\phi = \frac{5\pi}{4}$, **g** $\phi = \frac{3\pi}{2}$, **h** $\phi = \frac{7\pi}{4}$

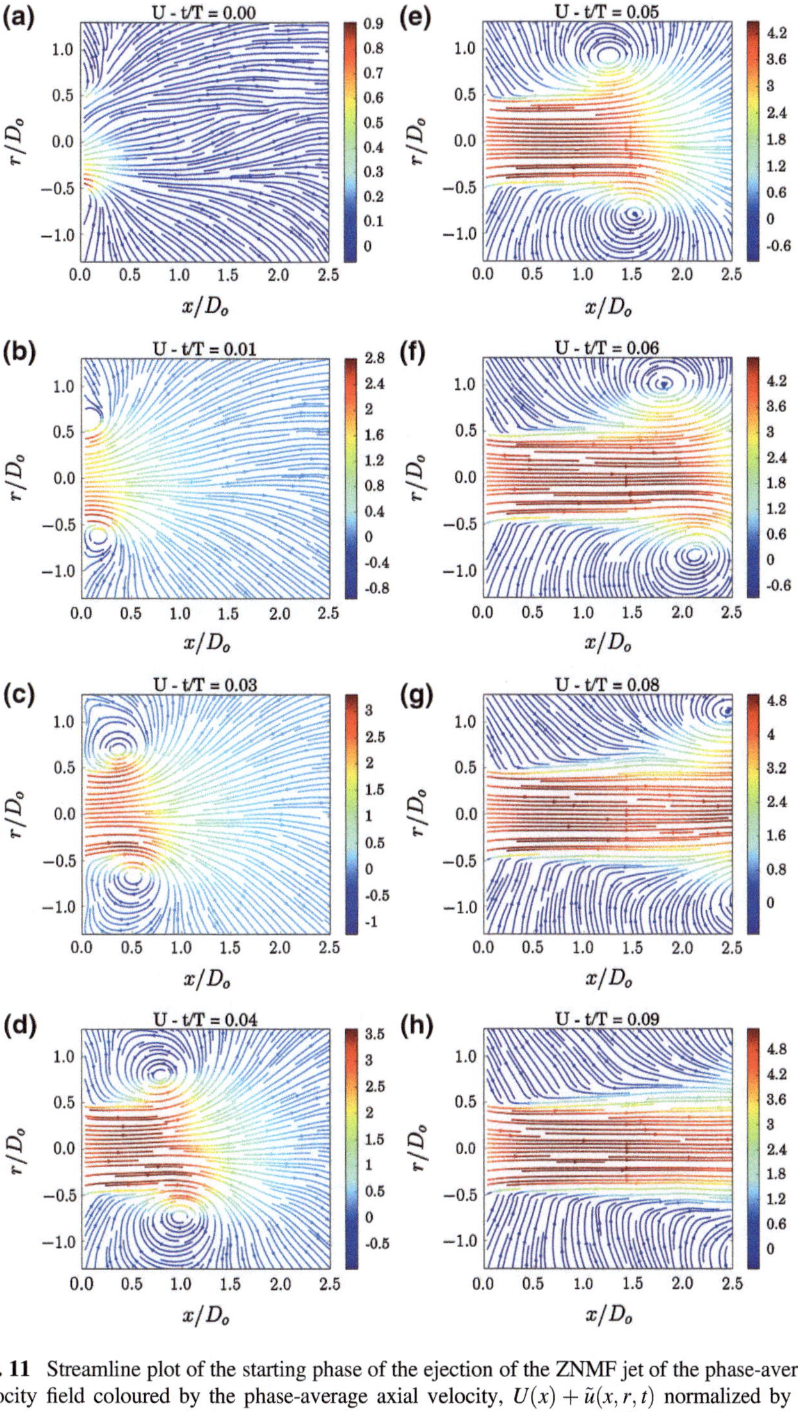


**Fig. 11** Streamline plot of the starting phase of the ejection of the ZNMF jet of the phase-average velocity field coloured by the phase-average axial velocity, $U(x) + \tilde{u}(x, r, t)$ normalized by $U_c$. The time evolution is shown for the 8 phases $\phi_n = \frac{n\pi}{78}$. **a** $\phi = 0, 2\pi$, **b** $\phi = \frac{3\pi}{78}$, **c** $\phi = \frac{5\pi}{78}$, **d** $\phi = \frac{7\pi}{78}$, **e** $\phi = \frac{9\pi}{78}$, **f** $\phi = \frac{11\pi}{78}$, **g** $\phi = \frac{13\pi}{78}$, **h** $\phi = \frac{15\pi}{78}$

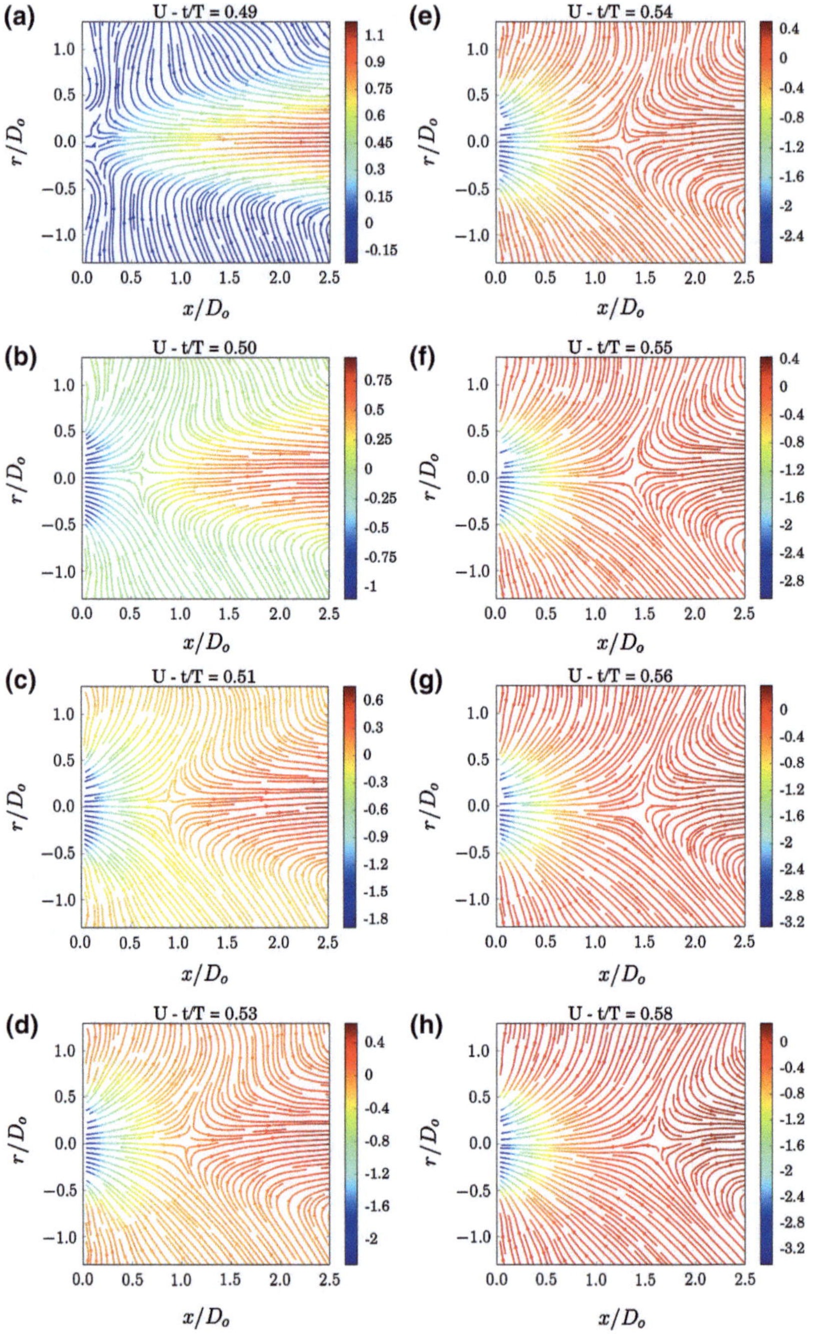


**Fig. 12** Streamline plot of the starting phase of the suction of the ZNMF jet of the phase-average velocity field coloured by the phase-average axial velocity, $U(x) + \tilde{u}(x, r, t)$ normalized by $U_c$. The time evolution is shown for the 8 phases $\phi_n = \frac{n\pi}{78}$. **a** $\phi = \frac{77\pi}{78}$, **b** $\phi = \frac{79\pi}{78}$, **c** $\phi = \frac{81\pi}{78}$, **d** $\phi = \frac{83\pi}{78}$, **e** $\phi = \frac{85\pi}{78}$, **f** $\phi = \frac{87\pi}{78}$, **g** $\phi = \frac{89\pi}{78}$, **h** $\phi = \frac{91\pi}{78}$

axial velocity region gets pushed downstream ahead of the saddle point and the parabolic shape of the edge of the positive axial velocity region increases as shown in Fig. 12a–d. During the remainder of this initial suction phase the axial velocity is found to be quite small downstream of the saddle point with predominately radial velocity as shown in Fig. 12e–h. The upstream region from where fluid can be entrained into the ZNMF jet orifice increases as the saddle point moves downstream, engulfing into the ZNMF jet cavity fluid which originates radially from a region adjacent to the rigid wall from where the ZNMF jet emanates. Figure 10e–g shows the entire suction phase of the ZNMF jet at a coarser level, the saddle point has a maximum downstream excursion which reaches up to $x \approx 2.8D_o$.

The phase-average streamline pattern shows that the saddle point is at $x \approx 2.5D_o$ when $\phi = 1.78\pi$ (i.e. $t = 0.89T$) and continues retreating towards the orifice along the jet axis (in an ideal case), reaching the orifice just before $\phi = 2\pi$ when the ejection phase of the ZNMF jet starts all over again.

The corresponding phase-correlated azimuthal vorticity field evolution is shown in Fig. 13. At the initial phase position of the ZNMF jet ejection phase shown in Fig. 13a, the phase correlated vorticity field is made up of a positive azimuthal vorticity sheet originating from the lip of the ZNMF jet orifice spreading in the axial direction with a negative vortex ring located at the ZNMF jet orifice lip on the outer side of this sheet and surrounding it. This is consistent also with the phase correlated axial velocity profiles shown in Fig. 9a for $x < 0.5D_o$. As already mentioned the initial transient phase of the ejection phase of ZNMF jet appears to be completed by $\phi = 15\pi/78$, thus Fig. 13b shows a typical azimuthal vorticity distribution of a laminar jet with high negative azimuthal vorticity in the shear layer close to the ZNMF jet orifice consistent with the top-hat phase-correlated axial velocity profile observed in Fig. 9b in the ZNMF jet orifice neighbourhood which has a fairly uniform axial velocity in the jet core with no vorticity. Downstream this is followed by spreading of the shear layer with the corresponding lowering of the phase correlated azimuthal vorticity which is consistent with the development of the Gaussian-like phase correlated axial velocity profile in Fig. 9b. This phase-correlated vorticity distribution during the remainder of the ejection phase of the ZNMF jet remains fairly unchanged analogous to the streamline pattern between $\pi/4 \leq \phi < \pi$. When the suction phase begins at $\phi = \pi$, the shear layer phase-correlated vorticity switches in sign but otherwise no appreciable difference is observed except that the magnitude of the phase-correlated vorticity in the shear layer during the suction phase is only about 0.5 of the magnitude of the phase-correlated vorticity in the shear layer during the ejection phase after the transients have abated.

A more detailed description of the phase correlated azimuthal vorticity during the starting transient of the ejection phase is shown in Fig. 14. At the beginning of the ejection phase the correlated azimuthal vorticity field in the neighbourhood of the orifice lip consists of a positive azimuthal vorticity sheet originating from the lip of the ZNMF jet orifice spreading radially in the axial direction with a negative vorticity vortex ring located at the ZNMF jet orifice lip on the outer side of this sheet and surrounding it. At $\phi = 3\pi/78$ the azimuthal vorticity field has evolved to one where the vorticity within the sheet for $x < 0.5D_o$ has intensified with the

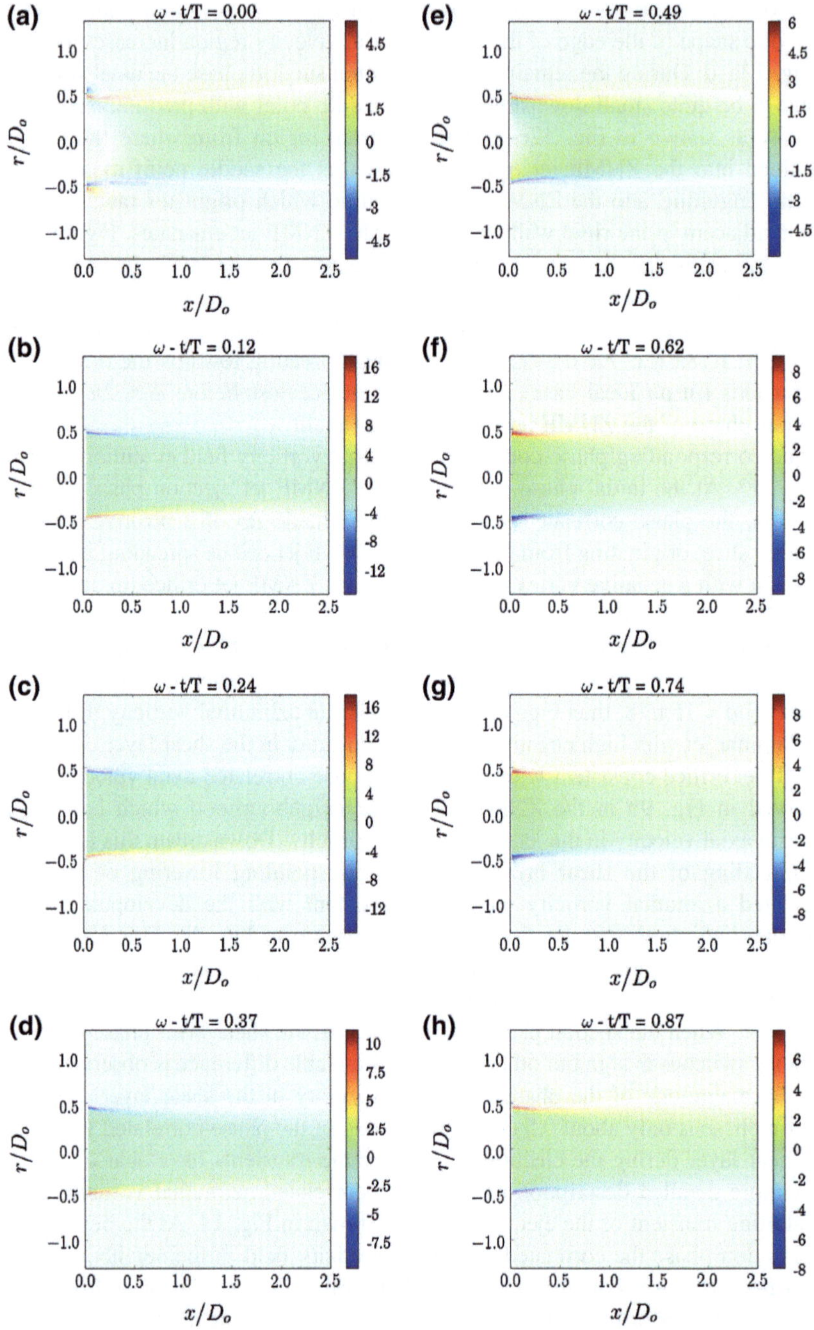


**Fig. 13** Phase correlated velocity vector field $(\tilde{u}, \tilde{v})$ coloured by the phase correlated non-dimensional azimuthal vorticity. The time evolution is shown for the 8 phases $\phi_n = \frac{n\pi}{4}$. **a** $\phi = 0, 2\pi$, **b** $\phi = \frac{\pi}{4}$, **c** $\phi = \frac{\pi}{2}$, **d** $\phi = \frac{3\pi}{4}$, **e** $\phi = \pi$, **f** $\phi = \frac{5\pi}{4}$, **g** $\phi = \frac{3\pi}{2}$, **h** $\phi = \frac{7\pi}{4}$

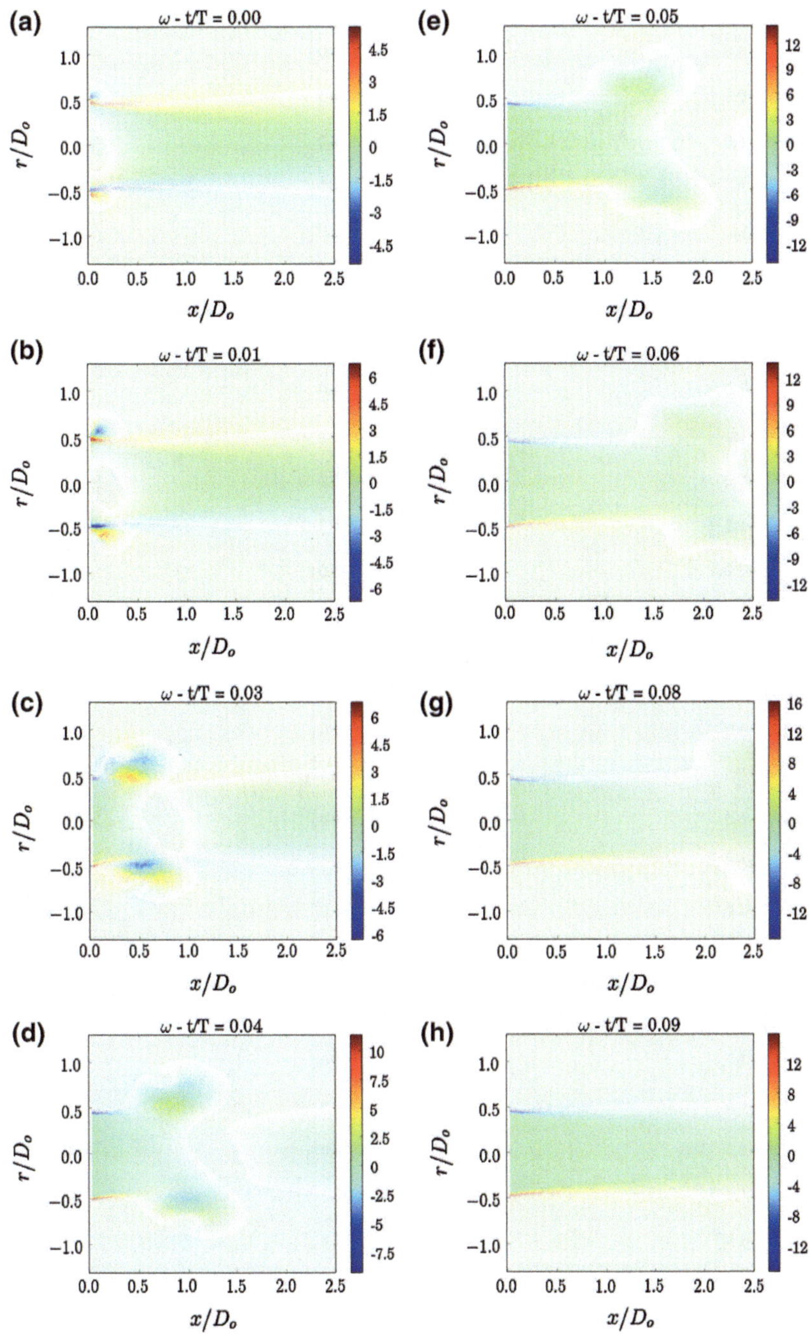


**Fig. 14** Starting phase of the ejection of the ZNMF jet shown by the phase correlated velocity vector field $(\tilde{u}, \tilde{v})$ coloured by the phase correlated non-dimensional azimuthal vorticity. The time evolution is shown for the 8 phases $\phi_n = \frac{n\pi}{78}$. **a** $\phi = 0, 2\pi$, **b** $\phi = \frac{3\pi}{78}$, **c** $\phi = \frac{5\pi}{78}$, **d** $\phi = \frac{7\pi}{78}$, **e** $\phi = \frac{9\pi}{78}$, **f** $\phi = \frac{11\pi}{78}$, **g** $\phi = \frac{13\pi}{78}$, **h** $\phi = \frac{15\pi}{78}$

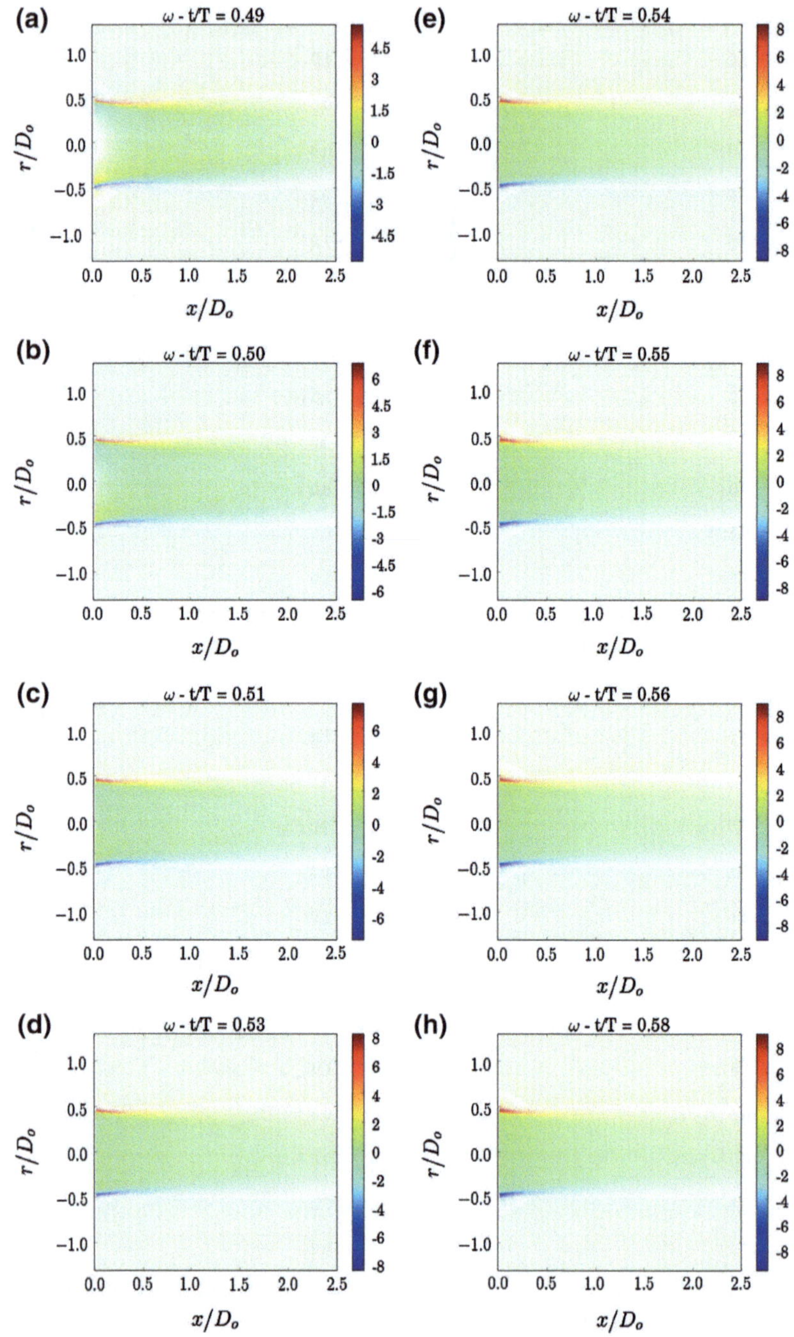


**Fig. 15** Starting phase of the suction of the ZNMF jet shown by the phase correlated velocity. **a** $\phi = \frac{77\pi}{78}$, **b** $\phi = \frac{79\pi}{78}$, **c** $\phi = \frac{81\pi}{78}$, **d** $\phi = \frac{83\pi}{78}$, **e** $\phi = \frac{85\pi}{78}$, **f** $\phi = \frac{87\pi}{78}$, **g** $\phi = \frac{89\pi}{78}$, **h** $\phi = \frac{91\pi}{78}$

negative vorticity vortex ring having grown as it moves away from the orifice lip while also amplifying as shown in Fig. 14b possibly due to azimuthal vortex stretching. The phase correlated positive vorticity of the shear layer diffuses and interacts with the negative vorticity of the vortex ring as both evolve downstream as shown in Fig. 14c–g, continuously annihilating each other as evidenced by the decrease of the vorticity in the interacting structure as it is followed downstream. Upstream of this interaction a phase correlated shear layer of negative vorticity develops which trails the vortex ring/shear layer vorticity interaction. This latter phase correlated negative vorticity layer is what remains during the ejection phase of the ZNMF jet as shown in Fig. 13b–d.

Figure 15 shows the detailed evolution of the phase-correlated vorticity from the beginning of the suction phase of the ZNMF jet period starting at $\phi = 77\pi/78$. Figure 15a shows that there is what appears to be an instant within which the sign of the phase-correlated azimuthal vorticity of the shear layer has an opposite vorticity region on the inner side of the layer in vector field $(\tilde{u}, \tilde{v})$ coloured by the phase correlated non-dimensional azimuthal vorticity. The time evolution is shown for the 8 phases $\phi_n = n\pi/78$.

The core of the jet where the fluid is now flowing into the orifice. This opposite vorticity region inside the core of the jet reduces in magnitude very rapidly and by $\phi = 79\pi/78$ the phase-correlated azimuthal vorticity field in the near field of the ZNMF jet consists of a thin sheet of positive vorticity which decays from the orifice lip in axial direction as it diffuses. This vorticity sheet starts with an effective length of approximately $1.5D_o$ which shortens to less than $D_o$ from the orifice lip in $x$ by the time the phase reaches $\phi = 91\pi/78$ as shown in Fig. 15g, and is found to remain so for most of the suction phase of the ZNMF jet as shown in Fig. 13f–h.

## 5 Summary

This chapter presents a study of the near field structure and evolution of a turbulent ZNMF jet using high-spatial resolution time-resolved MCCDPIV measurements. The spectral analysis of the axial velocity fluctuations show that near the orifice lip a dominant peak is found at the forcing frequency of 4 Hz, with the next dominant peak at the 3rd harmonic. The spectra develop along the shear layer with the 2nd harmonic becoming more prominent with downstream development and non-linear interactions developing from sum and difference components of the harmonics, suggesting a filling of the spectra as the flow evolves in the downstream direction and becomes a fully turbulent ZNMF jet in the far field as has been observed for these conditions [6].

The triple decomposition shows that all first order and second order statistics representing the mean flow in the former and the phase correlated and turbulent components in the latter evolve into self-similar profiles when scaled by the local half-width of the ZNMF jet and the centreline mean axial velocity. The self-similarity is clearly observable for $x > 2D_o$, this is remarkable as the ZNMF jet is found to develop self-similarity significantly further upstream and closer to the jet orifice than the equivalent continuous jet. However what is even more remarkable is that the phase correlated component of the velocity and the azimuthal vorticity also approach self-similar profiles $x > 2D_o$ for every phase of the piston motion, which suggests that each phase behaves like a quasi-mean jet when scaled by the ZNMF jet half width and the mean centreline velocity.

The analysis of the topology of the near field of the phase-correlated motion of the ZNMF jet shows that after the initial transient of the ejection phase of the ZNMF jet has completed, the streamline pattern remains that of a classical continuous laminar jet for the remainder of the ejection phase. The initial transient of the ZNMF ejection phase consists of a streamline pattern that is akin to a starting vortex ring. The corresponding phase correlated azimuthal vorticity evolution begins with a positive azimuthal vorticity sheet originating from the ZNMF jet orifice lip, spreading radially in the axial direction with a negative vorticity vortex ring located at the ZNMF jet orifice lip on the outer side of this sheet and surrounding it. The azimuthal vorticity field evolves during the initial stages of the ejection transient to one where the vorticity within the sheet first intensified with the negative vorticity vortex ring growing in size as it moves away from the orifice lip while the vorticity amplifies possibly due to azimuthal vortex stretching. This is followed by a stage where the phase-correlated positive vorticity of the shear layer diffuses and interacts with the negative vorticity of the vortex ring as both evolve downstream and start annihilating each other as evidenced by the decrease of the vorticity in the interacting structure. Upstream of this interaction a phase correlated shear layer of negative vorticity develops which trails the vortex ring/shear layer vorticity interaction. This latter phase-correlated negative vorticity layer is what remains during the remainder of the ejection phase of the ZNMF jet.

The suction phase of the ZNMF jet is characterized by the development of a saddle point at the centre on the ZNMF jet orifice at the beginning of this phase. The edge of the positive axial velocity region downstream of the saddle point is found to have a parabolic shape with the radial separatrices of the saddle marking a demarcation surface between fluid that can be drawn into the ZNMF jet through the orifice and fluid that will be entrained into the downward flowing jet during the suction phase of the ZNMF jet. As the suction phase proceeds, the saddle point moves along the jet axis in the downstream direction initially accelerating and then slowing down at a rate that is possibly directly proportional to the acceleration of the piston during the suction stroke. The positive axial velocity region gets pushed downstream ahead of the saddle point and the parabolic shape of the edge of the positive axial velocity region is observed to increase. During the rest of this initial suction phase the axial velocity is found to be quite small downstream of the saddle point with predominately radial velocity The potential upstream region from where fluid can be drawn into the ZNMF jet orifice increases as the saddle point moves downstream. The phase-average streamline pattern suggests that the saddle point moves to $x \approx 2.75D_o$ before it starts retreating back towards the orifice along the jet axis, reaching the orifice just before $\phi = 2\pi$ when the ejection phase of the ZNMF jet starts all over again.

**Acknowledgments** The support of this research by the Australian Research Council through facilities provided by an ARC LIEF grant are gratefully acknowledged. The MCCDPIV analysis was carried out on the NCI MASSIVE high performance computer, use of the NCI infrastructure for this purpose is gratefully acknowledged. The author is also very grateful to and would also like to acknowledge his colleagues Drs Nicholas Buchmann, Omid Amili and H.M Cui from LTRAC, who undertook the ZNMF jet experiments.

## References

1. Arnaud, J., Tomar, S., & Soria, J. (2004). Investigation of the mean flow pattern in zero-net-mass-flux elliptical-jets in cross-flow using planar-laser-induced fluorescence. In *15th Australasian Fluid Mechanics Conference*.
2. Buchmann, N. A., Atkinson, C., & Soria, J. (2013). Influence of ZNMF jet flow control on the spatiotemporal flow structure over a NACA-0015 airfoil. *Experiments in Fluids, 54*(3), 1485.
3. Cater, J., & Soria, J. (2002). The evolution of round zero-net-mass flux jets. *Journal of Fluid Mechanics, 472*, 167–200.
4. Cater, J., Soria, J., & Lim, T. T. (2004). The interaction of the piston vortex with a piston-generated vortex ring. *Journal of Fluid Mechanics, 499*, 327–343.
5. Cui, H. M., Buchmann, N., Amili, O., & Soria, J. (2011). Investigating zero-net-mass-flow jets with high Reynolds number using time-resolved PIV. In *9th International Symposium on Particle Image Velocimetry*, Kobe, Japan.
6. da Andrade, E. N. C. (1931). On the circulations caused by the vibration of air in a tube. In *Proceedings of the Royal Society of London, Series A: Containing Papers of a Mathematical and Physical Character, 134*(824), 445–470.
7. Ellenrieder, K.V., Kostas, J., & Soria, J. (2001). Measurements of a wall-bounded turbulent, separated flow using HPIV. *Journal of Turbulence, 2*, 1–15.
8. Fouras, A., & Soria, J. (1998). Accuracy of out-of-plane vorticity measurements using in-plane velocity vector field data. *Experiments in Fluids, 25*, 409–430.
9. Glezer, A. (2011). Some aspects of aerodynamic flow control using synthetic-jet actuation. *Philosophical Transactions of the Royal Society A: Mathematical, Physical and Engineering Sciences, 369*(1940), 1476–1494.
10. Gordon, M., Cater, J., & Soria, J. (2004). Investigation of the mean passive scalar field in zero-net mass-flux in cross-flow using PLIF. *Physics of Fluids, 16*(3), 794–808.
11. Gordon, M., & Soria, J. (2002). PIV measurements of a zero-net-mass-flux jet in cross flow. *Experiments in Fluids, 33*, 863–872.
12. Hart, D. P. (2000). PIV error correction. *Experiments in Fluids, 29*(1), 13–22.
13. Hussain, A. K. M. F., & Reynolds, W. C. (1970). The mechanics of an organized wave in turbulent shear flow. *Journal of Fluid Mechanics, 41*(02), 241–258.
14. Ingård, U., & Labate, S. (1950). Acoustic circulation effects and the nonlinear impedance of orifices. *The Journal of the Acoustical Society of America, 22*, 211.
15. Kitsios, V., Cordier, L., Bonnet, J. P., Ooi, A., & Soria, J. (2010). Development of a nonlinear eddy viscosity closure for the triple-decomposition stability analysis of a turbulent channel. *Journal of Fluid Mechanics, 664*, 74–107.
16. Lighthill, S. J. (1978). Acoustic streaming. *Journal of Sound and Vibration, 61*(3), 391–418.
17. Mathis, R., Duke, D., Kitsios, V., & Soria, J. (2008). Use of zero-net-mass-flow for separation control in diffusing S-duct. *Experimental Thermal and Fluid Science, 33*(1), 169–172.
18. Parekh, D. E., Kibens, V., Glezer, A., Wiltse, J. M., & Smith, D. M. (1996). Innovative jet flow control: Mixing enhancement experiments. *AIAA Journal, 308*, 96.
19. Raffel, M., Willert, C., & Kompenhans, J. (1998). *Particle image velocimetry*. Berlin: Springer.

20. Smith, B. L., & Glezer, A. (1998). The formation and evolution of synthetic jets. *Physics of Fluids*, *10*(9), 2281–2297.
21. Smith, B. L., & Glezer, A. (2002). Jet vectoring using synthetic jets. *Journal of Fluid Mechanics*, *458*, 1–34.
22. Soria, J. (1994). Digital cross-correlation particle image velocimetry measurements in the near wake of a circular cylinder. In *International Colloquium on Jets, Wakes and Shear Layers* (pp. 25.1–25.8). Melbourne, Australia: CSIRO.
23. Soria, J. (1996). An adaptive cross-correlation digital PIV technique for unsteady flow investigations. In A. Masri & D. Honnery (Eds.), *1st Australian Conference on Laser Diagnostics in Fluid Mechanics and Combustion* (pp. 29–45). NSW, Australia: University of Sydney.
24. Soria, J. (1996). An investigation of the near wake of a circular cylinder using a video-based digital cross-correlation particle image velocimetry technique. *Experimental Thermal and Fluid Science*, *12*, 221–233.
25. Soria, J. (1998). Multigrid approach to cross-correlation digital PIV and HPIV analysis. In *13th Australasian Fluid Mechanics Conference* (pp. 381–384). Melbourne: Monash University.
26. Soria, J. (2006). Particle image velocimetry—application to turbulence studies. In *Turbulence and Coherent Structures in Fluids, Plasmas and Nonlinear Media* (pp. 309–347). Hackensack, NJ: World Scientific Publication.
27. Soria, J., Cater, J., & Kostas, J. (1999). High resolution multigrid cross-correlation digital PIV measurements of a turbulent starting jet using half frame image shift film recording. *Optics and Laser Technology*, *31*, 3–12.
28. Tuck, A., & Soria, J. (2008). Separation control on a NACA 0015 airfoil using a 2D micro ZNMF jet. *Aircraft Engineering and Aerospace Technology*, *80*(2), 175–180.
29. Westerweel, J., Dabiri, D., & Gharib, M. (1997). The effect of a discrete window offset on the accuracy of cross-correlation analysis of digital PIV recordings. *Experiments in Fluids*, *23*(1), 20–28.

# Simulation and Modeling of Synthetic Jets

H. Tang and S. Zhong

**Abstract** This chapter gives a comprehensive review on the state-of-the-art numerical simulation and modeling of synthetic jet (SJ), an oscillatory, zero-net-mass-flux jet that has great potential in flow control. The numerical techniques dealing with the SJ oscillation, laminar-turbulent transition, and turbulence in quiescent conditions are introduced first. From the CFD simulations, the flow physics associated with SJs issuing from offices of different shapes, mainly high-aspect-ratio rectangular slots and low-aspect-ratio rectangular or circular orifices, are then summarized and discussed. Low-dimensional models that are used for fast prediction of the performance of SJ actuators in quiescent conditions are also introduced. The review then focuses on the CFD simulations of SJ interaction with cross flows, including both attached and separated flows. Different numerical methods are introduced and compared. The flow physics associated with SJs interacting with laminar boundary layers, turbulent boundary layers, different types of separated flows, as well as flows around an airfoil are also summarized and discussed. In the end, research about the coupling of CFD simulation or low-dimensional modeling with optimization and control algorithms for SJ-based flow control is briefly reviewed. We hope this chapter provides a better picture of the state-of-the-art numerical simulation and modeling of SJs for the flow separation control applications and future researchers in this field can benefit from reading it.

**Keywords** Synthetic jet · Active flow separation control · CFD simulation · Low-dimensional modelling

H. Tang (✉)
Department of Mechanical Engineering, The Hong Kong Polytechnic University, Kowloon, Hong Kong SAR, China
e-mail: h.tang@polyu.edu.hk

S. Zhong
School of Mechanical, Aerospace and Civil Engineering, University of Manchester, Manchester M13 9PL, UK



D.T.H. New and S.C.M. Yu (eds.), *Vortex Rings and Jets*,
Fluid Mechanics and Its Applications 111, DOI 10.1007/978-981-287-396-5_4

# 1 Introduction

Synthetic jets (SJ) as a promising flow control means have been a hot research topic for many years. A typical synthetic-jet actuator (SJA), as illustrated in Fig. 1, consists of a cavity bounded by rigid walls, with an orifice at one end and an oscillating diaphragm at the other. During the downward motion of the diaphragm, the ambient fluid is drawn into the cavity. As the diaphragm moves back up towards the orifice plate, the fluid is pushed out of the cavity and separates from the edges of the orifice. If the jet is sufficiently strong, the resulting free shear layer then rolls up to form a vortex ring at the exit of the orifice and as this cycle of suction and blowing is repeated periodically, a jet is "synthesized" from a train of vortex rings that propagate away from the orifice. Since the mass flow being ingested into the cavity is exactly equal to that being expelled during an SJ actuation cycle, this jet is also called a "zero-net-mass-flux" (ZNMF) jet. Nevertheless, it is able to inject non-zero momentum into a fluid system from which it is synthesized. And due to this unique ability, an SJ offers many desirable advantages over a conventional jet. These include a better effectiveness than steady jets due to its unsteady nature, the elimination of the compressed air supply and associated piping system, and the ease for micro-electro mechanical systems (MEMS) applications.

Since 1990s, SJs have been extensively investigated in laboratory experiments, aimed primarily at various aeronautical applications including flow separation control [2, 9, 83], mixing enhancement [92, 102], turbulence control [41, 59, 61], and jet vectoring [82]. As for flow separation control, significant separation delay or elimination has been demonstrated on various aerodynamic bodies such as airfoils and circular cylinders with characteristic Reynolds numbers up to $10^6$ and beyond.

In order to design SJAs for better or optimal flow control, a comprehensive understanding of the behavior of SJs and the interaction between SJs and attached or separated cross flows is required. Although many experimental investigations

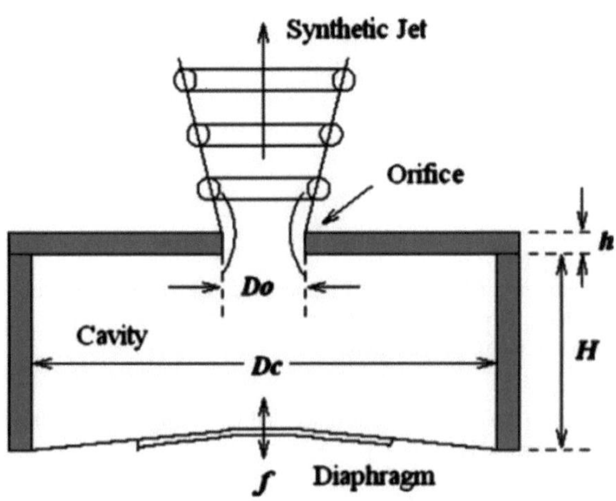


**Fig. 1** Schematic of a typical SJA

have been conducted in this aspect, due to the large number of parameters, multi-physics nature, and the limitation of existing measurement technologies, they can be costly and sometimes impractical. Hence both computational and theoretical capability has been actively sought to predict the characteristics of SJs and their interaction with attached or separated flows. Many types of computational fluid dynamics (CFD) and low-dimensional models have been applied, including steady or unsteady Reynolds averaged Navier-Stokes (RANS or URANS), large eddy simulations (LES), blended RANS-LES, and direct numerical simulations (DNS). Various degrees of success have been reported. However, most researchers today would probably agree that we are still a long way away from consistently predicting these types of flows with a high level of sustained accuracy and confidence.

To this end, how to predict the performance of SJAs and the resulting SJ flow with high level of fidelity is a major concern. Therefore in this chapter the start-of-the-art numerical simulations of SJs will be reviewed, including both CFD simulations and low-dimensional modeling, with the successes and remaining challenges for computational methods being pointed out.

This chapter is organized as follows: in Sect. 2 the status of both the CFD simulation and low-dimensional modeling of SJs in quiescent flows is reviewed. In Sect. 3 the simulations and modeling of SJs in cross flows are presented, with a focus on the interaction between SJs and attached or separated flows. In Sect. 4 the coupling of SJ simulations with control algorithms is discussed. In the end a summary of this chapter is given.

## 2 Synthetic Jets in Quiescent Flows

Although the practical relevance of an isolated SJ in quiescent conditions is relatively limited, it is regarded as an important baseline case deserving a separate study. To choose suitable SJAs for flow separation control, accurate and fast prediction of the performance of SJAs in quiescent conditions is the first step. The dependency of SJ quantities, such as the jet velocity and vortex structure, on the geometric and operating conditions of SJAs needs to be fully understood. In this section, therefore, the state-of-the-art CFD simulation and low-dimensional modeling of SJAs and SJs in quiescent conditions are reviewed, and major results from these numerical studies are presented. Before that, a dimensional analysis is briefly introduced.

### 2.1 Dimensional Analysis

Consider a typical SJA shown in Fig. 1, which consists of a cavity, an orifice and an oscillating diaphragm. The behavior of the SJ depends on eight parameters: (1) diaphragm oscillation amplitude $\Delta$, (2) diaphragm oscillation frequency $f$, (3) diaphragm area $A_d$, (4) orifice area $A_o$, (5) orifice depth or thickness $h$, (6) cavity volume $V_c$, (7) fluid kinematic viscosity $v$, and (8) fluid density $\rho$.

According to the Buckingham–Pi theorem, these parameters can be condensed into the following five independent, non-dimensional parameters that describe the actuator geometric and operating conditions:

$$\pi_1 = \Delta/A_o^{\frac{1}{2}}, \pi_2 = A_d/A_o, \pi_3 = V_c/(A_d A_o^{\frac{1}{2}}), \pi_4 = h/A_o^{\frac{1}{2}}, \pi_5 = fA_o/v \qquad (1)$$

The SJ performance parameters expressed in their non-dimensional form are expected to be functions of these non-dimensional parameters.

If the fluid is assumed incompressible, it is easy to obtain that

$$\begin{aligned} L &\triangleq \bar{U}_o/fA_o^{\frac{1}{2}} \propto \Delta A_d/A_o^{\frac{3}{2}} = \pi_1\pi_2 \\ Re &\triangleq \bar{U}_o A_o^{\frac{1}{2}}/v \propto \Delta f A_d/vA_o^{\frac{1}{2}} = \pi_1\pi_2\pi_5 \end{aligned} \qquad (2)$$

where $\bar{U}_o$ is the spatial- and time-averaged jet blowing velocity over an actuation cycle, $L$ the non-dimensional stroke length representing the length of the fluid column, and $Re$ the Reynolds number based on the jet velocity and the size of SJA orifice. Another two frequently used non-dimensional parameters, i.e., the Strouhal number $St$ and the Stokes number $S$, are functions of $L$ and $Re$

$$\begin{aligned} St &\triangleq fA_o^{\frac{1}{2}}/\bar{U}_o \propto 1/L \\ S &\triangleq (fA_o/v)^{\frac{1}{2}} \propto (Re/L)^{\frac{1}{2}} \end{aligned} \qquad (3)$$

If the orifice size $A_o^{1/2}$, diaphragm oscillating frequency $f$, and fluid density $\rho$ are chosen as the repeating parameters, the non-dimensional jet performance parameters is related to $L$ through

$$\begin{aligned} &\bar{U}_o/fA_o^{\frac{1}{2}} \propto L \\ &\bar{\dot{Q}}_o/fA_o^{\frac{3}{2}} \propto L \\ &\bar{\dot{M}}_o/\rho f^2 A_o^2 \propto L^2 \\ &I_o/\rho f A_o^2 \propto L^2 \\ &\Gamma_o/fA_o \propto L^2 \end{aligned} \qquad (4)$$

where $\bar{\dot{Q}}_o$ is the time-averaged volume flow rate, $\bar{\dot{M}}_o$ the time-averaged momentum flux rate, $I_o$ the total momentum flux or impulse discharged during a cycle, and $\Gamma_o$ the total vortex circulation produced in a cycle. Alternatively, if the orifice size $A_o^{1/2}$, fluid viscosity $v$, and fluid density $\rho$ are chosen as the repeating parameters, the above jet performance parameters can be related to $Re$ as well. The expressions can be easily derived and are not presented here for brevity. It is noteworthy that the qualitative relations described in Eq. 4 are valid no matter whether the fluid inside SJAs is incompressible or compressible. This statement has been supported by Tang and Zhong [87].

In this chapter, since many simulation and experimental cases of SJs are involved, if possible, all the corresponding non-dimensional stroke length $L$ and Reynolds number $Re$ will be converted to the values calculated based on the definitions given in Eq. 2. This will make the flow and SJA operating conditions comparable among different cases.

## 2.2 CFD Simulations

As an SJ consists of a train of vortex structures, its CFD simulations in quiescent flows usually focus on two aspects, i.e., the production and evolution of vortex structures. Depending on the shape of orifice, the vortex structures produced are different: a high-aspect-ratio rectangular orifice (2D slot) produces vortex pairs, a square orifice produces square vortex structures, and a circular orifice produces vortex rings. In recent years, SJs produced through 2D slots in quiescent and cross flows have been studied intensively due to their predominant 2D nature in the near field, which makes the data analysis much easier. In comparison, fewer investigations have been carried out on low-aspect-ratio rectangular or circular SJs. However, these SJs are attracting more and more attentions in flow control applications, especially when microelectronic mechanical systems (MEMS) based SJAs are used, because they are space saving and easily arrayed. In this section, the CFD simulations on SJs from 2D slots and circular orifices are reviewed. Before doing so, some common issues on the SJA CFD simulations are introduced.

### 2.2.1 Mathematical Representation of SJA Perturbation

For SJ simulations, it is crucial to correctly capture the spatial and temporal variation of the velocity at the orifice exit. For simplicity, in some investigations the SJA is not included in the computational domain. Instead, the orifice exit is treated as a suction/blowing or transpiration boundary. Kral et al. [39] introduced a sinusoidal variation of the jet velocity with different spatial distributions at the orifice exit

$$u_o(\xi = 0, \eta, t) = U_o f(\eta) \sin(\omega t) \tag{5}$$

where $\xi$ denotes the jet streamwise direction, $\eta$ the cross-stream direction, and $u_o$ the instantaneous jet velocity. If necessary, higher harmonics can be added into Eq. 5 to introduce jet velocities with multiple frequencies. Different spatial variations over the orifice can be considered, including

$$f(\eta) = \begin{cases} 1 \\ \sin(\pi\eta) \\ \sin^2(\pi\eta) \end{cases} \tag{6}$$

In Eq. 6, $f(\eta) = 1$ represents a "top hat" distribution, which, among the three expressions, matches the most with what was observed in experiments. Mallinson et al. [46] applied Eq. 5 together with this distribution in their simulations.

The velocity boundary condition can be replaced by a pressure gradient boundary condition, with the application of the streamwise momentum equation by ignoring the viscous term [39]

$$\frac{\partial p}{\partial \xi} = -\rho \frac{\partial u_o}{\partial t} \tag{7}$$

Although simple and useful, the above simplification may bring in significant discrepancies since the velocity variation and distribution are significantly affected by the flow inside the SJA. Rizzeta et al. [65] simulated the cavity flow and the external flow separately using the chimera overset zonal mesh method, and applied the orifice exit velocity information obtained from the cavity flow to the external flow simulations. They concluded that the resultant jet velocity profiles as well as the exterior flow fields are different from the analytical expressions described in Eqs. 5–7.

To model the cavity flow, the movement of the diaphragm has to be considered. Most of practical SJAs used diaphragms with a piezoelectric material bonded onto a brass disc. Considering a circular diaphragm and assuming a uniform piezoelectric force applied on the diaphragm, an analytical expression for the out-of-plane displacement of the diaphragm $w(r, t)$ can be given as [34, 59, 60]

$$w(r,t) = \sum_{n=1}^{\infty} [A_n J_n(kr) + C_n I_n(kr)] \cos(n\theta) \cos(\omega_n t) \tag{8}$$

where the subscript $n$ denote the mode number of the oscillation, $\omega_n$ is the nth natural frequency, the expression in the bracket describes the radial variation of displacement, and $\cos(n\theta)$ describes the circumferential variation of displacement. The radial components are given in terms of Bessel function $J_n$ and modified Bessel functions of the first kind, $I_n$. $A_n$, $C_n$ and $k$ are determined by the supporting conditions of the diaphragm. Mallinson et al. [47] used the 0th mode of Eq. 8, i.e., quasi-static deformation of the diaphragm, in their simulations.

Tang and Zhong [84] described the deformation of their mechanical driven diaphragm (a stainless steel disc clamped around the edge and driven by a rod at the center) at the 0th mode using the following formula

$$w(r,t) = \frac{\Delta}{2}\left[1 - \left(\frac{r}{a}\right)^2 + 2\left(\frac{r}{a}\right)^2 \ln\left(\frac{r}{a}\right)\right] \cos(\omega t) \tag{9}$$

where $a$ is the radius of the diaphragm. In their simulations, the SJA operating frequency is only 50 Hz, far below the first natural frequency of the diaphragm. Using this formula, they obtained simulated jet velocity close to that from experimental measurements.

For diaphragms other than circular shape, no analytical expression has been used for their deformation even it is available for some simple shapes, such as rectangular. Past investigations just assumed a piston-type motion [65] or a simple sinusoidal deformation motion. They adjusted the oscillation magnitude until the simulated mean jet velocity information matches with the corresponding experimental data.

To model the oscillation of the diaphragm, two major numerical techniques were applied. In one method, the mesh in the cavity deforms with the deformation of the diaphragm [47, 65]. In the other method, i.e., the immersed boundary method, a localized body force along desired points in the computational mesh is imposed to drive the fluid at a specified velocity so that the force has the same effect as a moving boundary [42, 52, 62]. Both techniques involve complicated numerical treatment and are time consuming.

To reduce the computational costs, velocity or transpiration boundary conditions were applied at the neutral position of the diaphragm. Tang and Zhong [84] used the time derivative of Eq. 9 along with the diaphragm amplitude measured in experiments as the velocity inlet boundary condition at the diaphragm neutral position, and obtained simulated jet velocity close to the experimental measurements. However, in general cases, the amplitude of velocity oscillation may need to be adjusted to match the experimental velocity information at the orifice exit.

For accurate simulations of the production and evolution of SJs from an SJA, replicating the flow conditions at the orifice exit is more important than the accurate modeling of the flow in the cavity. Utturkar et al. [89] examined the sensitivity of the jet to the design of cavity. Changes in the cavity aspect ratio as well as in the location of the oscillating diaphragm were considered. They found the changes in the cavity design have a limited effect on the jet produced by the cavity. Vatsa and Turkel [91] used simplified cavity geometry to simulate the cavity flow inside an SJA. They also applied a simplified transpiration boundary condition on the bottom side instead of the actual diaphragm position on the sidewall in the simulation, as shown in Fig. 2. A uniform velocity profile with multiple frequency components is applied at this transpiration boundary. The frequency components are determined by curve fitting the measured velocity at the orifice exit with the fast Fourier transform to reflect the proper mode shape and to ensure zero net mass transfer. The amplitude of this transpiration velocity is determined numerically to match the simulated peak velocity at the orifice exit with experimental measurements. With this treatment, they managed to obtain the time history of jet velocity and the streamwise and spanwise distribution of time-averaged jet velocity close to the experimental data (see Fig. 3). Zhang and Wang [109] adopted similar techniques and also obtained very good agreements in the velocity variation history at three different stations between the simulated results and experimental data. Based on these studies, it is suggested that the jet flow can be computed/modeled with

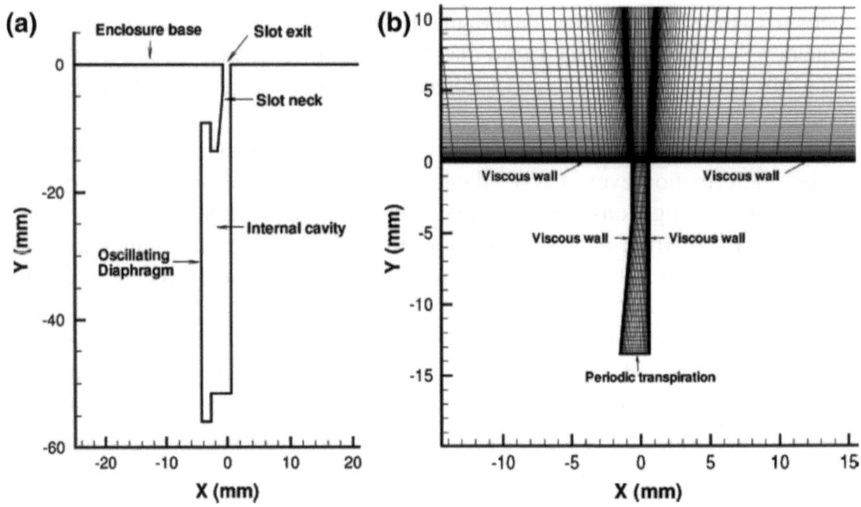


**Fig. 2** Simplification of the cavity geometry and boundary conditions: **a** actual cavity and **b** simplified cavity (From Vatsa and Turkel [91]. Reproduced with kind permission from the authors)

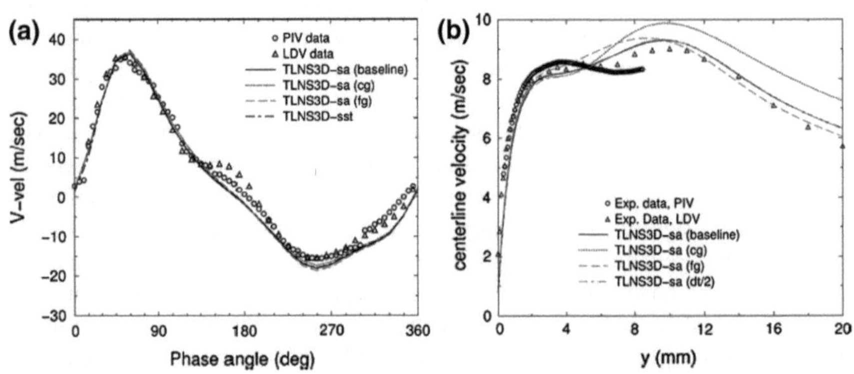


**Fig. 3** Comparison of **a** simulated time history of jet velocity near orifice exit and **b** simulated time-averaged velocity along the jet centerline with experimental data (From Vatsa and Turkel [91]. Reprinted with kind permission from the authors)

sufficient fidelity without precisely matching the cavity design. However, it is noted that compressibility effects were not included in these simulations and for this reason the conclusion regarding flow insensitivity to cavity design is only limited to incompressible flows.

### 2.2.2 Laminar Turbulent Transition and Turbulence Modeling

SJs produced through an orifice by the oscillation of diaphragm may undergo a transition from laminar to turbulent flow. Depending on the frequency and amplitude of the diaphragm oscillation, the transition may occur at different streamwise locations. For low-frequency and low-amplitude oscillations, a laminar SJ may be produced near the orifice exit and it then becomes turbulent further downstream, like what was observed by Smith and Glezer [81] and Yao et al. [105]. With high-frequency and high-amplitude oscillations, turbulent SJs may be produced right at the orifice exit, although the flow inside the cavity could still be laminar.

Kotapati et al. [38] investigated the turbulence transition inside and outside the cavity in their numerical study on a 2D-slot SJ. It was found that, outside the slot exit, turbulent fluctuations occur almost exclusively during the blowing cycle, and center around the phase where the local phase-averaged velocity is at its maximum (shown in Fig. 4a, b). During the suction cycle, the flow accelerates towards the slot and this tends to stabilize the flow. Consequently, the fluctuation level during this phase of the cycle is small. The behavior of the velocity fluctuation inside the slot and cavity, on

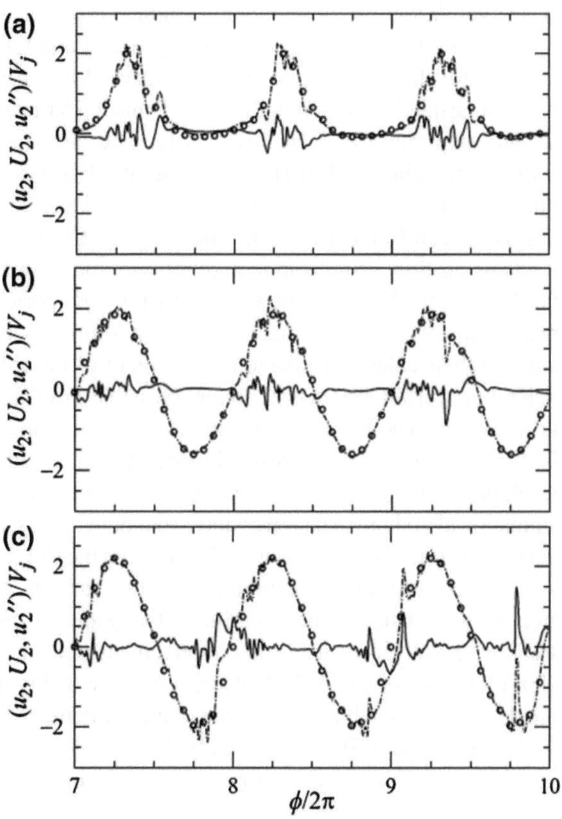


**Fig. 4** Temporal variation of instantaneous streamwise velocity $u_2/V_j$, phase-averaged streamwise velocity $U_2/V_j$ and turbulent fluctuation $u_2''/V_j$ inside the slot and cavity along the *centerline*: $x_2/d$ = **a** 1.93, **b** 0, **c** −2.0, *dashed lines*, $u_2/V_j$; *solid lines*, $u_2''/V_j$; *open circles* $U_2/V_j$ (From Kotapati et al. [38]. Reproduced with permission)

the other hand, is quite different. The turbulent fluctuations occur primarily during the acceleration phase of the cycle, i.e., during the phase of the cycle when the bulk velocity increases from its minimum value to its maximum (shown in Fig. 4c). They suspected the reason is that the turbulence fluctuations inside the slot are initiated in the cavity and convected into the slot during the expulsion.

For more accurate SJ simulations, therefore, numerical methods that can account for laminar-turbulent transition are preferred. Different numerical methods have been employed in simulating turbulent SJs, including direct numerical simulations (DNS), large eddy simulations (LES), unsteady Reynolds-Averaged Navier-Stokes equations (unsteady RANS or URANS), blended RANS-LES or detached eddy simulations (DES), and even non-Navier-Stokes equation based methods, such as lattice Boltzmann method [36, 49, 92].

### 2.2.3 Production and Evolution of Synthetic Jets from 2D Slots

SJs produced through high-aspect-ratio slots in quiescent and cross flows have been studied intensively due to their predominant 2D nature in the near field. Spanwise homogeneity makes the study more attractive in a laboratory setting and, even more so, for validation of computational studies, especially for the RANS simulations.

The experimental results obtained by Smith and Glezer [80, 81] became the first database for the validation of many numerical investigations. In their experiments, they synthesized a nominally 2D jet (0.5 mm × 75 mm) by the oscillatory motion of a circular diaphragm in a shallow cavity. The circular diaphragm was driven at its resonant frequency (1140 Hz) by a centrally bonded piezo-ceramic disk. The non-dimensional stroke length $L$ ranges between 5.3 and 25, and the Reynolds number $Re$ varies between 104 and 489. They investigated the near-field formation and evolution of SJs using flow visualization and hot-wire measurements, and observed that, near the slot exit, the SJ flow is dominated by the time-periodic formation, advection, and interactions of discrete vortex pairs, which ultimately become turbulent, slow down, and lose their coherence due to rapid growth of instabilities.

Kral et al. [39] reported 2D incompressible flow simulations of both laminar and turbulent SJs. The latter was obtained by the solution of URANS equations with the Spalart–Allmaras one-equation turbulence model. Simulation of a laminar SJ shows the development of a train of vortex pairs (as shown in Fig. 5a). While the turbulent simulation also shows a pair of counter-rotating vortices, the vortex pair quickly diffuses and is not distinct beyond one wavelength from the slot exit (as shown in Fig. 5b), which is in agreement with the experimental observations. The simulation results for the streamwise variation of the centerline velocity and the cross-stream distributions of the streamwise velocity were in good agreement with the measurements of Smith and Glezer [80]. However, the lack of three- dimensionality in the laminar flow simulations failed to capture the breakup of the vortex train observed in the experiments.

Rizzetta et al. [65] investigated both 2D and 3D flow fields of finite aspect-ratio (30) SJs using DNS of the compressible N-S equations. The exterior flow field was constructed using high-order compact-difference schemes in conjunction with

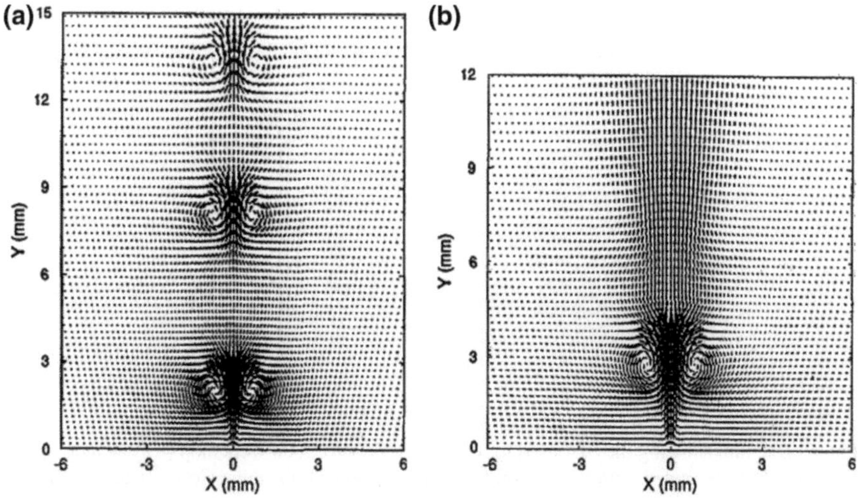


**Fig. 5** Simulated velocity vectors of **a** a laminar synthetic jet and **b** a turbulent synthetic jet at similar operating conditions (From Kral et al. [39]. Reprinted by permission of the American Institute of Aeronautics and Astronautics, Inc.)

non-dispersive spatial filters by enforcing the symmetry condition along the jet centerline that precluded asymmetric interactions between vortices across the centerline. A comparison between time-averaged velocity distributions downstream of the slot of both 2D and 3D simulations and the measurements of Smith and Glezer [80] yielded a reasonable agreement near the jet centerline. Three dimensionality in their simulations helped capture the breakdown of external flow into a turbulent jet, a phenomenon observed in the experiments but missed in any strictly 2D calculations.

Other numerical simulations on similar SJA settings include Lee and Goldstein [42] who reported 2D computations of an array of two slot SJs performed using an immersed boundary method. For a comprehensive review of both the experimental and computational investigations on these high-aspect-ratio SJs, see Glezer and Amitay [24].

To further assess the CFD capability in simulating SJ flow as well as turbulent separation control, a workshop on CFD validation of three selected test cases was held at the NASA Langley Research Center in March 2004 (CFDVAL2004 Workshop). These three test cases were selected and ordered in an increasing level of complexity with respect to the flow physics involved: (1) a flow field generated by a plane SJ in a quiescent condition, (2) a circular oscillatory jet in a cross flow, and (3) the control of a turbulent separated flow over a wall-mounted hump model. A number of CFD simulations were conducted for each of the test cases. And by comparing the simulation results with the complete experimental data sets provided by NASA Langley Research Center, the capability of the state-of-the-art CFD techniques in simulating SJ flows and turbulent separation control were evaluated. All the test data and a list of submitted CFD simulations undertaken are available to

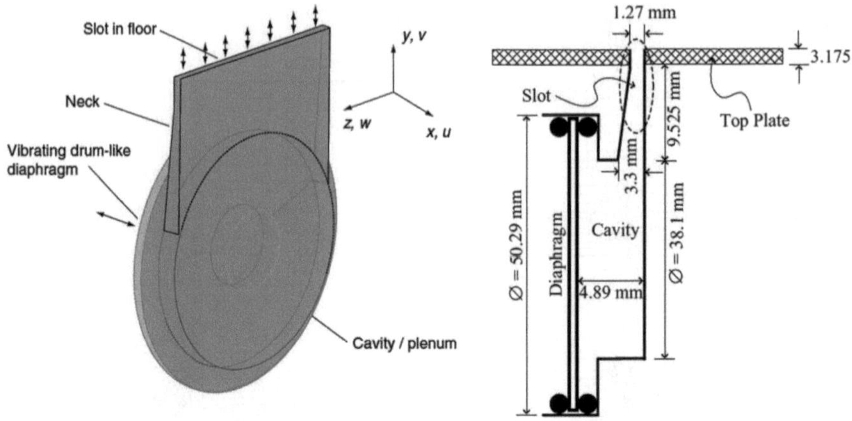


**Fig. 6** Diagram showing the SJA used for Case1 in CFDVAL2004 workshop (*left*) isometric view (Rumsey [68]. Reprinted with permission from Multi-Science Publishing Co Ltd.) and (*right*) side view (From Kotapati et al. [38]. Reproduced with permission)

the public through NASA's website http://cfdval2004.larc.nasa.gov/ (cited in Sep 2014). A summary of these numerical simulation efforts can be found in Rumsey et al. [69, 70] and Rumsey [67].

In this section, we only focus on the simulations related to Case 1, i.e., a plane SJ produced through a rectangular slot in quiescent air. The slot is 1.25 mm wide and 35.6 mm long, giving an aspect ratio of about 28. The jet was driven by a circular piezoelectric diaphragm of 50.8 mm in diameter, mounted on one sidewall of the cavity, as shown in Fig. 6. The diaphragm was driven at 445 Hz. The SJ peak velocity is around 40 m/s, according to which the non-dimensional stroke length is estimated to be about $L = 23$ and the Reynolds number about $Re = 1075$. In this experiment, Yao et al. [105, 106] measured the flow velocity outside the slot using three different techniques, i.e., hot-wire anemometry, laser-Doppler velocimetry (LDV) and particle image velocimetry (PIV). The time-averaged contours of streamwise jet velocity from two orthogonal views are shown in Fig. 7, in which a significant three dimensionality can be observed. In the view along the slot length (Fig. 7a), the jet is narrow near the slot exit, and turns into a wider and diffusive jet about 12*h* above the slot, where *h* is the slot width. In the view along the slot width, a more complicated plane jet structure is developed (Fig. 7b). The jet flow emanates uniformly near the jet exit. At a height of about 4*h*, the end vortices roll toward the center part of the jet. The homogeneous central region does not appear to extend much above 8*h* or so. From the experimental data the jet flow was believed to be transitional [68].

The jet produced through the slot of aspect ratio 28 in this case is categorized as a rectangular or finite-span SJ. Sahni et al. [73] classified SJs as plane 2-D or very-high-aspect-ratio slots (with aspect ratio of 75 or above), circular, elliptic or low-aspect-ratio slots (with aspect ratio below 5) and rectangular or finite-span slots (with aspect ratio between 5 and 75), and they found the behaviours of these types of SJs are quite different.

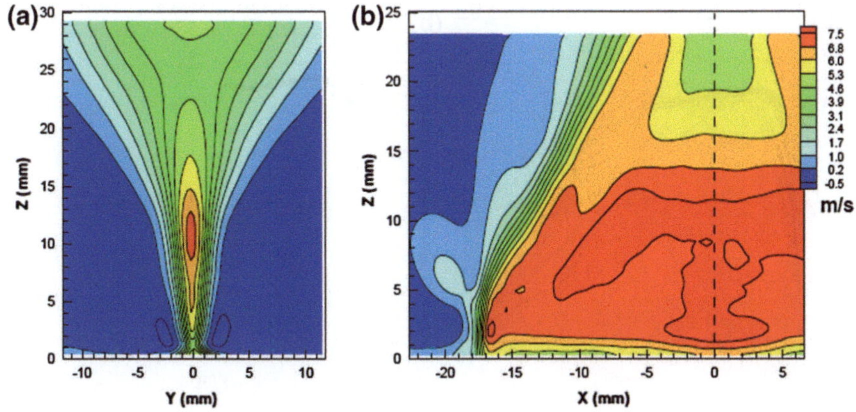


**Fig. 7** Contours of mean streamwise velocity of a plane synthetic jet: **a** view along slot length, **b** view along slot width (From Yao et al. [106]. Reprinted by permission of the American Institute of Aeronautics and Astronautics, Inc.)

The submitted simulations for this case used a broad range of computational methods, including DNS [38], DES [11, 100, 101], URANS [1, 7, 58, 90, 91, 109], and even a 1-D model [104].

Since as shown in Fig. 7 the end-wall effects caused significant three dimensionality in the SJ flow, Yao et al. [105, 106] and Rumsey et al. [67, 68] suggested that 2D CFD simulation is only limited to the region within $8h$ distance from the slot exit. However, most submitted simulations treated the flow as a 2D flow. The side-mounted piezoelectric driver and complex cavity geometry were difficult to simulate, so all computations made approximations inside the cavity or simply applied transpiration boundary conditions directly on the wall from which the SJ emanated. As a result, the various computations often did not even start off with the same boundary conditions at the slot exit as in the experiment.

Most of the submitted simulations did not faithfully model the entire cavity and used a simple sine wave transpiration boundary condition at the diaphragm position. The actual drum-like diaphragm, combined with the complex 3D cavity shape, likely produced additional modes that affected the SJ emanating from the slot. For these reasons, results from these simulations did not appear to match the experimental data at the slot exit very well. Figure 8a shows the comparison of the computed and the measured time-averaged velocity along the centerline. A very wide range of results is observed, with some simulations being in better agreement with experiment than others. The computed turbulence quantities also did not compare well at all with the experimental data. After the workshop, Vatsa and Turkel [91] improved their original transpiration boundary condition through the process of curve-fitting the measured velocity at the slot exit using the fast Fourier transform (FFT), and they achieved better agreement with the experimental data, as shown in Fig. 8b. This FFT technique was later adopted by Zhang and Wang [109] and a good agreement was also achieved.

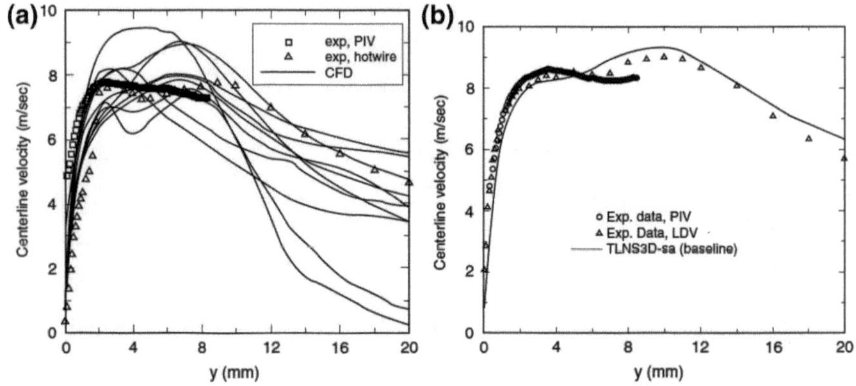


**Fig. 8** Time-averaged velocity along the SJ centerline for Case1 in CFDVAL2004, comparing experimental data with **a** simulation results from 10 workshop submissions and **b** simulation results from Vatsa and Turkel [91] (From Rumsey [68]. Reprinted with kind permission from Multi-Science Publishing Co Ltd.)

To investigate the laminar-turbulent transition of the SJ flow, Kotapati et al. [38] introduced a small spanwise sinusoidal perturbation of 1 % of jet velocity for a short period of time in the first cycle of their 3D DNS simulations, and the three dimensionality was then developed on its own through the inherent instability in the flow. With this operation, they were able to capture the breakdown of the counter-rotating vortex pairs and the transition to turbulence due to spanwise instabilities. They also found the turbulence was not well-developed within the slot, but according to the spectra the flow breaks down rapidly into a well-developed turbulent jet a short distance away from the slot exit, as shown in Fig. 9. However, the edge of the slot was not included in their simulations and hence the end-wall effects were not captured.

Through the above plane SJ simulations based on the experimental data from both Smith and Glezer [80, 81] and Yao et al. [105, 106], one of the main advancements has been the development/recognition of techniques to better match the boundary conditions of the experiment at the slot exit. When the simulations matched experiments better at the slot exit, they tended to match better in the exterior flow field. Although the cavity shape itself does not appear to be a key factor, the jet exit flow field is a priori knowledge in these cases in order to determine appropriate boundary conditions. However, Xia and Qin [100] managed to simulate the drum-like motion of the piezoelectric membrane that did not make use of this knowledge at the slot exit. Such a truly predictive method is more desirable, because one may then be able to predict actuator performance with only knowledge of the characteristics of the vibrating driver. Another advancement has been the recognition of the importance of end effects for this flow field. Even with a slot aspect ratio as large as 28, the end effects become significant in a relatively short distance from the slot exit [105, 106].

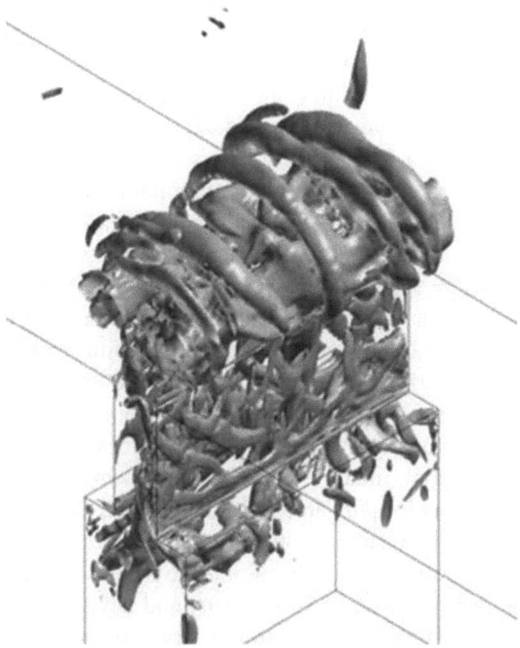

**Fig. 9** Visualization of vertical structures in the near field for Case1 in CFDVAL2004 from DNS results (From Kotapati et al. [38]. Reproduced with permission)

Many different methodologies were seen to be able to yield reasonable results in the near field, including low-order URANS, DES, and direct N-S simulations (note: not DNS). For the direct N-S simulations, solely laminar results were generally poor, but with sufficient resolution and small enough numerical dissipation that allowed the N-S solver to develop turbulent structures, results were good in the near field. It is unclear whether the URANS methods are fully adequate for predicting mean-flow quantities, and if so, which models work and which do not. With high-fidelity boundary conditions, the linear Spalart–Allmaras and SST $k$-$\omega$ models appeared to do reasonably well for many features of interest in the near field. Nevertheless, Carpy and Manceau [7] suggested that linear models are missing out key physics and more advanced nonlinear models or even full Reynolds stress model may be required. Challenges that remain also include the need to perform computations that accurately capture the end effects.

### 2.2.4 Production and Evolution of Synthetic Jets from Orifice of Circular or Other Shapes

Compared to the investigations on high-aspect-ratio plane SJs, the number of numerical investigations on low-aspect-ratio (circular or square) SJs is relatively fewer. However, low-aspect-ratio SJs are capable of producing more effective separation control per orifice area by generating strong streamwise vorticity that enhances the mixing within the boundary layer. This type of SJs is attractive in flow control applications also because they are space saving and can be easily arrayed.

Unlike plane SJs, there is no widely used experimental database for CFD validations for low-aspect-ratio SJs. Mallinson et al. [46] measured the SJ flow field outside an SJA with a circular orifice of 0.75 mm diameter using a hot-wire probe. The SJA operates at the diaphragm's resonance frequency 1450 Hz. The SJ peak velocity is about 30 m/s and the non-dimensional stroke length and Reynolds number are estimated as about $L = 8.8$ and $Re = 480$, respectively. At these conditions, the SJ was found to be turbulent, which was confirmed by comparing their laminar and turbulent (URANS using the standard $k$-$\varepsilon$ model) simulation results with the test data. The variation of the centerline velocity obtained from both experiments and simulations starts to follow the self-similar trend ($1/y$) from approximately 10 orifice diameters downstream, much less than continuous jets for which the self-similar flow is not established until approximately $40D_o$. Similar observations were also made by Rathnasingham and Breuer [60] for circular, and Smith and Glezer [80, 81] for planar, SJs. In Mallinson et al.'s laminar simulations, a train of vortex rings can be captured, as shown in Fig. 10a. However, when the turbulence is considered, a large vortex ring is trapped near the orifice as shown in Fig. 10b, and newly produced vortex rings dissipate just near the orifice, "feeding" to this large vortex ring. This is similar to what was captured in the simulations by Karl et al. [39, see Fig. 5].

Tang and Zhong [84] simulated circular SJs at two different diaphragm oscillation settings. Their SJA had an orifice of 5 mm diameter and was operated at a fixed frequency 50 Hz, which was far below its resonant frequency of the diaphragm and actuator. An incompressible flow solver was used in this simulation. In the first case the non-dimensional stroke length is $L = 2.0$ and the Reynolds number $Re = 170$, whereas in the second case $L = 3.2$ and $Re = 280$. Based on smoke visualization observations, the jet flow was laminar in the first case and became turbulent after about $2D_o$ from the orifice exit in the second case. For the second case, four different turbulence models, i.e., the standard $k$-$\varepsilon$, re-normalization group (RNG) $k$-$\varepsilon$, standard $k$-$\omega$ and Reynolds stress models were applied. It was found that the simulation results for the laminar case agreed well with the experimental data, and for the turbulence simulations the RNG $k$-$\varepsilon$ and standard $k$-$\omega$ models produced the best match with the experimental data.

For SJAs with an orifice of small diameter, typically less than 1 mm, the compressibility and viscosity effects in the SJA become significant causing the SJs behave differently from large-scale SJs. Tang and Zhong [87] simulated circular SJs produced through an orifice of diameter 0.5 mm. For this simulation the compressible flow solver was applied. The non-dimensional stroke length varies between 1.4 and 4.8 and the Reynolds number varies between 10 and 50. The compressibility inside the SJA makes the simulated jet velocity deviate from the prediction based on impressibility assumption. Figure 11 shows the vortex rings produced from SJAs of $D_o = 5$ mm and 0.5 mm at the same $L$ value. It is seen that the vortex structures of the two SJs have identical shapes and convective velocities.

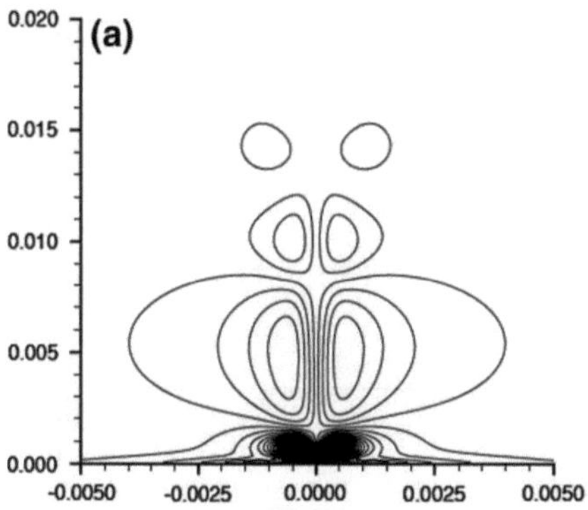


**Fig. 10** Vorticity contours for **a** laminar flow simulation, and **b** turbulent flow simulation at the same phase (From Mallinson et al. [46]. Reprinted with kind permission from the authors)

Nevertheless, the value of the highest vorticity level in the vortex center for the $D_o = 0.5$ mm case (Fig. 11a) decreases more rapidly than that of the $D_o = 5$ mm case (Fig. 11b), indicating a higher dissipation rate. The simulation also reveals that for small-scale SJs, while the linear relations predicted by the incompressible flow model [85] between the non-dimensional jet flow parameters (non-dimensional $\bar{U}_o$ and $\bar{\dot{M}}_o$ et al.) and the actuator operating parameters ($\Delta$ and $f$) are no longer valid, the linear relations between the non-dimensional jet performance parameters and the non-dimensional jet flow parameters ($L$ and $Re$) still hold.

Using DNS, Ravi et al. [62] investigated the production and evolution of SJs from rectangular slots of three different aspect ratios, i.e., 1, 2, and 4, in quiescent air. The non-dimensional stroke length varies from $L = 6.2$–$12.2$, and the Reynolds number is fixed at $Re = 227$. As shown in Fig. 12, for the square slot a distinct vortex ring emerges, which forms a nearly circular vortex ring right after leaving the slot. For the slots of aspect ratio 2 and 4, an "axis-switching" phenomenon occurs to the rectangular vortex ring, which rotates in its own plane by 90° as it propagates downstream and becomes perpendicular to the slot. The jet velocity profile at the slot exit was also examined. It was found that with the increase of the aspect ratio, the jet velocity profile along the slot width tends to be more parabolic.

Later Ravi and Mittal [63] extended the above numerical work and simulated the jet flow from a rectangular slot with a higher aspect ratio 8. The non-dimensional stroke length is high at about $L = 40$, and the Reynolds number is fixed at $Re = 300$. Ample flow details were captured in this simulation. Figure 13 shows the evolution of the produced vortex structures at four instants. In the earlier phase of blowing shown in Fig. 13a, the rectangular vortex ring emerging from the slot forms a nearly circular vortex ring that is connected to the slot by a rectangular shear layer. At a later stage in the expulsion shown in Fig. 13b this circular vortex is still visible although it has propagated a significant distance from the slot so that it nearly catches up with the vortices from the previous cycle. A number of other vortices form upstream of the circular vortex, including a strong elongated vortex that has

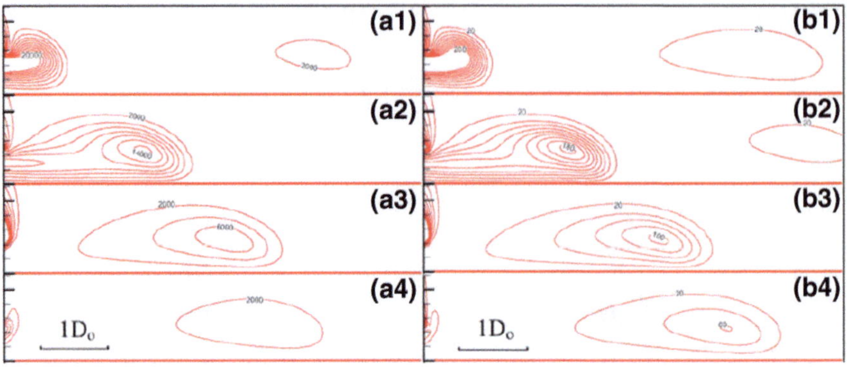


**Fig. 11** Time sequences of vorticity contour for synthetic jets from an orifice of **a** $D_o = 0.5$ mm and **b** $D_o = 5$ mm. The non-dimensional stroke lengths of both synthetic jets are the same (Reprinted from Tang and Zhong [87]. With kind permission from Springer Science and Business Media)

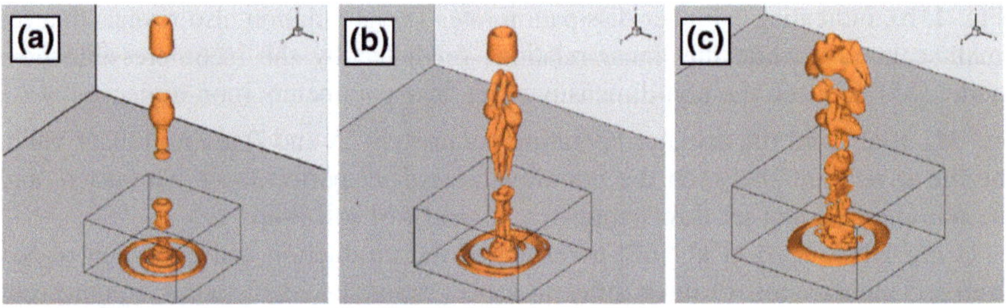


**Fig. 12** Vortex structures produced from an SJA with a slot of aspect ratio of **a** 1, **b** 2, and **c** 4. $Re = 227$ (From Ravi et al. [62]. Reprinted by permission of the American Institute of Aeronautics and Astronautics, Inc.)

rotated 90 degrees from the slot, indicating the occurrence of "axis switching". At a later stage in the cycle when flow is ingested into the cavity (Fig. 13c) the expelled vortices evolves into a complex vortex cluster containing a sequence of hairpin-like vortices. These vortices are formed due to the stretching of vortex filaments by the underlying strain field induced by the jet. At a later stage in the cycle (Fig. 13d) the vortex cluster propagates further downstream and the vortex strength starts to diminish due to the influence of viscosity.

Wu and Leschziner [94, 95] used LES to simulate a single SJ issuing into quiescent surroundings as well as into a zero-pressure-gradient turbulent boundary layer over a flat plate. For the former simulation, they re-produced the vortex structures generated by an SJA with a 4.4 mm by 4.4 mm square orifice. The SJA's operating conditions match with those in the experiment by Garcillan et al. [21] at $L = 11.5$ and $Re = 240$. The simulated evolution of the produced vortex rings is presented in Fig. 14. During the blowing stroke, the detached shear layer rolls up to

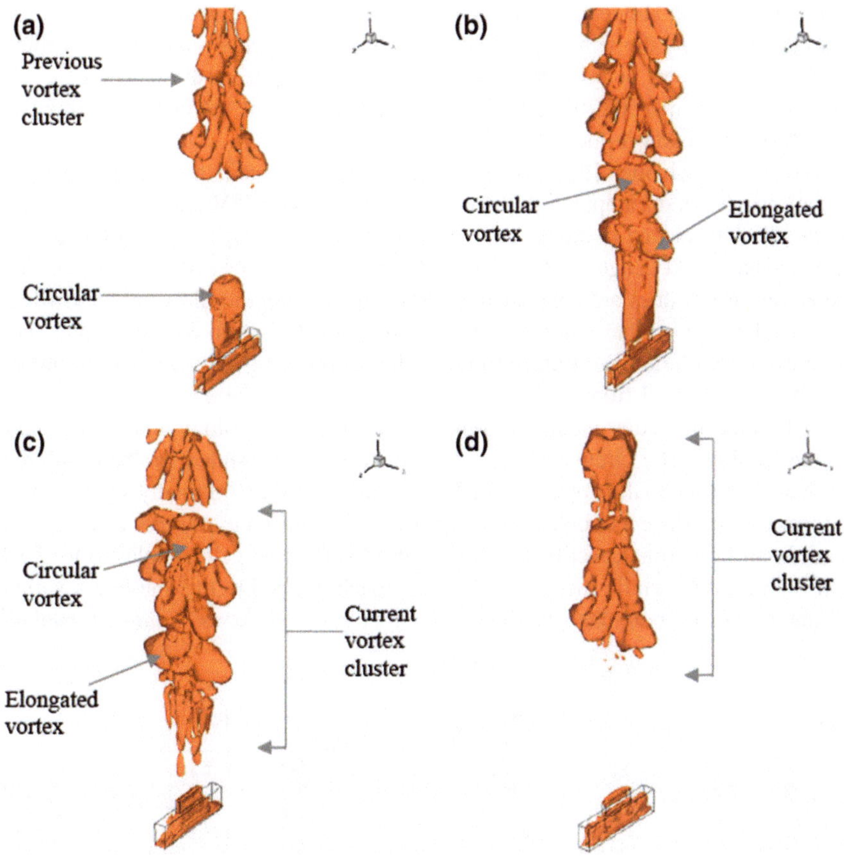


**Fig. 13** Evolution of vortex structures produced from an SJA with a slot of aspect ratio 8. *Re* = 300 (From Ravi and Mittal [63]. Reprinted with kind permission from the authors). **a** Maximum expulsion (Phase $\phi$ = 90). **b** Maximum volume (Phase $\phi$ = 180). **c** Maximum ingestion (Phase $\phi$ = 270). **d** Maximum volume (Phase $\phi$ = 360)

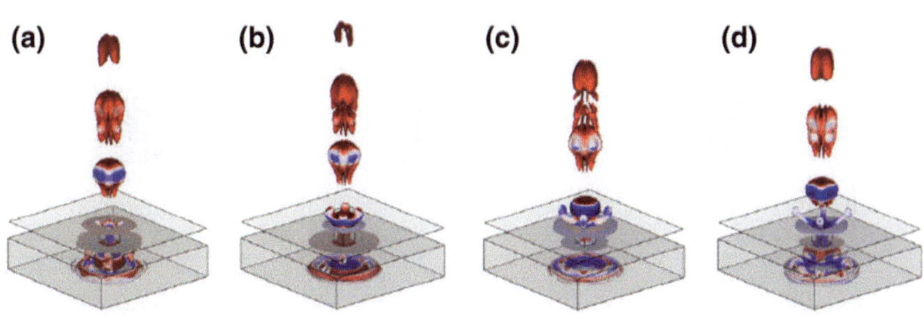


**Fig. 14** Evolution of vortex rings produced by a square-orifice SJA operating at $L$ = 11.5 and *Re* = 240. The structures are determined by iso-surfaces of vorticity magnitudes and colored by streamwise velocity. Four phases are showed: **a** $\phi = 0$; **b** $\phi = \pi/2$; **c** $\phi = \pi$; **d** $\phi = 3\pi/2$ (Reprinted from Wu and Leschziner [95], with permission from Elsevier)

form a vortex ring above the orifice. Upon formation, the ring then propagates away from the orifice by the self-induced velocity and subsequently merges with vorticity patches that are left behind by the break-up of the ring from the preceding cycle. This process of merging triggers, in turn, the break-up of the current vortex ring prior to its diffusion in the far field.

Wu and Leschziner [95] also simulated the SJs issued through circular and square orifices and compared the simulation results with PIV data from Cater and Soria [8]. The orifice diameter is $D_o$ = 2 mm. The SJs were operated at a non-dimensional stroke length of $L^*$ = 333 and a Reynolds number of $Re^*$ = 5000. Note the $L^*$ and $Re^*$ values are calculated based on the momentum flow velocity, which is defined based on the non-zero mean momentum flux of an SJ during one actuation cycle. At these operating conditions, the resulting SJ flow is highly turbulent, which was observed from both the experiments and simulations. Wu and Leschziner [95] paid particular attention to the effects of mesh topology and orifice shape on the SJ flow. The circular shape of the orifice is represented by three different meshes, as shown in the upper row of Fig. 15. In mesh a, an originally square grid is deformed to produce a circular geometry, which is referred to as *deformed mesh.* In mesh b, an orthogonal, rectilinear mesh is used with solid cells employed to block non-fluid regions within the orifice. The approach applied to this mesh is a variant of the immersed-boundary method, which is referred to as *solid-cell method.*

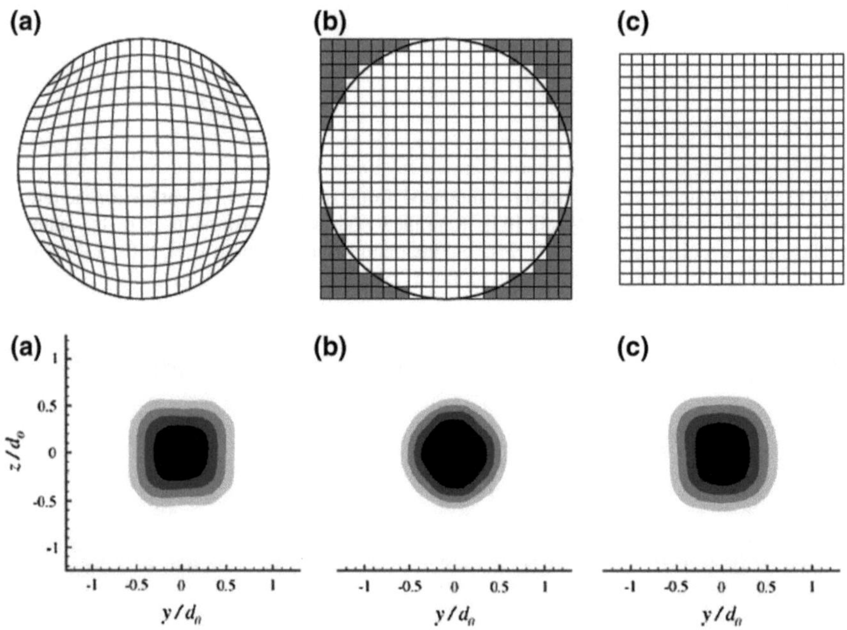


**Fig. 15** *Upper* Approaches to mesh the circular orifice. *Lower* Contour plots of time-averaged, streamwise velocity at the cross-flow plane $x/D_o$ = 2. **a** Deformed mesh; **b** solid cell; **c** square orifice (Reprinted from Wu and Leschziner [95], with permission from Elsevier)

In mesh c, a square orifice is used to simply replace the original circular orifice, which is referred as *square jet*. The cross-sectional area of this square orifice is kept the same as that of the circular orifice. The lower row of Fig. 15 shows the contours of the mean streamwise velocity in a cross-stream plane $2D_o$ away from the orifice. It was found that the *deformed mesh*, although providing a representation of the circular geometry, returned a near-orifice velocity that is not markedly different from that produced with the square orifice. The reason is because the strong deformation of the mesh exerts an undesirable influence on the flow. In contrast, the fully rectilinear mesh of the *solid-cell method* returned a much better representation of the axisymmetric flow. The *square jet* reflects, as expected, the shape of the underlying orifice. Further comparison indicates that the *deformed mesh* and *solid-cell method* led to only minor deviations in the time-averaged flow and turbulent-stress profiles. The *square jet*, however, produced higher levels of turbulence, due to the increased straining at the corners. Based on these findings, Wu and Leschziner [95] concluded that the *solid-cell method* is the best meshing method for their code, and they adopted this method in their flow-control simulations.

## 2.3 Low Dimensional Modeling

Due to the large parameter space involved in SJA design, the task of optimizing and selecting actuators with the desired performance using time-consuming CFD simulations is not practical. Hence, reduction in the complexity of the full CFD model described by the 2D or 3D N-S equations is necessary, and low-dimensional or reduced-order models are sought. By solving the simplified governing equations, a low-dimensional model enables the key performance parameters of an SJA, such as the jet velocity and resonant frequency, with an acceptable accuracy.

By assuming an incompressible flow inside the cavity and an inviscid flow in the orifice duct, Tang and Zhong [85] developed a zero-dimensional, incompressible model for SJAs. This model directly links the oscillatory displacement of the diaphragm to the jet velocity and other jet performance parameters, such as the mass flow rate, momentum flux rate, jet impulse, and vortex circulation. It gives reliable prediction of the characteristics of SJs for an actuator with a large orifice. However it significantly over-predicts the jet velocity when the orifice is small such that the effect of viscosity in the orifice duct becomes dominant and the compressibility in the cavity cannot be ignored.

Rathnasingham and Breuer [60] developed a zero-dimensional, compressible flow model for SJAs, which consists of five coupled, first-order, nonlinear ordinary differential equations. The diaphragm made of a thin circular plate is considered and resembled by a piston that moves in the cylindrical cavity. The SJA is treated as a second-order dynamic system. The flow in the orifice is modeled using the unsteady Bernoulli equation, so that the SJ velocity can be calculated from the pressure difference across the orifice. The air in the cavity is assumed isothermal and its density and pressure are determined by applying the equation of state of ideal gases.

In the end, equations for the diaphragm position and velocity, the fluid density and pressure, and the jet velocity, are derived. With this model, they characterized the viscous effects by the Stokes number and found an optimal operating Stokes number by matching the viscous flow solution to an ideal, inviscid flow solution. Compressibility and flow unsteadiness are shown to lead to nonlinear fluid effects that dominate when the orifice size is small. The model was shown to predict the trends of the actuator behavior reasonably well. Later Crook et al. [9] adopt this model in their analysis and added one more equation which incorporates the force produced by the piezo-ceramic material attached on the diaphragm to the above system of equations. Their model did not perform well, however, since the steady Bernoulli equation was applied to the oscillating orifice flow.

Lockerby and Carpenter [45] further extended Rathnasingham & Breuer's model, and developed a one-dimensional, compressible flow model. In their model, the movement of the diaphragm is modeled based on the classic thin-plate theory. It is assumed that the thermodynamic process occurring in the cavity is isothermal and instantaneous, and the air can be modeled as a perfect gas. The flow through the orifice is modeled as an unsteady pipe flow to which the N-S equations for an axisymmetric, parallel flow are applied. This approach is based on the assumption that streamlines in the orifice exit are parallel to its axis, which is an adequate approximation if the orifice length-to-diameter ratio is much larger than unity. The model is able to predict the jet velocity of SJAs from an input voltage applied on the diaphragm. Despite over-predicting the magnitude of the jet velocity, this model yields optimal cavity height and orifice depth that are close to the experimental observations [9].

Tang and Zhong [86] modified Lockerby & Carpenter's model to enable the diaphragm oscillation frequency and amplitude to vary independently. With this modified model (named the static compressible model), the effect of changing SJA geometrical and operating parameters on the jet peak velocity was investigated. It was found, depending on the flow conditions inside the orifice duct, the SJA may operate in two distinct regimes, i.e., the Helmholtz resonance regime and the viscous flow regime. In the Helmholtz resonance regime, the resultant SJ is generated by the mass flow physically displaced by the oscillating diaphragm coupled with the Helmholtz resonance in the actuator. In the viscous flow regime, the Helmholtz resonance is completely damped by the viscous effect such that the jet is produced by the diaphragm oscillation alone.

By representing the characteristics of a speaker-driven SJA using an equivalent electrical circuit, McCormick [48] made the first attempt to model SJAs using the lumped element model (LEM). This model was further developed by Gallas et al. [19, 20] to model a piezoelectric-driven SJA. The main assumption employed in the LEM is that the characteristic length scale of the governing physical phenomena is much larger than the largest geometric dimension. If this assumption holds, the temporal and spatial variations in an acoustic system can be decoupled, and the governing partial differential equations for the distributed system can be "lumped" into a set of coupled ordinary differential equations. In the LEM, the individual components of an SJA are modeled as elements of an electrical circuit (such as

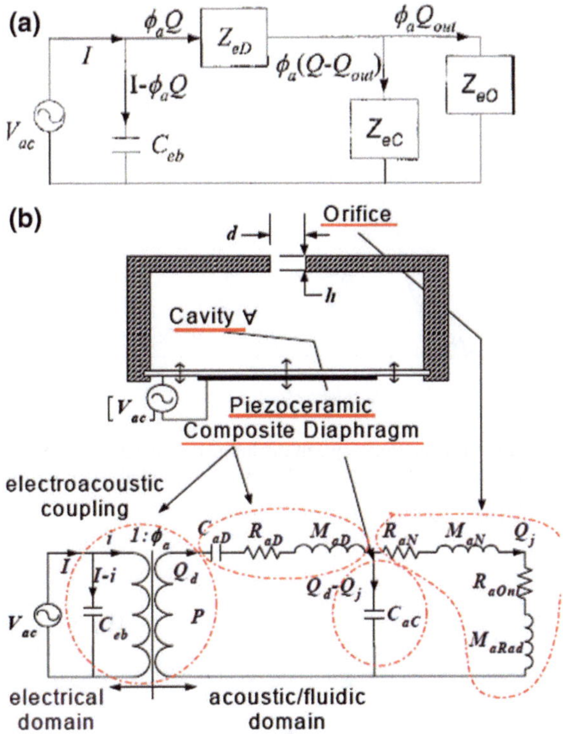


**Fig. 16** Equivalent circuit of a piezoelectric-driven synthetic jet actuator **a** schematic of the model (From Gallas et al. [19]. Reprinted with kind permission from the authors) **b** detailed electric representation of the equivalent circuit (From Gallas et al. [20]. Reprinted with kind permission from the authors)

capacitors, inductors or resistors) with the volume velocity $Q$ acting as the "electrical current" and the pressure difference $\Delta P$ between the inside and the outside of the actuator as the "electrical voltage" (see Fig. 16). The linear composite plate theory is used to calculate the net volume velocity of the diaphragm, such that the electric input is transformed into the diaphragm deformation. The acoustic mass and resistance in the orifice are obtained with the incompressible, fully developed laminar pipe flow theory. As a result, the LEM neglects the compressibility effects inside the orifice duct. The model was applied to two prototypical SJAs and, by artificially tuning the damping coefficient of the diaphragm, it provides a good agreement with the measured peak jet velocity (Fig. 17). One of the conclusions of this study is that a better model is required to model the unsteady flow in the orifice duct, including the entrance and exit effects.

Based on the model developed by Gallas et al. [19, 20], Tang and Zhong (2009) proposed a new LEM for SJAs in which the mechanical movement of diaphragm is decoupled from the fluid phenomenon in the actuator cavity to allow the modeling of fluid mechanics aspect of the SJA to be investigated and validated separately. They also incorporated the "minor losses" occurring at the orifice exit into the model to improve the modeling fidelity. They compared the modeling results for an SJA from three different models, i.e., the incompressible model [85], the static compressible model [86], and the LEM, and concluded that the LEM is capable of

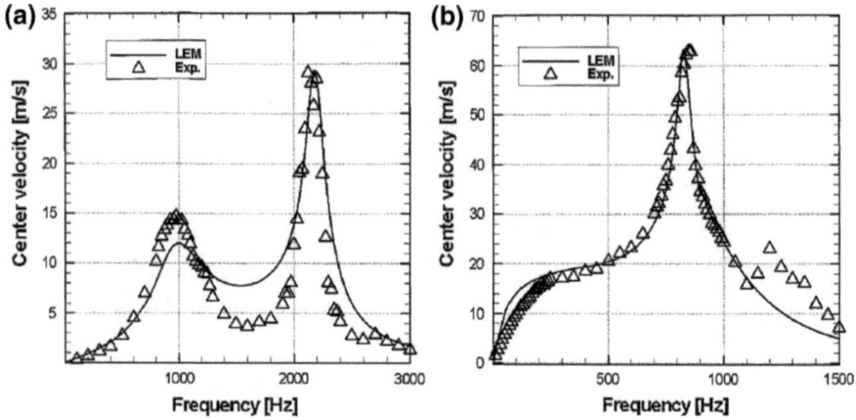


**Fig. 17** Comparison between the lumped element model and experiment for two cases **a** damping coefficient $\zeta = 0.03$, **b** $\zeta = 0.02$ (From Gallas et al. [19]. Reprinted with kind permission from the authors)

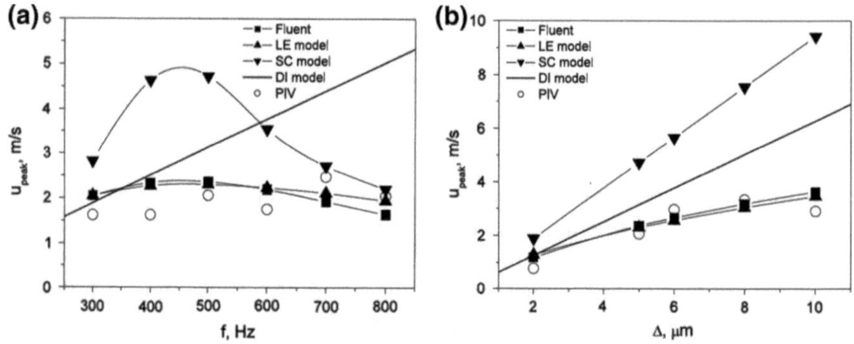


**Fig. 18** Comparison of the SJ peak velocity predicted by three low dimensional models (Reprinted from Tang and Zhong [89]. With kind permission from Springer Science and Business Media)

providing a reliable prediction of the jet peak velocity required for designing SJAs for practical applications (see Fig. 18). They also showed that the LEM is capable of predicting the temporal variation of SJs both in phase and magnitude when the actuator is operating away from the SJA's Helmholtz resonance frequency. Although the model fails to predict the correct phase information at the Helmholtz resonance frequency, it can still produce the jet peak velocity that is in good agreement with experimental data.

Yamaleev et al. [103] proposed a quasi-one-dimensional model for SJA simulations, in which the one-dimensional Euler equations are used to compute the cavity flow, and the 2D N-S equations are used to compute the flow inside and outside the orifice. The oscillating diaphragm is simulated as a quasi-one-dimensional moving boundary. Compared to 2D or 3D simulations, this low-dimensional model has demonstrated its computational efficiency. However, compared to the aforementioned low-dimensional models it is still not cheap since the 2D full CFD simulation is still needed in the outer portion of the orifice. They found that if the Reynolds number based on the slot width $d$ is larger than 500 and the interface between the 1D Euler equations and the 2D N-S equations is located more than $2d$ away from the slot exit inside the orifice, this model provides high accuracy in the exterior flow field. Yamaleev and Carpenter [104] applied this model to a realistic SJA listed in Case1 of CFDVAL2004 Workshop. They provided the conditions at which their model is applicable: (1) Similar to the constraint in the LEM, the actuator size should be much smaller than the wavelength of diaphragm oscillations. (2) The cavity volume, orifice length, slot exit size, diaphragm area, frequency, and amplitude of the quasi-one-dimensional model should be equal to those of the 3D actuator, whereas the quasi-one-dimensional cavity profile may not exactly coincide with that of the realistic actuator.

It should be noted that the efficiency obtained by using low-dimensional models is offset by a reduced accuracy that is resulted from the simplification. Therefore, a balance between the efficiency and the desirable level of accuracy has to be made in the development of any low-dimensional model.

## 3 Synthetic Jets in Cross Flows

In the flow-separation-control applications, single or an array of SJs are issued into a cross flow. The interaction of SJs with cross flows has been an interesting topic associated with boundary layer flows, vortex dynamics, and flow separation. However, the simulation of this type of flow phenomenon also poses a number of challenges. First, the scale disparity between the small SJA orifice and the large body of fluid being controlled by the SJ may cause a serious grid-resolution problem. In addition, the need to resolve the near-wall region of the boundary layer, especially at high Reynolds numbers, substantially aggravates the resource problem. Second, when the fluid is compressible, the acoustic characteristics of the cavity are important. This introduces a time-scale that is much smaller than the SJA actuation period. Third, if the flow is turbulent, getting statistically converged time-averaged and phase-averaged results requires many actuation cycles to be included in the simulation. Also, it is not practical to simulate a turbulent boundary layer approaching the SJ from its inception, and the prescription of unsteady, turbulent conditions across an inflow plane fairly close to the SJ is not an easy job. And finally, if flow separation is involved, how to accurately predict the separation point, the re-attachment point, and the dynamics associated with flow separation has always been very challenging. With all these challenges in mind, therefore, the state-of-the-art techniques in simulating SJs in cross flows will be reviewed and discussed in this section.

## 3.1 Dimensional Analysis

For an SJ issuing into a cross flow, the list of physical parameters described in Sect. 2.1 on which the SJ flow depends should be extended to include the freestream velocity $U_\infty$, the local boundary-layer thickness $\delta$, and the wall shear stress $\tau_w$. Based on the dimensional analysis, the characteristics of the SJ at its orifice exit plane will depend on the following non-dimensional parameters (Zhong et al. 2007)

$$\begin{aligned} &\pi_1 = \Delta/A_o^{\frac{1}{2}}, \pi_2 = A_d/A_o, \pi_3 = V_c/(A_d A_o^{\frac{1}{2}}), \pi_4 = h/A_o^{\frac{1}{2}}, \\ &\pi_5 = fA_o/\nu, \pi_6 = \delta/A_o^{\frac{1}{2}}, \pi_7 = fA_o^{\frac{1}{2}}/U_\infty, \pi_8 = \tau_w/\rho U_\infty^2 \end{aligned} \tag{10}$$

Therefore in comparison to SJs in quiescent conditions (Eq. 1), three additional non-dimension parameters, i.e., the ratio of boundary-layer thickness to orifice size $\pi_6$, the Strouhal number $St = \pi_7$ (different from the Strouhal number for SJs in quiescent conditions, defined in Eq. 3), and the non-dimensional shear stress $\pi_8$, also play important roles in determining the characteristics of an SJ as it enters a cross flow. A velocity ratio *VR*, is often introduced to represent the relative strength between the SJ velocity and the freestream velocity of the cross flow

$$VR \triangleq \bar{U}_o/U_\infty \tag{11}$$

Alternatively, a momentum coefficient $C_\mu$ is also frequently used in literature to characterize the relative strength of single SJs or an array of SJs

$$C_\mu \triangleq n\bar{U}_o^2 A_o/U_\infty^2 A_{ref} \tag{12}$$

where $n$ is the total number of SJs in operation, and $A_{ref}$ the characteristic area corresponding to the cross flow.

The cross-flow boundary layer is usually characterized by a momentum thickness based Reynolds number

$$Re_\theta \triangleq U_\infty \theta/\nu \tag{13}$$

If the cross flow is applied on an object of finite dimension, such as an airfoil or a hump, a reduced frequency is defined as

$$F^+ \triangleq fc/U_\infty \tag{14}$$

where $c$ is the characteristic length of the object.

## 3.2 Synthetic Jets in Attached Cross Flows

Mittal et al. [52] simulated the interaction of a plane SJ with a laminar flat plate boundary layer using an incompressible N-S solver. On the upstream side of the outer flow domain, a velocity inlet boundary condition corresponding to a Blasius boundary layer profile is applied. Four different cases corresponding to different velocity ratios, i.e., $VR = +\infty$ (quiescent external flow), 0.95, 0.21, and 0.10, were simulated and compared. Figure 19 compares the vortex structures at the end of blowing. In comparison to a synthetic jet in a quiescent flow, at the velocity ratio of 0.95 the initially symmetric vortex pair becomes inclined towards the downstream of the cross flow. Due to the velocity gradient in the boundary layer, the upstream branch of the vortex pair convects faster than its downstream counterpart and even catches up with the downstream branch of the preceding vortex pair (shown in Fig. 19b). As the velocity ratio decreases to 0.21, the vortex pair is unable to penetrate the boundary layer.

Its upstream branch is cancelled out by the boundary layer, which has opposite-signed vorticity, and its downstream branch entrains fluid from freestream and grows in size. The entrainment of high-momentum freestream fluid into the boundary layer by this spanwise vortex is beneficial to flow separation control. As the velocity ratio decreases further, the vortex pair is too weak to penetrate the boundary layer, and no direct entrainment is observed.

The change in velocity ratio also changes the flow inside the cavity. As the velocity ratio decreases, the initially symmetric, large area of almost stagnant fluid near the two sidewalls of the cavity shown in Fig. 19a shrinks and becomes asymmetric because the suction of higher momentum fluid energizes the fluid inside the cavity.

The influence of the cross flow on the SJ velocity profile at the slot exit was also investigated by Mittal et al. [52]. When there is no cross flow, as shown in Fig. 20a, the velocity profiles at all phases are symmetric about the centerline and they are quite different during the blowing and suction strokes. The profile exhibits more of a "jet–like" profile during blowing whereas it is more "plug–like" during suction. When a cross flow is introduced, the profile becomes skewed towards the downstream during the blowing stroke, while it is relatively unchanged during the suction stroke. As the cross-flow velocity increases or the velocity ratio decreases, the

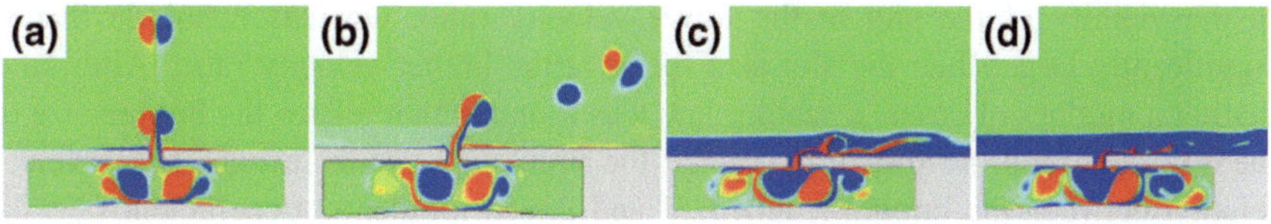


**Fig. 19** Vorticity contours at the end of blowing for a plane SJ with $VR =$ **a** $+\infty$; **b** 0.95; **c** 0.21; **d** 0.10 (From Mittal et al. [52]. Reprinted by permission of the American Institute of Aeronautics and Astronautics, Inc.)

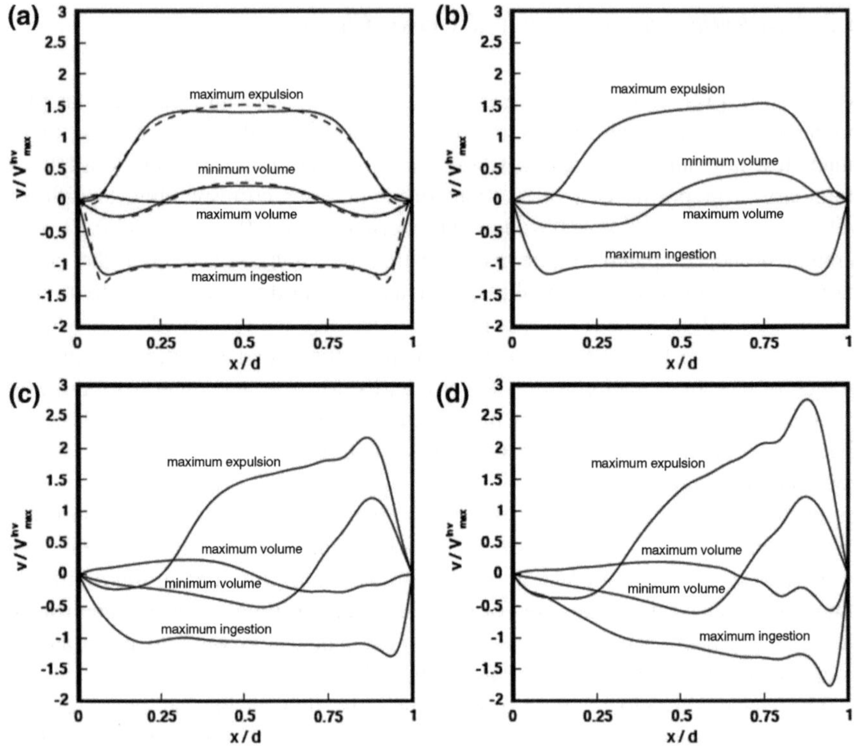


**Fig. 20** Velocity profiles at the orifice exit for a plane SJ with $VR$ = **a** $+\infty$; **b** 0.95; **c** 0.21; **d** 0.10 (From Mittal et al. [52]. Reprinted by permission of the American Institute of Aeronautics and Astronautics, Inc.)

skewness of the profile further increases. At the smallest velocity ratio of 0.10, the profile during the suction stroke becomes skewed too.

The so-called "virtual aeroshaping" effect caused by a plane SJ in a cross flow was also investigated using the same numerical framework [51]. Due to the zero-net-mass-flux constraint, SJs are capable of forming recirculation bubbles in the mean exterior flow, which can be significantly larger in size than the SJA slot size. Figure 21 shows the streamlines of mean velocity field for two different cases, in which the non-dimensional boundary thickness is fixed at $\delta/d = 5$, and the velocity ratios are $VR = 0.63$ and 2.5, respectively. Only a small recirculation bubble forms in the vicinity of the slot at the lower velocity ratio, whereas a bubble that is significantly larger than the slot size forms at the higher velocity ratio. After simulating a number of cases, a scaling law was found which relates the bubble size to the flow condition. As shown in Fig. 22, the simulation results reveal a nearly linear relation between the bubble length $L_r/d$ and the SJ momentum coefficient $C_\mu$.

The 3D flow field of low-aspect-ratio SJs in cross flow was also simulated by the same group [62]. The interactions of a laminar boundary layer (LBL) with SJs from slots of three different aspect ratios, i.e., $w/d = 1$, 2, and 4, are shown in Fig. 23.

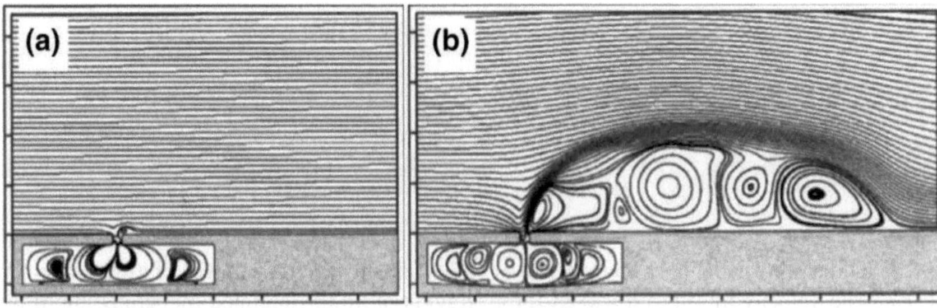


**Fig. 21** Virtual aeroshaping sketched by streamline of mean flow with $\delta/d$ = 5. **a** $VR$ = 0.63; **b** $VR$ = 2.5 (Reprinted with permission from Mittal and Rampuggoon [51], AIP Publishing LLC.)

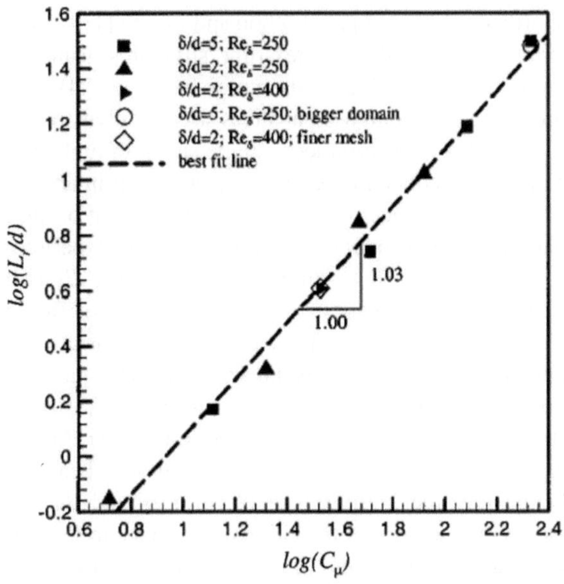


**Fig. 22** Correlation between normalized bubble length ($L_r/d$) and momentum coefficient ($C_\mu$) (Reprinted with permission from Mittal and Rampuggoon [51], AIP Publishing LLC.)

In these three cases the non-dimensional boundary layer thickness is kept at $\delta/d = 2$ and the velocity ratio is held at $VR = 0.25$. For all the three cases, the formation of a sequence of vortex-loop type structures that convect downstream in the boundary layer was observed. Clear vortex rings in the streamwise direction are found at about a cavity length distance from the slot center.
These vortex rings penetrate the boundary layer and emerge out of the boundary layer indicating that 3D SJs have more momentum than their 2D counterpart.

For low-aspect-ratio, especially circular SJs issuing into an LBL, the interaction results in the generation of streamwise vortex structures, which is more capable of entraining high-momentum flow into the boundary layer. As the magnitudes of the non-dimensional stroke length and the velocity ratio increase, the resulting vortex structures appear in different shapes, i.e., hairpin vortices that attached to the wall,

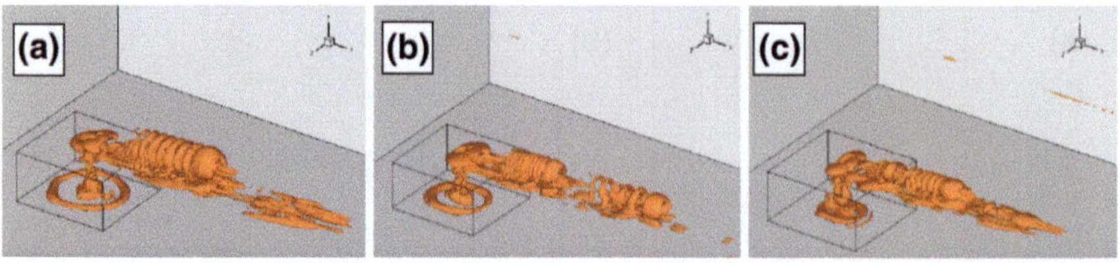


**Fig. 23** Interaction of a laminar boundary layer with synthetic jets from an SJA of aspect ratio **a** 1; **b** 2; **c** 4. The velocity ratio is $VR = 0.25$ (From Ravi et al. [62]. Reprinted by permission of the American Institute of Aeronautics and Astronautics, Inc.)

stretched vortex rings that gradually move towards the edge of the boundary layer, and tilted or distorted vortex rings that penetrate the edge of the boundary layer shortly downstream [111]. Zhou and Zhong [114] examined the formation of hairpin structures and tilted-vortex-ring structures using their simulated results. They reproduced the primary structures (see Fig. 24), confirming that hairpin vortices produce a pair of high shear stress streaks downstream of the SJA orifice whereas the tilted vortex rings produce a single streak of high shear stress. They also revealed a hierarchy of coherent vortices consisting of primary, secondary and tertiary structures (see Fig. 25) hence providing a further insight of the formation of the two different patterns of wall shear stress observed in the experiment by Jabbal and Zhong [33].

By using the same simulation framework, Zhou and Zhong [115] further did a systematic investigation on how the nature of the vortex structures produced by circular SJs and their associated surface shear stress patterns vary with the jet operating condition. They presented a *VR-L* parameter map, in which the transitional boundary separating the one-streak and two-streak pattern of surface shear stress is shown as a red dash line (see Fig. 26). Above the line the tilted vortex ring type of vortices and the one-streak surface shear stress pattern are observed, and below the line hairpin type of vortices and the two-streak surface shear stress pattern are seen. The transitional boundary in the *VR-L* parameter map is nearly a vertical line at around $L = 1.6$ and it turns sharply into a horizontal line at $VR \approx 0.32$. By investigating the increase of the space- and time-averaged skin friction coefficient, they concluded that in order to maximize the impact on the near-wall flow while keeping the energy expenditure low, it is wise to maximize the accumulated effect of hairpin vortices by keeping the spacing between consecutive hairpin vortices similar to the local boundary layer thickness upstream of the separated flow instead of producing stronger individual structures.

When a turbulent boundary layer (TBL) is involved, the simulation and analysis become more complicated, as described in the beginning of this section. Using LES, Wu and Leschziner [95] simulated a circular SJ issuing into a fully turbulent flat-plate boundary layer at a momentum-thickness Reynolds number $Re_\theta = 2380$. The non-dimensional stroke length of the SJ is $L = 29$, the velocity ratio is $VR = 2.0$, and the ratio of boundary-layer thickness to orifice diameter is $\delta/D_o = 10$. The computational grid resolves the boundary layer down to $\Delta y^+ \approx 0.5$, with both the streamwise and spanwise resolution being $\Delta x^+$, $\Delta z^+ \approx 24$. The incoming TBL is

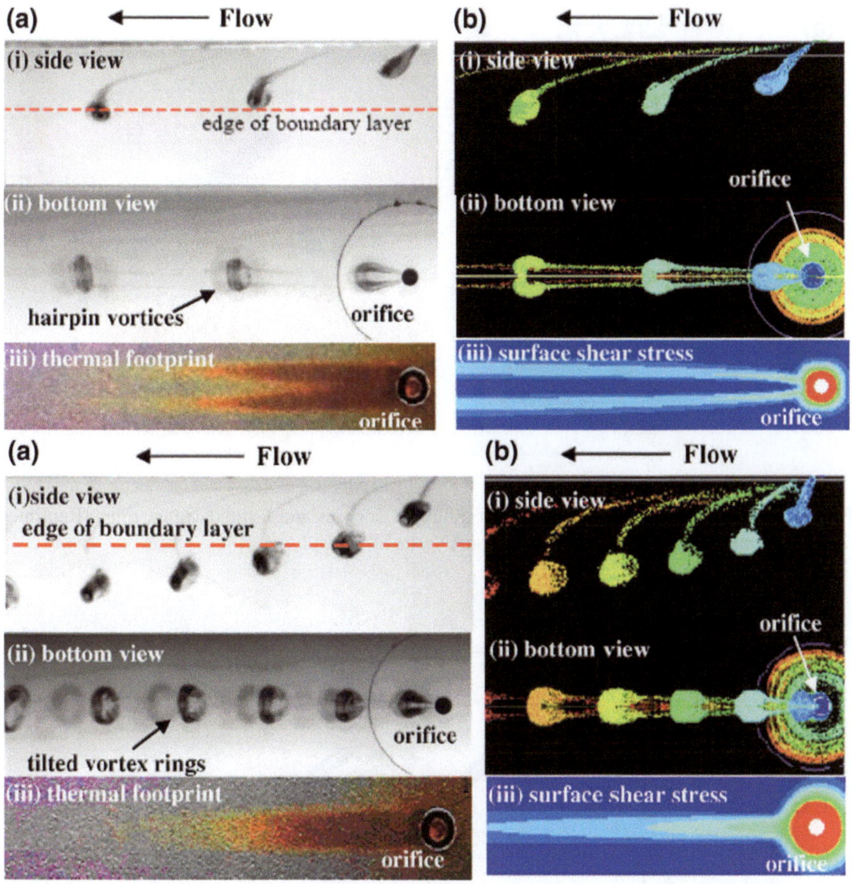


**Fig. 24** Comparison of (*left*) dye and liquid crystal visualization from experiments and (*right*) particle trace and surface shear stress from simulations for (*upper*) hairpin vortices and (*lower*) tilted vortex rings (Reprinted from Zhou and Zhong [114], with permission from Elsevier)

realized by an unsteady representation of the inlet conditions, through the use of 5000 instantaneous flow realizations extracted from a separate precursor simulation for a canonical boundary layer at the same Reynolds number, generated with a rescaling method. The circular geometry of the nozzle is represented by means of the solid-cell technique, as depicted in Fig. 15b. Spanwise periodicity is imposed on the lateral sides. As shown in Fig. 27, the streamwise vortex structures produced by the interaction of the SJ with the TBL are not as distinguishable as in LBLs. These vortex structures break up into turbulent patches very soon.

Wen and Tang [93] compared the generation and evolution of SJ-induced hairpin vortices and the accompanying vortex structures in two different boundary layers, one is an LBL with a momentum-thickness Reynolds number of $Re_\theta = 180$ at the SJA orifice exit and the other is a TBL with a Reynolds number of $Re_\theta = 420$.

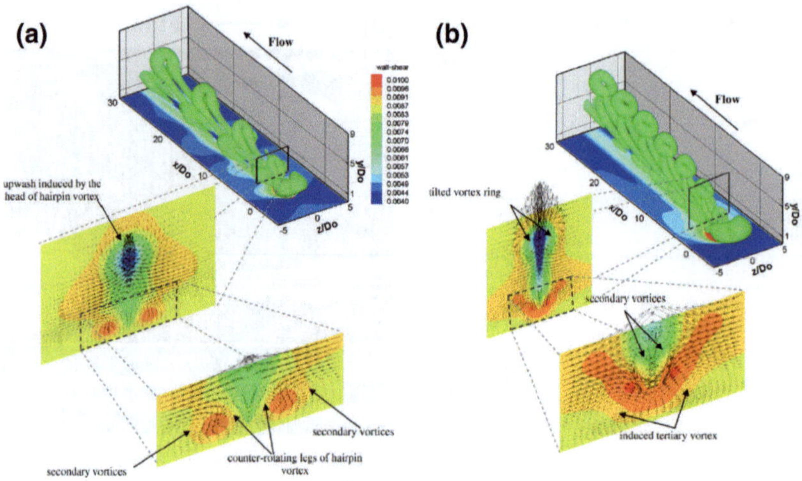


**Fig. 25** Instantaneous vortex structures of **a** hairpin vortices and **b** tilted vortex rings (Reprinted from Zhou and Zhong [114], with permission from Elsevier)

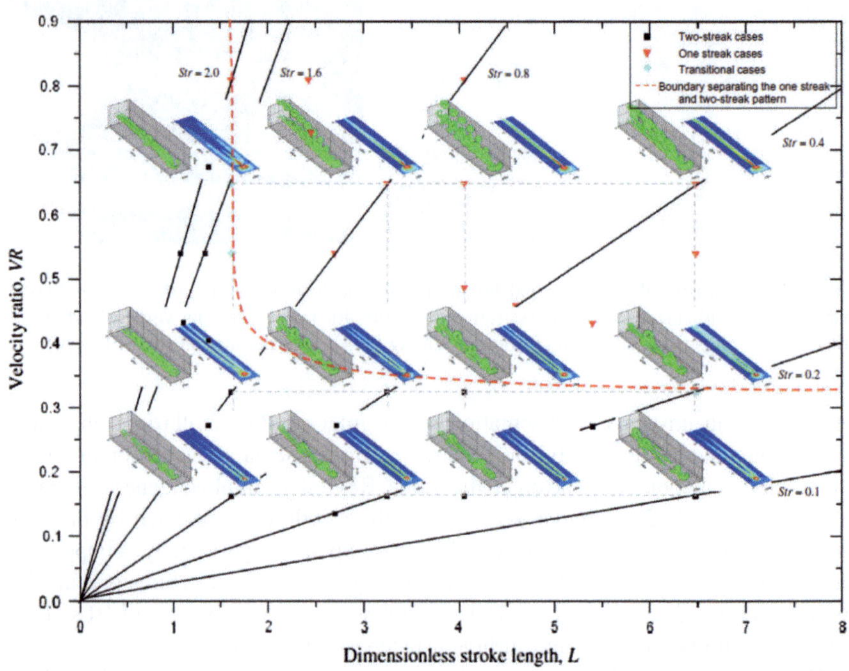


**Fig. 26** *VR-L* parameter map of different vortex structures and their corresponding time-averaged wall shear stress (Reprinted from Zhou and Zhong [115], with permission from Elsevier)

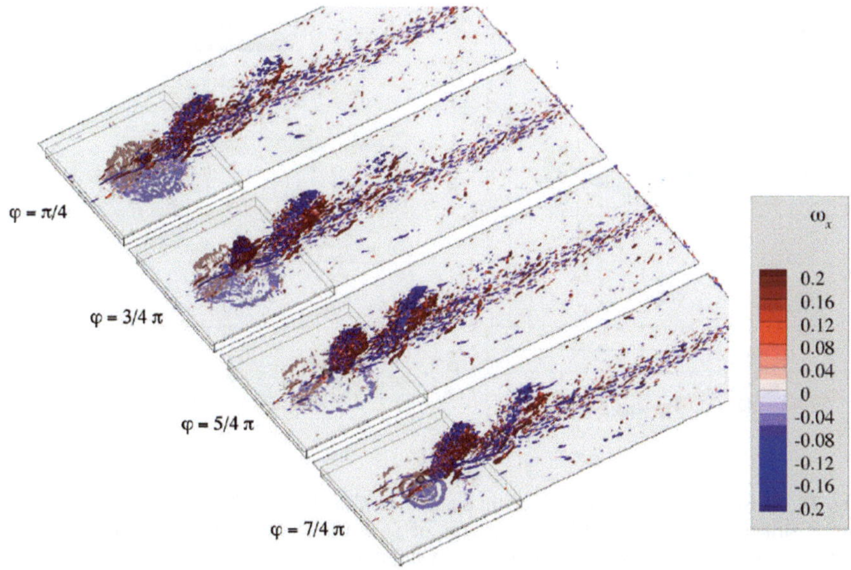


**Fig. 27** A circular SJ in a turbulent crossflow (Reprinted from Wu and Leschziner [95], with permission from Elsevier)

Both boundary layers were developed in the same freestream velocity and the SJA operates under the same conditions, i.e., the velocity ratio $VR = 0.17$ and the non-dimensional stroke length $L = 1.7$. From the *VR-L* parameter map Fig. 26, hairpin vortices shall be produced, which was also confirmed by the simulation results in this study. For the TBL simulation, LES was used and the incoming TBL was realized by employing a mean velocity profile superimposed with perturbations from a fluctuating vorticity field. Figure 28 demonstrates the time-averaged hairpin vortex in both boundary layers at the same phase, from the top and side views. It was found that in the TBL the hairpin vortex travels faster and is in a more bent shape than in the LBL. The hairpin vortex in the TBL is asymmetric, oscillates laterally, and dissipates fast, whereas it is symmetric and persistent in the LBL. Secondary vortices were observed in both cases. It was found the SJA suction also contributes to the generation of secondary vortices, and plays a major role in the TBL. The influence of hairpin vortices on both boundary layers was also examined in the context of flow separation control, through the investigation of the boundary layers' velocity profile, wall shear stress, and shape factor. The change of velocity and velocity gradient profiles due to the passage of hairpin vortices reveals that the influence of the hairpin vortex on the TBL is wider but much weaker than on the LBL. It was suggested that a much stronger SJ is required for the TBL in order to achieve similar flow separation control effect to that in the LBL.

In Case 2 of CFDVAL2004, SJs from a circular orifice of 6.35 mm diameter issued into a TBL with a frequency of 150 Hz are simulated. The Mach number of the cross flow is $M = 0.1$. The approximate ratio of boundary-layer thickness to

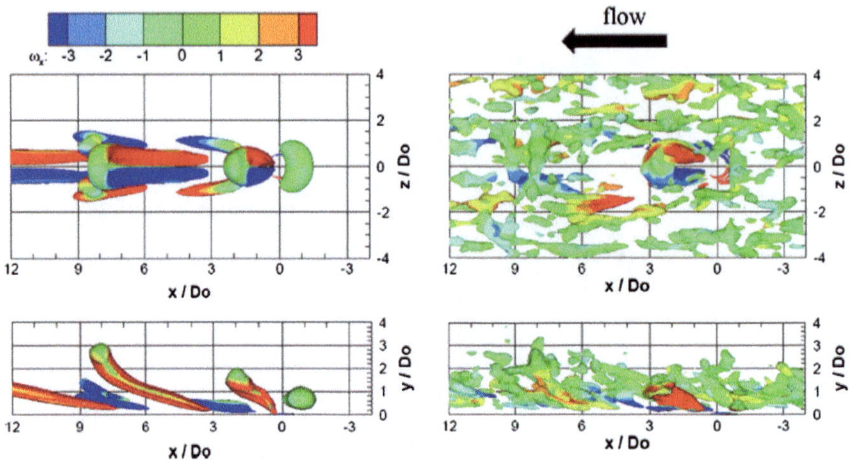


**Fig. 28** Phase-averaged hairpin vortices in the LBL (*left*) and in the TBL (*right*) at $t = 3T/4$. The vortex structures are identified by iso-surfaces of $Q$ = 10. The contours represent streamwise vorticity (Reprinted from Wen and Tang [93], with permission from Elsevier)

orifice diameter is $\delta/d = 3.3$. The velocity ratio can be estimated as $VR = 0.4$ and the non-dimensional stroke length is estimated as about $L = 15$. More details about this benchmark test can be found in Schaeffler and Jenkins [75, 76]. Simulations on this case were conducted from different research groups. Most of them used URANS [5, 10, 12, 32, 66, 71, 72], two used DES [10, 99], and one used LES [12]. Rumsey et al. [69, 70] and Rumsey [67, 68] have made very good summaries on these simulations. They concluded that both URANS and LES could achieve equally reasonable results in terms of mean flow quantities (both time-averaged and phase-averaged). LES is able to achieve better agreement with turbulence data than URANS. Qualitative agreement of CFD with experiment was reasonably good, but quantitative comparisons showed significant variations. Different turbulence models were found to have less impact than different grids, codes, or other solution variants. The benefits of using appropriate inflow boundary conditions (seeded unsteady turbulent fluctuations instead of steady-RANS-like turbulent profile) for LES to insure adequate turbulent eddy content were also demonstrated [12]. The reason why the CFD simulations did not perform well in quantitative comparisons may be because the quality of the experimental data. Rumsey [67, 68] suggested it would be worth revisiting the experiment or possibly establishing new jet-in-cross-flow benchmark experiments for validation.

### 3.3 Synthetic Jets in Separated Cross Flows

When the cross-flow flow encounters an adverse pressure gradient, such as over a hump or an airfoil, it tends to separate from the wall. An accurate prediction, not to

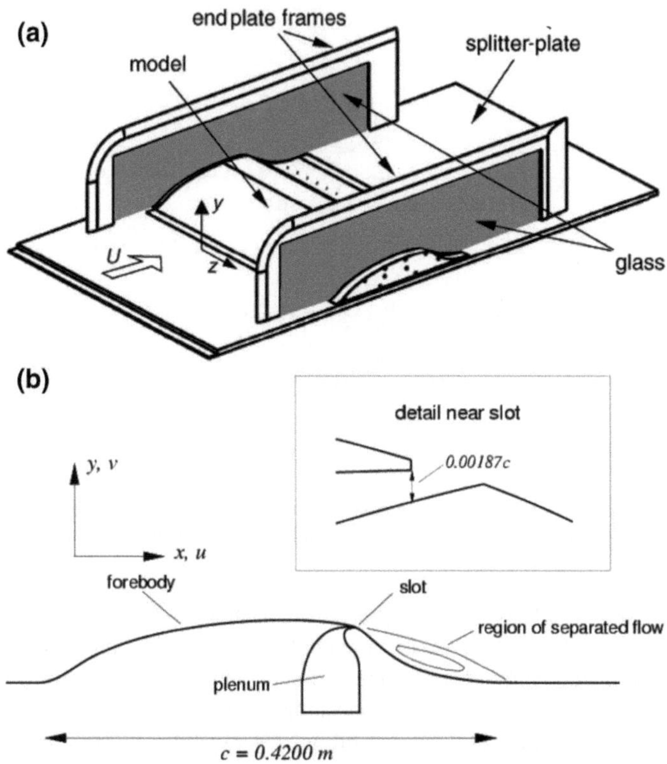


**Fig. 29** Case 3 hump model for CFDVAL2004 **a** isometric view of the model mounted on the splitter-plate with endplates in place (From Greenblatt et al. [25]. Reprinted with kind permission from the authors) and **b** close view of the model (From Rumsey [67]. Reprinted with kind permission from the author)

mention control, of flow separation and, if applicable, re-attachment at high Reynolds numbers is a challenging task for CFD simulations. The simulation of interaction of SJs with such flows and the resulting flow separation control is even more challenging.

To help assess the capability of the state-of-the-art CFD techniques in simulating the interaction of SJs with separated cross flows and the resulting flow separation control, Case 3 of CFDVAL2004 provided a benchmark experiment for CFD validation. As shown in Fig. 29, in this case a cross flow flows over a nominally 2D wall-mounted hump of chord $c$ = 420 mm and maximum height of 53.7 mm. The hump configuration is similar to that in Seifert and Pack [78]. The cross-flow Mach number is $M = 0.1$, with a TBL of a thickness $\delta/h = 0.46$ and a Reynolds number $Re_\theta$ = 5,700 developed about two chords before the hump model, where $h$ is the height of the hump. Over the hump, the uncontrolled flow separates near 65 % chord, and reattaches downstream behind the end of the hump. A plane SJ that forms through a 0.79-mm slot and spans the entire hump model locates near the

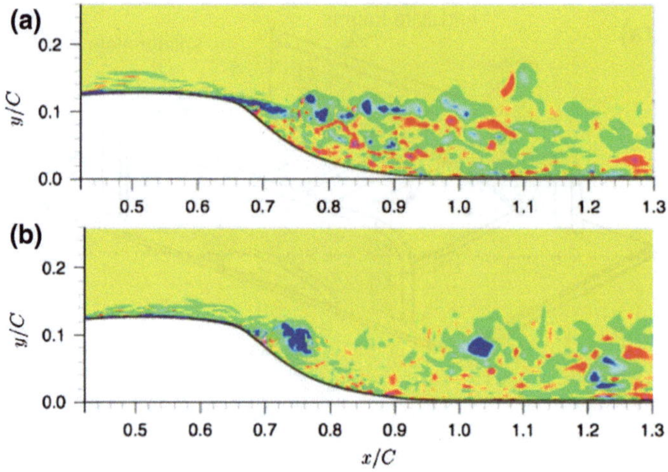


**Fig. 30** Instantaneous spanwise vorticity contours: **a** uncontrolled case and **b** SJ-controlled case (Adapted from You et al. [108]. Reprinted with kind permission from the authors)

separation point. The SJA was driven electromechanically by a rectangular-shaped, rigid piston mounted on an elastic membrane deep inside the cavity chamber. It operated at a frequency of 138.5 Hz and achieved a peak velocity of 26.6 m/s, roughly corresponding to the non-dimensional stroke length $L = 77$ and velocity ratio $VR = 0.25$. At this actuation condition, the resulting SJ was able to significantly lessen the size of the separation bubble, but was not strong enough to fully attach the separated flow. It was intentionally set in order to increase the challenge for the CFD simulations.

Gross features of the flow separation in the uncontrolled and controlled cases are visualized using the spanwise vorticity contours as shown in Fig. 30. The incoming TBL separates due to the strong adverse pressure gradient near the cavity slot and produces abundant small-scale vortices between the separated shear layer and bottom wall (Fig. 30a). In the SJ-controlled case shown in Fig. 30b, a periodic production and convection of large-scale vortices is observed. The repeated process of vortex rollup and shedding is also found to reduce effectively the separation bubble size.

Because of the ample physics involved and its challenges to CFD, this hump case has attracted more than 20 different research groups, the most among the three cases in CFDVAL2004. Rumsey et al. [69, 70] and Rumsey [67, 68] have made good summaries on these simulations at different points of time. It was concluded that RANS/URANS turbulence models fail in general: the eddy viscosity in the separated shear layer region was significantly under-predicted in magnitude, leading to too little mixing and hence too late a re-attachment downstream. DES methods have also been used, but have shown mixed success. They require careful considerations on grid spacing and definition of the region where the LES is to be active.

Success has been made in the realm of LES. Using LES, several groups have obtained significant improvement over RANS/URANS methods. By resolving many of the 3D turbulent eddies in the flow field, the mixing due to turbulence in the separated region can be better predicted. Šarić et al. [74] performed LES with a constant-coefficient Smagorinsky SGS model, and reported good predictions in the baseline case, but significant deviations of mean velocity profiles from the experimental data provided by Greenblatt et al. [25, 26] in the SJ-controlled case. The numerical scheme employed in their LES is based on the deferred-correction approach, and the numerical dissipation can severely impact the effectiveness of the SGS model. Morgan et al. [53] also obtained good agreement with experimental data for the uncontrolled case using implicit LES (ILES), which relies on numerical dissipation from 6th-order low-pass Pade-type spatial filtering to play the role of a SGS model. However, the Reynolds number employed in the ILES is only about one-fifth of the value of the experiments. Although much better than RANS/URANS simulation results, their SJ-controlled case predicted too much separation. Franck and Colonius [18] applied ILES with the same type of spatial filter of even higher order scheme (eighth order) but found little difference between results using LES and ILES.

You et al. [108] produced the best simulation results using LES with a dynamic Smagorinsky SGS model. In contrast to the other LES techniques that employ upwind-biased schemes [74] or the ILES [17, 18, 53], they used an energy-conservative second-order central-difference scheme on a staggered mesh. Aliasing errors were controlled by enforcing kinetic energy conservation as opposed to numerical dissipation or filtering, thus avoiding artificial damping of small scales. A comparison of the time-averaged re-attachment location predicted by the representatives of different methods is given in Table 1. Similar comparison was also conducted for the mean surface pressure coefficients (Fig. 31), and for the mean and turbulence quantities (Fig. 32). It is showed that the LES of You et al. [108] generally provides the best results. Surprisingly, in this simulation they did not include the SJA cavity flow. Instead, they only applied a simple transpiration boundary condition at the SJA slot exit, which has a parabolic velocity profile and varies in a sinusoidal manner.

**Table 1** Predicted reattachment locations ($x/c$) for Case 3 of CFDVAL2004

| | Uncontrolled case | SJ-controlled case |
|---|---|---|
| Typical RANS/URANS [70] | 1.24 | 1.22 |
| DES [74] | 1.12 | 1.11 |
| LES [74] | 1.11 | 1.05 |
| LES [108] | 1.09 | 1.01 |
| ILES [53] | 1.14 | 1.10 |
| Experiment [25, 26] | 1.11 ± 0.003 | ≈0.98 |

Adapted from Rumsey [68]. Reprinted with kind permission from Multi-Science Publishing Co Ltd.

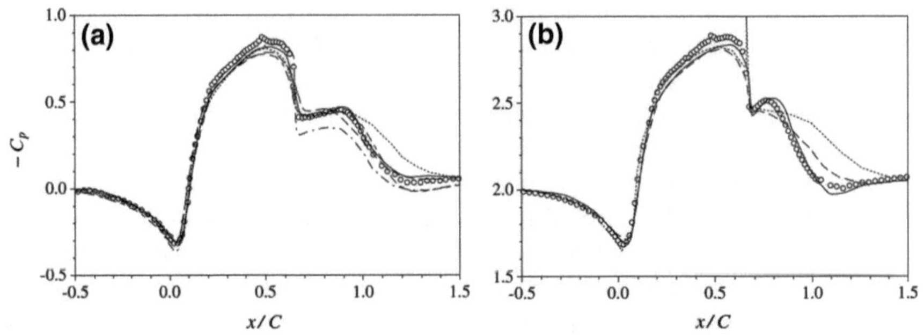


**Fig. 31** Surface pressure coefficient for **a** uncontrolled and **b** SJ-controlled cases for Case 3 of CFDVAL2004. *Solid line* LES of You et al. [108]; *dash–dotted line* LES of Šarić et al. [74]; *dashed line* ILES [53]; *dash–short-dashed line* DES [40]; *dotted line* URANS [6]; *open circles* experiment [25, 26] (Adapted from You et al. [108]. Reprinted with kind permission from the authors)

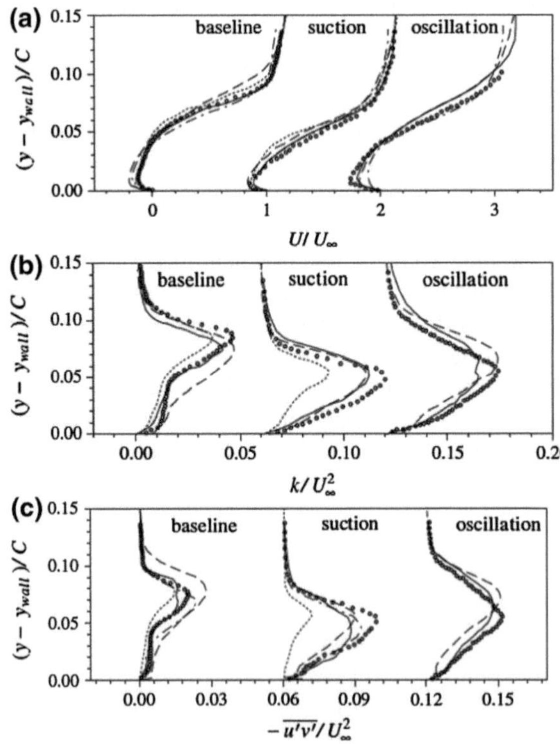


**Fig. 32** Profiles of **a** mean streamwise velocity, **b** turbulent kinetic energy, and **c** Reynolds shear stress at x/c = 0.8. *Solid line* LES of You et al. [108]; *dash–dotted line* LES of Šarić et al. [74]; *dashed line* ILES [53]; *dash–short-dashed line* DES [40]; *dotted line* URANS [6]; *open circles* experiment [25, 26]. The profiles for the suction and oscillation cases in **a** are shifted by 1 and 2, respectively. The profiles for the suction and oscillation cases in **b**, **c** are shifted by 0.06 and 0.12, respectively (Adapted from You et al. [108]. Reprinted with kind permission from the authors)

Franck and Colonius [17, 18] investigated the effect of compressibility on the uncontrolled and controlled flow over the wall-mounted hump. They explored the uncontrolled flow of the Mach numbers ranging from 0.1 to 0.6 and compared with experimental data from both Greenblatt et al. [25, 26] and Seifert and Pack [78]. The SJ control was also implemented at Mach numbers of the same range. It was found that the control at the same condition is less effective at higher Mach numbers in terms of the drag and separation bubble length reduction.

In addition to the CFDVAL2004 efforts, Garnier et al. [22] also evaluated the URANS capabilities for separation flow control by means of SJs. They simulated the control of a separated flow over a rounded backward-facing step. The step height is $h = 20$ mm and the maximum slope of the step is 35°. The Mach number is $M = 0.33$ and the freestream velocity is $U_\infty = 109.6$ m/s. The boundary layer thickness and momentum thickness at $x/h = -1$ are $\delta = 0.5h$ and $\theta = 0.05h$, respectively. The step height based Reynolds number is $Re_h = 30{,}000$ and the momentum thickness based Reynolds number is $Re_\theta = 1350$ at the separation point. The SJ orifice consists of a 2D slot of 0.55 mm width. Its windward edge is located near the mean separation point of the uncontrolled flow. The momentum coefficient of the SJ is $C_\mu = 3.3 \times 10^{-3}$ based on the root-mean-square value of the SJ velocity and separation length. The evaluation was carried out by comparing the URANS results with the LES results (validated by their DNS results, [13]), in the context of investigating the frequency response of the separated flow to the periodic excitation of an SJ. For URANS, four turbulent models were assessed, including the Spalart–Allmaras model, the *k*-*ω* model of Wilcox, the *k*-*ω* model of Menter with SST correction, and a *k*-*l* variant of the EARSM model. It was found that although the two variants of *k*-*ω* models and the *k*-*l* EARSM model are able to identify properly the frequency at which the length of the separation bubble is minimized, the amplitude of the reduction is clearly underestimated. They concluded that these URANS models suffer from two important limitations: firstly, they are not able to account for the bubble length reduction at high frequencies, and secondly they are not even able to reproduce accurately the uncontrolled flow.

Using LES, the same group [12, 13, 22, 57] has done a series of studies on active control of separated flows over a rounded backward-facing step, focusing on the effects of SJ frequency and strength. In Dandois et al. [12, 13], two different reduced frequencies were considered, i.e., $F^+ = 0.5$ and 4, where $F^+$ is based on the separation length of the uncontrolled flow. The momentum coefficient was fixed at $C_\mu = 0.01$, corresponding to a velocity ratio of $VR = 0.5$. $F^+ = 0.5$ is close to the natural shedding frequency of the step, at which the SJA operates in the “SJ mode” (non-zero-momentum injection) with non-dimensional stroke length $L = 6.3$. Under this condition, a series of downstream-traveling large vortices forms in the separated region as shown in Fig. 33a. High-momentum fluid is carried from the freestream toward the wall by the entrainment of these vortices, and hence the level of turbulent kinetic energy production is doubled compared to in the uncontrolled case. As a result, the separation length reduces by half. As shown in Fig. 33b, however, at $F^+ = 4$ the SJA generates large-amplitude sound waves and operates in the “acoustic streaming mode” (zero-momentum injection) with $L = 0.78$, which is not sufficient to form an SJ according to the criteria of Shuster and Smith [77] and

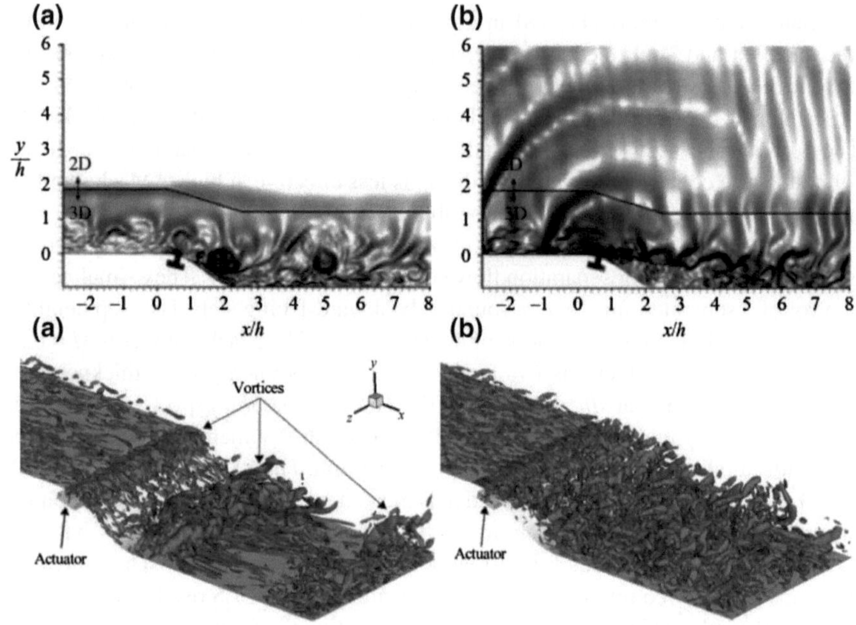


**Fig. 33** Instantaneous flow visualization at **a** $F^+ = 0.5$ and **b** $F^+ = 4$. *Upper* Pseudo-schlieren contours; *lower* Q-criterion iso-surface structures (From Dandois et al. [13]. Reproduced with permission)

Holman et al. [30]. The sound waves significantly modify the stability of the mean streamwise velocity profile, causing a reduced turbulent kinetic energy level. As a result, the separation length increases by 43 %.

Pamart et al. [57] extended the above work and studied the performance of the SJs at 11 different frequencies (varying between $F^+ = 0.1$ and 10) and 3 different momentum coefficients ($C_\mu = 0.0033, 0.0066$, and 0.01). They used different criteria to determine the optimal SJ frequency and strength, including the length of separation bubble, wall pressure coefficient, integral of turbulence level, maxima of turbulent kinetic energy, and pressure drag etc. They concluded that the best frequency is relative to the chosen criteria, and no frequency is optimal for every chosen criterion. Nevertheless, it was found that most of individual optimal frequencies fall in the vicinity of the reduced frequency range $F^+ = 1.2$–1.5. This frequency range corresponds to a range of Strouhal number based on the uncontrolled bubble height, i.e., $St_{bh} = 0.20$–0.25. As for the influence of SJ strength, it was found a higher momentum coefficient enhances the control effect without changing the optimal frequency.

By controlling a laminar separation bubble in an adverse-pressure-gradient boundary layer induced by a faring above a flat plate, Ozawa et al. [56] and Hong [31] investigated the effects of the forcing frequency and amplitude of a circular SJ on eliminating the separation bubble. By doing so, the effect of pressure gradient on the separation bubble is decoupled from the effect of curvature. For simplification, in their LES simulations the fairing-induced adverse pressure gradient was generated by defining the static pressure values obtained from their wind tunnel tests on the upper boundary of the computational domain, and the circular orifice was replaced by a square orifice. By studying the variation of shape and size of the separation bubble as well as the variation of maximum fluctuating velocity at three different frequencies, they confirmed that the optimal reduced frequency $F^+$ should be in the order of O(1), and the SJA is effective when the forcing frequency is in the lower range of the Tollmien–Schlichting instability. They concluded that the forcing frequency had significantly stronger influence on SJA effectiveness than the forcing amplitude.

To explore the roles of three natural frequency scales in the SJ-based separation control, i.e., the frequency corresponding to the roll-up of shear layer $f_{SL}$, the frequency corresponding to the separation bubble $f_{sep}$, and the wake shedding frequency $f_{wake}$, Mittal et al. [50] and Kotapati et al. [37] designed a novel numerical test rig. Figure 34 shows the test rig in 2D, which consists of an elliptic airfoil with chord $c$ and thickness $t = 0.02c$ at 0° angle of attack (AOA) in a freestream. A separation bubble is induced at a desired location on the upper surface of the airfoil by applying an adverse pressure gradient through blowing and suction on the upper boundary of the computational domain. The blowing and suction are applied through a zero-vorticity-flux boundary condition derived by Na and Moin [54]. This simple configuration facilitates the study of all the features of a canonical separated flow such as the leading edge boundary layer inception, open/closed suction-side separation, and a wake that includes vortices from the suction and pressure sides. In addition, like in the studies by Ozawa et al. [56] and Hong [31], this numerical test rig also permits independent description of the adverse pressure gradient without the confounding effect of curvature. The flow at a chord Reynolds number of $Re = 60{,}000$ with and without the SJ actuation was calculated using a 2D N-S simulation, and the response of the flow to the systematic variation of actuation

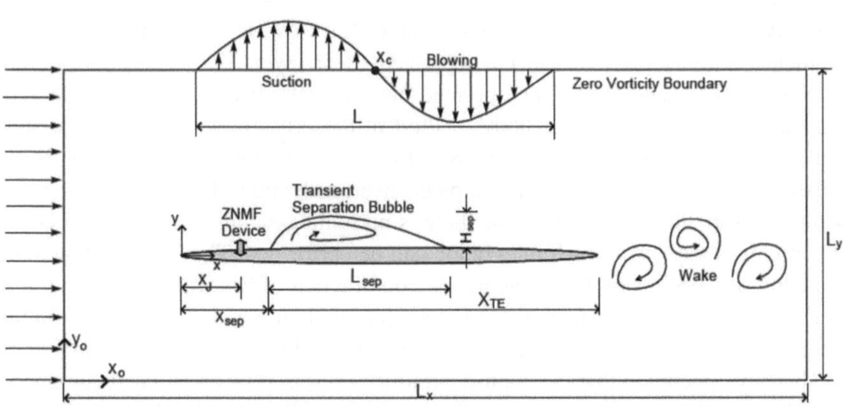


**Fig. 34** Schematic of the numerical test rig used in Mittal et al. [50], Kotapati et al. [37], and Aram et al. [4] (From Kotapati et al. [37]. Reprinted with kind permission from the authors)

frequency was investigated by focusing on the resonant interaction between the SJ forcing and the natural time scales of the flow. It was found that at low forcing frequencies ($f <= f_{SL}$) both the shear layer and the separation zone are able to lock-on to the forcing frequency or its sub-harmonics, whereas high actuation frequency ($f > f_{SL}$) does not effectively couple with the frequency of either the shear layer or the separation zone. Furthermore, the actuation at the separation bubble frequency $f_{sep}$ (or its sub-harmonics) is more effective in flow separation suppression. Positive control of the separation zone was also found to lower the wake shedding frequency. Aram et al. [4] later extended the problem from 2D to 3D, and modified the test rig a little bit by replacing the 2 % thick elliptical airfoil with a 13 % thick flat plate connected with a 4:1 elliptic leading edge and blunt trailing edge. LES was applied. The simulation results of uncontrolled flow at a chord Reynolds number of $Re = 10^5$ show that the shear layer and separation bubble are locked on to the same dominant frequency, i.e., $f_{SL} = f_{sep}$, whereas the wake region has lower frequency $f_{wake}$. It was found that operating the SJ at a frequency close to $f_{SL}$ or $f_{sep}$ causes a significant reduction in the size of separation bubble, whereas higher excitation frequency is not effective.

Although most of the above numerical investigations as well as some experimental studies suggested that the SJA optimal frequency is of the same order of magnitude as the natural frequencies of the separated flow, Amitay et al. [3] and Glezer and Amitay [24] suggested through their experiments that much higher (at least one order of magnitude higher) frequency actuation can also be helpful in flow separation control. This is because at frequencies of this level the SJA actuation is decoupled from that of the global flow instability and is able to produce a virtual surface modification downstream of the actuator, which displaces the streamlines and results in a localized favorable pressure gradient. This virtual surface modification leads to a thinner boundary layer that is more resistant to separation.

Leschziner and Lardeau [44] reviewed the state-of-the-art simulations of SJs in boundary-layer separation control, covering both plane and circular SJs. They confirmed the problem of URANS in simulating such complex flows, i.e., URANS often gives a poor representation of separated flows owing to insufficient level of turbulence activity predicted in the separated shear layer. For the scale-resolving simulations (LES and DNS), they pointed out the scale disparity between the very small jet orifice and the large body of fluid in the cross flow is the most serious challenge. It is especially so in the case of circular SJs. One of the reasons is the statistically converged time-averaged, phase-averaged, and turbulence quantities are extremely difficult to obtain because there is no homogeneous direction in circular-SJ configurations, over which integration can be performed. They demonstrated the capability of LES in simulating the interaction of a pair of SJs with a separated flow over a rounded backward-facing step by comparing their simulation results with the experimental data obtained from Zhang and Zhong [110] and Zhong and Zhang [113]. They concluded that the control authority exercised by an isolated circular SJ is modest, despite its high momentum, and that circular SJs offer less opportunity for resonance between the actuation frequency and 2D instability modes in the separated shear layer.

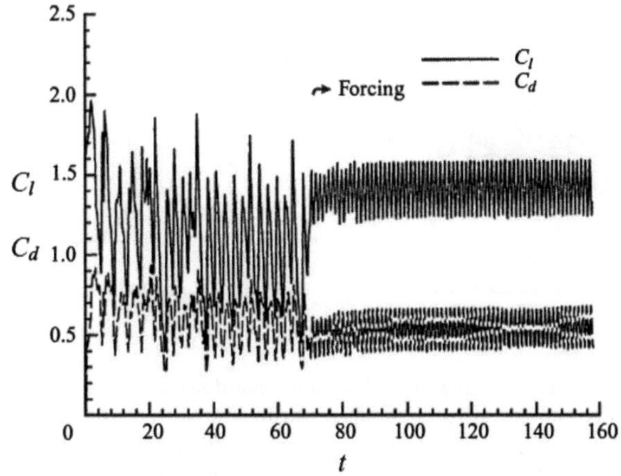


**Fig. 35** Time histories of $C_l$ and $C_d$ of a NACA0012 airfoil before and after the actuation of the SJA. $Re = 5 \times 10^5$, $M = 0.2$, AOA = 25°. The forcing frequency is two times of the natural shedding frequency of the airfoil (From Wu et al. [98]. Reproduced with permission)

When the SJ technology is applied on an airfoil, the control of airflow around the airfoil will affect the airfoil's aerodynamic performance, such as the lift, drag, and pitching moment. Since emerging, this flow control technology has been extensively tested on airfoils and even aircraft models, both in lab experiments and in simulations. Most of the numerical effects were made on the stall control or post-stall flow control of airfoils. Wu et al. [96–98] used an algebraic Baldwin-Lomax turbulence model based URANS solver to investigate post-stall lift enhancement for a NACA0012 airfoil at a Reynolds number of $Re = 5 \times 10^5$ and Mach number of $M = 0.2$ with the issuance of an SJ normal to the airfoil surface near the leading edge. They found that lift increase and drag reduction in the post-stall regime, i.e., AOAs between $\alpha = 18°$ and 35°, can be achieved. As an example, Fig. 35 shows the time history of lift coefficient and drag coefficient before and after the SJ actuation. It clearly indicates the lift increases and the drag decreases after actuating the SJ at a higher harmonic of the airfoil natural shedding frequency. The phase of the velocity at multiple locations were also investigated to show the control effect. As shown in Fig. 36a, the phase variation at a point downstream the separation point is chaotic before the SJ is actuated. After the SJ is actuated at a higher harmonic of the airfoil natural shedding frequency, the chaotic phase variation becomes very organized as shown in Fig. 36b.

Using a Spalart-Allmaras turbulence model based URANS solver, Donovan et al. [14] did a similar investigation on a NACA0012 airfoil at a much higher Reynolds number $Re = 8.5 \times 10^6$. Performance benefits were obtained by placing the actuators very close to the airfoil leading edge. A significant increase of 29 % in lift was obtained using the SJA with a momentum ratio $C_\mu = 0.0052$ and frequency $F^+ = 1$ in the post-stall regime ($\alpha = 22°$). Similar URANS based investigations on the stall or post-stall flow control of airfoils were also conducted by Ravindran [64], Nae [55],

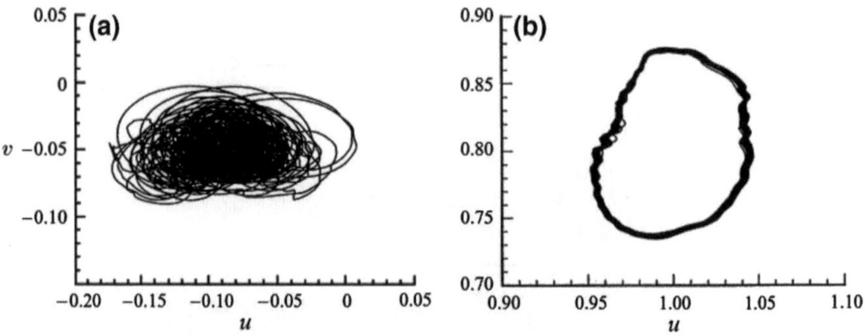


**Fig. 36** Phase diagram of instantaneous flow at a point downstream of the separation point and close to the suction surface of a NACA0012 airfoil, $Re = 5 \times 10^5$, $M = 0.2$, AOA = 25° **a** uncontrolled flow **b** SJ-controlled flow. The forcing frequency is two times of the natural shedding frequency of the airfoil (From Wu et al. [98]. Reproduced with permission)

Hassan [28], Duvigneau and Visonneau [16], Duvigneau et al. [15], and Kim and Kim [35], covering the chord Reynolds numbers range from $O(10^5)$ to $O(10^6)$.

From the discussion in the previous section about the simulation of SJ interaction with separated cross flows, it can be concluded here that, for the simulation of stall or post-stall control over airfoils that involves strong pressure gradients, flow separation, and confluence of boundary layers and wakes, traditional CFD techniques based on URANS are not able to provide results that quantitatively match experimental data. However, using the LES or DNS solver is very resource demanding for this kind of aerodynamics problem. You and Moin [107] utilized an unstructured-grid finite-volume LES solver to evaluate the effectiveness of a plane SJ over a NACA0015 airfoil at $\alpha = 16.6°$ and the chord Reynolds number of $Re = 8.96 \times 10^5$. The frequency of the SJ actuation was set $F^+ = 1.284$ and the momentum ratio $C_\mu = 0.0123$. Simulation results were compared with the experimental data from Gilarranz et al. [23], and qualitative and quantitative agreements were obtained for both uncontrolled and controlled cases. The delay of the onset of flow separation resulted in an about 70 % increase in the lift.

Instead of at stall or post-stall AOAs, the beneficial aerodynamic effects from the use of SJs on an airfoil at low AOAs were qualitatively demonstrated by Hassan and Janakiram [27]. In this study, an array of 10 SJs was placed on either the lower or the upper surface of a NACA0012 airfoil in a cross flow at the chord Reynolds number $Re = 3 \times 10^6$ and Mach number $M = 0.6$. Each of the SJs is numerically represented by only a single grid node, on which a sinusoidal transpiration boundary condition is applied. An algebraic Baldwin-Lomax turbulence model based URANS solver was used for this 2D simulation. The effects of the array location, jet velocity, and jet frequency on the lift, drag and pitching moment were investigated at two low AOAs, i.e., 0° and 5°. They found that although the SJ array can enhance the lift characteristics of the airfoil by carefully choosing the forcing amplitude and frequency, the mean drag increases as well.

The effectiveness of the SJ technology in flow control was also demonstrated using numerical simulations on more complicated configurations. For example, Höll et al. [29] deployed continuous and segmented slot SJs on the flap of a high-lift system and obtained increased lift in their numerical experiments at $Re = 7.5 \times 10^5$ using 3D $k$-$\omega$ turbulence model based URANS simulations. Sharma et al. [79] simulated the controlled transonic flow around a blend-wing-body aircraft equipped with an array of SJs at $Re = 5.5 \times 10^6$ and $M = 0.85$ using a 3D Spalart-Allmaras turbulence model based URANS solver. It was found that the aerodynamic efficiency of the aircraft ($C_L/C_D$) at $\alpha = 5°$ can be increased by 60 % with the operation the SJ array at a suitable frequency. However, the simulation results from URANS solvers for such complicated flows can only be qualitatively meaningful, and quantitative analysis requires advanced solvers with high fidelity, such as LES and DNS.

# 4 Coupling with Optimization and Control Algorithms

In previous sections, it has been introduced that many investigations aimed to find the optimal SJA geometrical or operating conditions for flow separation control. These "optimization" studies were conducted by manually choosing a series of discrete values in a certain range. For these studies, however, optimization algorithms were not really involved from the mathematics point of view. On the other hand, even when the optimal conditions are identified, how to effectively and efficiently achieve and maintain those conditions with the change of flow conditions is another challenge. This requires the integration of the SJ technology with suitable feedback control algorithms. With the development of computing hardware and numerical technologies, the combination of the flow solver (CFD or low-dimensional models) with advanced optimization or control algorithms for complex engineering problems, such as the SJ-technology based flow control, becomes more and more feasible. Although not much, some numerical efforts have been made in this aspect and will be briefly introduced in this section.

By incorporating a modified $k$-$\varepsilon$ model based URANS solver into the L-BFGS-B control algorithm, a limited-memory algorithm for nonlinear optimization problems subject to simple bounds on the control inputs, Nae [55] successfully minimized the drag experienced by a cylinder in a cross flow using the SJ technology. The control inputs were the forcing frequency and strength of two SJs, which were symmetrically located on the rear side of the cylinder (±45°), each covering a 10° arc length. The forcing frequency was bounded in the range of 0–2000 Hz, and the SJ peak velocity was bounded in the range of 0–20 m/s. At the Reynolds number $Re = 42{,}000$ and Mach number $M = 0.2$, the minimal mean drag was obtained, which is 28 % lower than that without control, using the forcing frequency of 1272 Hz and jet peak velocity of 20 m/s.

Duvigneau and Visonneau [16] coupled a SST $k$-$\omega$ model based URANS solver with an automatic optimization procedure, which relies on a derivative-free multi-directional search algorithm, for the control of the stall of a NACA0015 airfoil using an SJ that is located at 12 % of the chord. The objective function is the time-averaged lift, whereas the input variables are the SJ forcing frequency, velocity, and orientation. Their experimental parameters were chosen as the starting values for the optimization. It was found that, at the Reynolds number of $Re = 8.96 \times 10^5$, the efficiency of the control is significantly improved when optimal parameters are chosen. The maximum lift was increased by 34 % and the stall angle was delayed from 19° to 22°, compared to the starting values. The optimal input variables are characterized by an increase of the velocity amplitude and a decrease of the actuation frequency. Duvigneau et al. [15] further optimized the location of the SJ using the same methodology. To facilitate this optimization study, they implemented an unstructured grid refinement/coarsening method to automatically generate meshes adapted to the presence of the SJ at a prescribed location. The proposed methodology was applied to optimize the location of an SJ on the suction side of a NACA0012 airfoil at a Reynolds number of $Re = 2 \times 10^6$ and AOAs of 18° and 20°. Simulation results show that the jet location has a stronger effect at the AOA of 20°, at which lift increases by 57 % for the optimal location.

Recently Lee et al. [43] implemented SJs on light airfoils to realize rapid maneuvering (pitching and plunging). In the simulations, a closed-loop control scheme and vehicle dynamics are coupled with CFD simulations. To save the computational time while achieving enough accuracy for the CFD simulations, an Euler-URANS-LES hybrid method was used. The Euler equation was solved in regions with irrotational flow, especially upstream and freestream. The URANS equations are solved in regions close to the wall in which attached boundary layer is expected. And the LES-filtered N-S equations are solved in regions where substantial concentrations of vorticity are expected. The control design consists of a linear-quadratic regulator (LQR) control law and augmented adaptive control. This control framework was validated against their experiments for both pitch and plunge motions and good control results were achieved.

# 5 Summary

This chapter gives a comprehensive review on the state-of-the-art numerical simulation and modeling of SJs. The first part is the simulation and modeling of SJs in quiescent conditions. The numerical techniques dealing with the SJ oscillation, laminar-turbulent transition, and turbulence are briefly introduced. From the CFD simulations, the flow physics associated with SJs issuing from offices of different shapes, mainly high-aspect-ratio rectangular slots and low-aspect-ratio rectangular or circular orifices, are revealed. Low-dimensional models that are used for fast prediction of the performance of SJAs in quiescent conditions are also introduced. The second part of the review is about the CFD simulations of SJs interaction with cross flows, including attached and separated flows. For this type of simulations, similar to what Rumsey et al. [69, 70], Rumsey [67, 68], and Garnier et al. [22] have already concluded: RANS/URANS turbulence models fail in general, DES methods have shown mixed successes, and the most successful solver so far is LES. The flow physics associated with SJs interacting with laminar boundary layers, turbulent boundary layers, different types of separated flows, as well as flows around an airfoil are also revealed in the meantime when different numerical methods are introduced and compared. In the end, research about the coupling between CFD simulations or low-dimensional modeling with optimization and control algorithms for SJ-based flow control is reviewed. It is hoped that through this chapter, a better picture of the state-of-the-art numerical simulations and modeling of SJs for the flow separation control applications is provided.

## References

1. Adya, S., Han, D., & Hosder, S. (2010). Uncertainty quantification integrated to the CFD modeling of synthetic jet actuators. *AIAA Paper 2010-4411*.
2. Amitay, M., Kibens, V., Parekh, D., & Glezer, A. (1999). The dynamics of flow reattachment over a thick airfoil controlled by synthetic jet actuators. *AIAA Paper 1999-1001*.
3. Amitay, M., Smith, D. R., Kibens, V., Parekh, D. E., & Glezer, A. (2001). Aerodynamic flow control over an unconventional airfoil using synthetic jet actuators. *AIAA Journal, 39*, 361–370.
4. Aram, E., Mittal, R., Griffin, J., & Cattafesta, L. (2010). Towards effective ZNMF jet based control of a canonical separated flow. *AIAA Paper 2010-4705*.
5. Biedron, R., Vatsa, V., & Atkins, H. (2005). Simulation of unsteady flows using an unstructured Navier-Stokes solver on moving and stationary grids. *AIAA Paper 2005-5093*.
6. Capizzano, F., Catalano, P., Marongiu, C., & Vitagliano, P. L. (2005). URANS modelling of turbulent flows controlled by synthetic jets. *AIAA Paper 2005-5015*.
7. Carpy, S., & Manceau, R. (2006). Turbulence modelling of statistically periodic flows: synthetic jet into quiescent air. *International Journal of Heat and Fluid Flow, 27*(5), 756–767.
8. Cater, J. E., & Soria, J. (2002). The evolution of round zero-net-mass-flux jets. *Journal of Fluid Mechanics, 472*, 167–200.
9. Crook, A., Sadri, A. M., & Wood, N. J. (1999). The development and implementation of synthetic jets for the control of separated flow. *AIAA Paper 1999-3176*.
10. Cui, J., & Agarwal, R. K. (2005). 3-D CFD validation of an axisymmetric jet in cross-flow (NASA Langley workshop validation: Case 2). *AIAA Paper 2005-1112*.
11. Cui, J., & Agarwal, R. K. (2006). Three-dimensional computation of a synthetic jet in quiescent air. *AIAA Journal, 44*(12), 2857–2865.
12. Dandois, J., Garnier, E., & Sagaut, P. (2006). Unsteady simulation of synthetic jet in a crossflow. *AIAA Journal, 44*(2), 225–238.
13. Dandois, J., Garnier, E., & Sagaut, P. (2007). Numerical simulation of active separation control by a synthetic jet. *Journal of Fluid Mechanics, 574*, 25–58.
14. Donovan, J. F., Kral, L. D., & Cary, A. W. (1998). Active flow control applied to an airfoil. *AIAA Paper 1998-0210*.
15. Duvigneau, R., Hay, A., & Visonneau, M. (2007). Optimal location of a synthetic jet on an airfoil for stall control. *Journal of Fluids Engineering, 129*, 825–833.
16. Duvigneau, R., & Visonneau, M. (2006). Optimization of a synthetic jet actuator for aerodynamic stall control. *Computers & Fluids, 35*, 624–638.
17. Franck, J. A., & Colonius, T. (2008). Large-Eddy simulation of separation control for compressible flow over a wall-mounted hump. *AIAA Paper 2008-0555*.

18. Franck, J. A., & Colonius, T. (2010). Compressible large-eddy simulation of separation control on a wall-mounted hump. *AIAA Journal, 48*(6), 1098–1107.
19. Gallas, Q., Holman, R., Nishida, T., Carroll, B., Sheplak, M., & Cattafesta, L. (2003). Lumped element modeling of piezoelectric-driven synthetic jet actuators. *AIAA Journal, 41*(2), 240–247.
20. Gallas, Q., Holman, R., Raju, R., Mittal, R., Sheplak, M., & Cattafesta, L. (2004). Low dimensional modeling of zero-net mass-flux actuators. *AIAA Paper 2004-2413.*
21. Garcillan, L., Zhong, S., Pokusevski, Z., & Wood, N. J. (2004). A PIV study of synthetic jets with different orifice shape and orientation. *AIAA Paper 2004-2213.*
22. Garnier, E., Pamart, P. Y., Dandois, J., & Sagaut, P. (2012). Evaluation of the unsteady RANS capabilities for separated flows control. *Computers & Fluids, 61*, 39–45.
23. Gilarranz, J. L., Traub, L. W., & Rediniotis, O. K. (2005). A new class of synthetic jet actuators—part II: application to flow separation control. *Journal of Fluids Engineering, 127*, 377–387.
24. Glezer, A., & Amitay, M. (2002). Synthetic jets. *Annual Review of Fluid Mechanics, 34*, 503–529.
25. Greenblatt, D., Paschal, K. B., Yao, C. S., Harris, J., Schaeffler, N. W., & Washburn, A. E. (2006). Experimental investigation of separation control part 1: Baseline and steady suction. *AIAA Journal, 44*(12), 2820–2830.
26. Greenblatt, D., Paschal, K. B., Yao, C. S., & Harris, J. (2006). Experimental investigation of separation control part 2: Zero mass-efflux oscillatory blowing. *AIAA Journal, 44*(12), 2831–2845.
27. Hassan, A. A., & Janakiram, R. D. (1997). Effects of zero-mass synthetic jets on the aerodynamics of the NACA-0012 airfoil. *AIAA Paper 97-2326.*
28. Hassan, A. A. (2004). Oscillatory and pulsed jets for improved airfoil aerodynamics—a numerical simulation. *AIAA Paper 2004-227.*
29. Höll, T., Wassen, E., & Thiele, F. (2010). Active separation control on a high-lift configuration using segmented actuation slots. *AIAA Paper 2010-4249.*
30. Holman, R., Utturkar, Y., Mittal, R., Smith, B. L., & Cattafesta, L. (2005). Formation criterion for synthetic jets. *AIAA Journal, 43*, 2110–2116.
31. Hong, G. (2012). Numerical investigation to forcing frequency and amplitude of synthetic jet actuators. *AIAA Journal, 50*(4), 788–796.
32. Iaccarino, G., Marongiu, C., Catalano, P., & Amato, M. (2004). RANS modeling and simulations of synthetic jets. *AIAA Paper 2004-2223.*
33. Jabbal, M., & Zhong, S. (2008). The near wall effect of synthetic jets in a boundary layer. *International Journal of Heat and Fluid Flow, 29*, 119–130.
34. Joslin, R. D., Horta, L. G., & Chen, F. J. (1999). Transitioning active flow control to applications. *AIAA Paper 1999-3575.*
35. Kim, S. H., & Kim, C. (2009). Separation control on NACA23012 using synthetic jet. *Aerospace Science and Technology, 13*, 172–182.
36. Kim, J. W., Sankar, L., Min, B. Y., Yeshala, N., & Egolf, T. A. (2012). Multiscale modeling of active flow control for fuselage drag reduction. *AIAA Paper 2012-0075.*
37. Kotapati, R. B., Mittal, R., & Cattafesta, L. N. (2006). Numerical experiments in synthetic jet based separation control. *AIAA Paper 2006-0320.*
38. Kotapati, R. B., Mittal, R., & Cattafesta, L. N. (2007). Numerical study of a transitional synthetic Jet in quiescent external flow. *Journal of Fluid Mechanics, 581*, 287–321.
39. Kral, L. D., Donovan, J. F., Cain, A. B., & Cary, A. W. (1997). Numerical simulation of synthetic jet actuators. *AIAA Paper 1997-1824.*
40. Krishnan, V., Squires, K. D., & Forsythe, J. R. (2006). Prediction of separated flow characteristics over a hump. *AIAA Journal, 44*(2), 252–262.
41. Lee, C. Y., & Goldstein, D. B. (2001). DNS of microjets for turbulent boundary layer control. *AIAA Paper 2001-1013.*
42. Lee, C. Y., & Goldstein, D. B. (2002). Two-dimensional synthetic jet simulation. *AIAA Journal, 40*(3), 510–516.

43. Lee, S. K., Lopez Mejia, O. D., Moster, R. D., Muse, J. A., Kutay, A. T., & Calise, A. J. (2013). Simulation of rapidly maneuvering airfoils with synthetic jet actuators. *AIAA Journal, 51*(8), 1883–1897.
44. Leschziner, M. A., & Lardeau, S. (2011). Simulation of slot and round synthetic jets in the context of boundary-layer separation control. *Philosophical Transactions of the Royal Society A, 369*, 1495–1512.
45. Lockerby, D. A., & Carpenter, P. W. (2004). Modeling and design of microjet actuators. *AIAA Journal, 42*(2), 220–227.
46. Mallinson, S. G., Hong, G., & Reizes, J. A. (1999). Some characteristics of synthetic jets. *AIAA Paper 1999-3651.*
47. Mallinson, S. G., Kwok, C. Y., & Reizes, J. A. (2003). Numerical simulation of micro-fabricated zero mass-flux jet actuators. *Sensor and Actuators A, 105*, 229–236.
48. McCormick, D. C. (2000). Boundary layer separation control with directed synthetic jets. *AIAA Paper 2000-0519.*
49. Menon, S., & Soo, J. H. (2004). Simulation of vortex dynamics in three-dimensional synthetic and free jets using the large-eddy lattice Boltzmann method. *Journal of Turbulence, 5*, 032.
50. Mittal, R., Kotapati, R. B., & Cattafesta, L. N. (2005). Numerical study of resonant interactions and flow control in a canonical separated flow. *AIAA Paper 2005-1261.*
51. Mittal, R., & Rampuggoon, P. (2002). On the virtual aeroshaping effect of synthetic jets. *Physics of Fluids, 14*(4), 1533–1536.
52. Mittal, R., Rampuggoon, P., & Udaykumar, H. S. (2001). Interaction of a synthetic jet with a flat plate boundary layer. *AIAA Paper 2001-2773.*
53. Morgan, P. E., Rizzetta, D. P., & Visbal, M. R. (2006). High-order numerical simulation of turbulent flow over a wall-mounted hump. *AIAA Journal, 44*(2), 239–251.
54. Na, Y., & Moin, P. (1998). Direct numerical simulation of a separated turbulent boundary layer. *Journal of Fluid Mechanics, 370*, 175–201.
55. Nae, C. (2000). Numerical simulation of a synthetic jet actuator. In *Proceedings of 22nd International Congress of Aeronautical Sciences*, Harrogate, UK.
56. Ozawa, T., Lesbros, S., & Hong, G. (2010). LES of synthetic jets in boundary layer with laminar separation caused by adverse pressure gradient. *Computers & Fluids, 39*, 845–858.
57. Pamart, P. Y., Dandois, J., Garnier, E., & Sagaut, P. (2010). Large eddy simulation study of synthetic jet frequency and amplitude effects on a rounded step separated flow. *AIAA Paper 2010-5086.*
58. Park, S. H., Yu, Y. H., & Byun, D. Y. (2007). RANS simulations of a synthetic jet in quiescent air. *AIAA Paper 2007-1131.*
59. Rathnasingham, R., & Breuer, K. S. (1997a). System identification and control of a turbulent boundary layer. *Physics of Fluids, 9*, 1867–1869.
60. Rathnasingham, R., & Breuer, K. S. (1997b). Coupled fluid-structural characteristics of actuators for flow control. *AIAA Journal, 35*(5), 832–837.
61. Rathnasingham, R., & Breuer, K. S. (2003). Active control of turbulent boundary layers. *Journal of Fluid Mechanics, 495*, 209–233.
62. Ravi, B. R., Mittal, R., & Najjar, F. M. (2004). Study of three-dimensional synthetic jet flowfields using direct numerical simulation. *AIAA Paper 2004-0091.*
63. Ravi, B. R., & Mittal, R. (2006). Numerical study of large aspect-ratio synthetic jets. *AIAA Paper 2006-0315.*
64. Ravindran, S. S. (1999). *Active control of flow separation over an airfoil, TM-1999-209838.* Langley: NASA.
65. Rizzetta, D. P., Visbal, M. R., & Stanek, M. J. (1999). Numerical investigation of synthetic-jet flowfields. *AIAA Journal, 37*(8), 919–927.
66. Rumsey, C. L. (2004). Computation of a synthetic jet in a turbulent cross-flow boundary layer. *NASA/TM-2004-213273.*

67. Rumsey, C. L. (2008). Successes and challenges for flow control simulations (invited). *AIAA Paper 2008-4311*.
68. Rumsey, C. L. (2009). Successes and challenges for flow control simulations. *International Journal of Flow Control, 1*(1), 1–27.
69. Rumsey, C. L., Gatski, T. B., Sellers, W. L., Vatsa, V. N., & Viken, S. A. (2004). Summary of the 2004 CFD validation workshop on synthetic jets and turbulent separation control. *AIAA Paper 2004-2217*.
70. Rumsey, C. L., Gatski, T. B., Sellers, W. L., Vatsa, V. N., & Viken, S. A. (2006). Summary of the 2004 computational fluid dynamics validation workshop on synthetic jets. *AIAA Journal, 44*(2), 194–207.
71. Rumsey, C. L., Schaeffler, N. W., Milanovic, I. M., & Zaman, K. B. M. Q. (2005). Time-accurate computations of isolated circular synthetic jets in crossflow. *AIAA Paper 2005-5016*.
72. Rumsey, C. L., Schaeffler, N. W., Milanovic, I. M., & Zaman, K. B. M. Q. (2007). Time-accurate computations of isolated circular synthetic jets in crossflow. *Computers & Fluids, 36*(6), 1092–1105.
73. Sahni, O., Wood, J., Jansen, K., & Amitay, A. (2011). Three-dimensional interactions between a finite-span synthetic jet and a cross-flow. *Journal of Fluid Mechanics, 671*, 254–287.
74. Šarić, S., Jakirlić, S., & Tropea, C. (2005). Computational analysis of locally forced flow over a wall-mounted hump at high-Re number. In *Proceedings of Fourth International Symposium on Turbulence and Shear Flow Phenomena*, pp. 1189–1194.
75. Schaeffler, N. W., & Jenkins, L. N. (2004). The isolated synthetic jet in crossflow: a benchmark for flow control simulation. *AIAA Paper 2004-2219*.
76. Schaeffler, N. W., & Jenkins, L. N. (2006). Isolated synthetic jet in crossflow: experimental protocols for a validation dataset. *AIAA Journal, 44*(12), 2846–2856.
77. Shuster, J. M., & Smith, D. R. (2007). Experimental study of the formation and scaling of a round synthetic jet. *Physics of Fluids, 19*, 045109.
78. Seifert, A., & Pack, L. G. (2002). Active flow separation control on wall-mounted hump at high Reynolds numbers. *AIAA Journal, 40*(7), 1363–1372.
79. Sharma, A. S., Rawat, V. S., Kulkarni, V., & Saha, U. K. (2012). Numerical investigations on flow control over blended-wing-body transonic aircraft using synthetic-jet actuators. *Journal of Aircraft, 49*(2), 532–545.
80. Smith, B. L., & Glezer, A. (1997). Vectoring and small-scale motions effected in free shear flows using synthetic jet actuators. *AIAA Paper 1997-0213*.
81. Smith, B. L., & Glezer, A. (1998). The formation and evolution of synthetic jets. *Physics of Fluids, 10*(9), 2281–2297.
82. Smith, B. L., & Glezer, A. (2002). Jet vectoring using synthetic jets. *Journal of Fluid Mechanics, 458*, 1–34.
83. Smith, D., Amitay, M., Kibens, V., Parekh, D., & Glezer, A. (1998). Modification of lifting body aerodynamics using synthetic jet actuators. *AIAA Paper 98-0209*.
84. Tang, H., & Zhong, S. (2005). 2D numerical study of circular synthetic jets in quiescent flows. *Aeronautical Journal, 109*(1092), 89–97.
85. Tang, H., & Zhong, S. (2006). Incompressible flow model of synthetic jet actuators. *AIAA Journal, 44*(4), 908–912.
86. Tang, H., & Zhong, S. (2007). A static compressible flow model of synthetic jet actuators. *Aeronautical Journal, 111*(1121), 421–431.
87. Tang, H., & Zhong, S. (2008). Characteristics of small-scale synthetic jets—numerical investigation. In *IUTAM Symposium on Flow Control and MEMS*, pp. 135–140.
88. Tang, H., & Zhong, S. (2009). Lumped element modelling of synthetic jet actuators. *Aerospace Science and Technology, 13(6)*, 331-339.

89. Utturkar, Y., Mittal, R., Rampunggoon, P., & Cattafesta, L. (2002). Sensitivity of synthetic jets to the design of the jet cavity. *AIAA Paper 2002-0124*.
90. Vatsa, V. N., & Carpenter, M. (2005). Higher-order temporal schemes with error controllers for unsteady Navier-Stokes equations. *AIAA Paper 2005-5245*.
91. Vatsa, V. N., & Turkel, E. (2006). Simulation of synthetic jets using unsteady Reynolds-Averaged Navier-Stokes equations. *AIAA Journal, 44*(2), 217–224.
92. Wang, H., & Menon, S. (2001). Fuel-air mixing enhancement by synthetic microjets. *AIAA Journal, 39*(12), 2308–2319.
93. Wen, X., & Tang, H. (2014). On hairpin vortices induced by circular synthetic jets in laminar and turbulent boundary layers. *Computers & Fluids, 95*, 1–18.
94. Wu, D. K. L., & Leschziner, M. A. (2008). Large-eddy simulations of synthetic jets in stagnant surroundings and turbulent cross-flow. In *IUTAM Symposium on Flow Control and MEMS*, pp. 127–134.
95. Wu, D. K. L., & Leschziner, M. A. (2009). Large-eddy simulations of circular synthetic jets in quiescent surroundings and in turbulent cross-flow. *International Journal of Heat and Fluid Flow, 30*, 421–434.
96. Wu, J. M., Lu, X. Y., Denney, A. G., Fan, M., & Wu, J. Z. (1997a). Post-stall lift enhancement on an airfoil by local unsteady control, part I lift, drag and pressure characteristics. *AIAA Paper 1997-2063*.
97. Wu, J. M., Lu, X. Y., & Wu, J. Z. (1997b). Post-stall lift enhancement on an airfoil by local unsteady control, part II mode competition and vortex dynamics. *AIAA Paper 1997-2064*.
98. Wu, J. Z., Lu, X. Y., Denny, A. G., Fan, M., & Wu, J. M. (1998). Post-stall flow control on an airfoil by local unsteady forcing. *Journal of Fluid Mechanics, 371*, 21–58.
99. Xia, H., & Qin, N. (2005). Detached-eddy simulation for synthetic jets with moving boundaries. *Modern Physics Letters B, 19*(28–29), 1429–1434.
100. Xia, H., & Qin, N. (2006). Comparison of unsteady laminar and DES solutions of synthetic jet flow. *AIAA Paper 2006-3161*.
101. Xia, H., & Qin, N. (2008). DES applied to an isolated synthetic jet flow. In *Advances in Hybrid RANS-LES Modelling, Notes on Numerical Fluid Mechanics and Multidisciplinary Design*, Vol. 97, pp. 252–260.
102. Xia, Q., & Zhong, S. (2014). Enhancement of laminar flow mixing using a pair of staggered lateral synthetic jets. *Sensors and Actuators, A: Physical, 207*, 75–83.
103. Yamaleev, N. K., & Carpenter, M. H. (2005). A quasi-1-D model for arbitrary 3-D synthetic jet actuators. *AIAA Paper 2005-0643*.
104. Yamaleev, N. K., & Carpenter, M. H. (2006). Quasi-one-dimensional model for realistic three-dimensional synthetic jet actuators. *AIAA Journal, 44*(2), 208–216.
105. Yao, C. S., Chen, F. J., & Neuhart, D. (2006). Synthetic jet flowfield database for computational fluid dynamics validation. *AIAA Journal, 44*(12), 3153–3157.
106. Yao, C. S., Chen, F. J., Neuhart, D., & Harris, J. (2004). Synthetic jet flow database for CFD validation. *AIAA Paper 2004-2218*.
107. You, D., & Moin, P. (2008). Active control of flow separation over an airfoil using synthetic jets. *Journal of Fluids and Structures, 28*, 1349–1357.
108. You, D., Wang, M., & Moin, P. (2006). Large-eddy simulation of flow over a wall-mounted hump with separation control. *AIAA Journal, 44*(11), 2571–2577.
109. Zhang, P. F., & Wang, J. J. (2007). Novel signal wave pattern for efficient synthetic jet generation. *AIAA Journal, 45*(5), 1058–1065.
110. Zhang, S., & Zhong, S. (2011). Turbulent flow separation control over a two-dimensional ramp using synthetic jets. *AIAA Journal, 49*(12), 2637–2649.
111. Zhong, S., Millet, F., & Wood, N. J. (2005). The behaviour of circular synthetic jets in a laminar boundary layer. *Aeronautical Journal, 109*(1100), 461–470.

112. Zhong, S., Jabbal, M., Tang, H., Garcillan, L., Guo, F., Wood, N., & Warsop, C. (2007) Towards the design of synthetic jet actuators for full-scale flight conditions-part 1: the fluid mechanics of synthetic jet actuators. *Flow, Turbulence and Combustion, 78(3-4)*, 283-307.
113. Zhong, S., & Zhang, S. (2013). Further examination of the mechanisms of round synthetic jets in delaying turbulent flow separation. *Flow, Turbulence and Combustion, 91*(1), 177–208.
114. Zhou, J., & Zhong, S. (2009). Numerical simulation of the interaction of a circular synthetic jet with a boundary layer. *Computers & Fluids, 38*, 393–405.
115. Zhou, J., & Zhong, S. (2010). Coherent structures produced by the interaction between synthetic jets and a laminar boundary layer and their surface shear stress patterns. *Computers & Fluids, 39*, 1296–1313.

# Dynamics of Jets Issuing from Trailing-Edge Modified Nozzles

**T.H. New, D. Tsovolos and E. Tsioli**

**Abstract** This chapter will elaborate upon the fundamental flow behaviour associated with vortex-rings and jets issuing from nozzles, where the nozzle trailing-edges or lips are physically modified with selected geometries. This technique represents a passive but robust form of manipulating the underlying vortex-ring and jet circulation, such that improvements to their entrainment and mixing characteristics can be achieved. Implementations of simple inclined, hybrid inclined, notched, crown-shaped, chevron and stepped nozzles, as well as some of their implementations in circular and noncircular jets, single-stream or dual-stream coaxial jets, will be discussed as part of the overall understanding. In particular, recently observed influences of trailing-edge modifications upon the axis-switching behaviour of noncircular jets, as well as their relationships with coaxial jet flow parameters such as the velocity- and area-ratios will be presented. On top of key flow physics insights in terms of how the nozzle trailing-edge geometry will confer distortionary effects upon the basic vortex structures, the impact of such nozzles upon jet mixing efficacies will be discussed as well.

**Keywords** Trailing-edge modified nozzles · Indeterminate-origin nozzles · Jet-mixing · Noncircular jets · Coaxial jets

T.H. New (✉)
School of Mechanical and Aerospace Engineering, Nanyang Technological University, 50, Nanyang Avenue, Singapore 639798, Singapore
e-mail: dthnew@ntu.edu.sg

D. Tsovolos
School of Engineering, Brownlow Hill, University of Liverpool, Merseyside L69 5GH, UK

E. Tsioli
The Cyprus Institute, Energy, Environment and Water Research Center, Nicosia, Cyprus



D.T.H. New and S.C.M. Yu (eds.), *Vortex Rings and Jets*,
Fluid Mechanics and Its Applications 111, DOI 10.1007/978-981-287-396-5_5

# 1 Introduction

Jet flow is a ubiquitous flow phenomenon which can be found in many natural and manmade flow situations. From pulsating jets produced by squids to high-speed aircraft engine exhausts, questions on the roles of jet flows in biological flows, how they behave fundamentally and of course, how they may be manipulated to meet specific flow objectives for various engineering purposes, have been explored by many for decades. In seeking ways to alter jet flow behaviour, many passive and more recently, active strategies have been put forward and investigated by the scientific community. While active jet control techniques such as acoustic excitations, fluidic excitations, micro-jets, flap actuators and synthetic jets, amongst others, have been extensively studied, practical challenges in terms of their robustness, ease of implementation, cost and power requirements meant that they are at present limited to specialized applications. On the other hand, passive jet control techniques often proved to be more viable when these technical considerations are paramount, even if they may not possess the level of flexibility as compared to active jet control techniques. Passive jet control techniques typically involve geometrically modifying some physical aspects of the jet exit conditions. For instance, the jet exit azimuthal geometry may be changed from a conventional circular shape to a noncircular one, such as rectangular, square, elliptic, lobed or other more complex shapes. On the other hand, axial extents of the jet nozzles may be varied as yet another means of imposing geometrical modifications. Most of the earlier investigations paid great attention towards the first technique, where much of the focus was on the characterization of the vortex structures and their dynamics, as well as the comparative jet-mixing efficacies. It is only recently that more researchers are paying particular attention towards the second technique.

Before we proceed to understand how modifying the trailing-edges of jet nozzles may help to alter and improve the resultant jet-mixing characteristics by varying the initial flow and boundary conditions, it will be helpful to have an overview of fundamental jet flow dynamics. Note that this overview is not meant to be exhaustive but rather, aims to serve as a brief introduction to familiarize readers with some basic introduction to vortex dynamics associated with conventional non-modified circular jet flows. Circular jets have been very well-studied with extensive theoretical, experimental and numerical knowledge gathered over the past few decades. For a good appreciation on the classical theoretical framework governing circular jet flow behaviour, including those associated with more complex scenarios, the authors will refer readers to Bai [4], Abramovich [1], Rajaratnam [67] and Lee and Chu [33]. On the other hand, experimental and numerical work had shown that as an incompressible round jet exhausting from a flat jet exit, as is typically the case, into a quiescent environment at a moderate Reynolds number, its shear layer develops azimuthal instability waves which amplify as they convect further away from the jet exit. After some critical point, the instability waves eventually cause the jet shear layer to roll up into regular and coherent axisymmetric large-scale ring-vortices. As these ring-vortices travel downstream, naturally occurring but small-scale longitudinally-aligned counter-rotating vortices (or streamwise vortices, for the sake of brevity here) begin to form along the braid region between consecutive ring-vortices due to increased flow strains caused by the latter [34]. However, it should be noted that while these streamwise vortices form regularly around the jet azimuthal, their exact locations are by no means fixed spatially but instead in an arbitrary manner. Nonetheless, their presence and hence, subsequent interactions with the ring-vortices, aid in the breaking down of the large-scale vortex structures, in addition to that incurred under the actions of viscous effects. This phenomenon has been observed to promote mass and momentum exchange between the round jet and its quiescent surroundings, which in turn typically leads to improved jet-mixing levels.

Near-field behaviour of laminar noncircular jets is relatively similar in terms of the initiation of the wave instabilities which in turn lead to the rolling-up of the jet shear layer into coherent ring-vortices. In this case however, azimuthal outlines of the ring-vortices tend to resemble those of the noncircular jet exits only in their immediate vicinities. Such flat noncircular ring-vortices are inherently unstable and tend to deform under non-uniform self-induced velocity distribution along the entire circumference of every noncircular ring-vortex filament. This variation in self-induced velocity along the vortex filament circumference is caused by its non-uniform curvature resulting from the contour of the nozzle lip and produces an interesting variety of vortex behaviour. For instance, ring-vortices associated with elliptic and rectangular jets undergo axis-switching whereby their major and minor axes interchange as they convect away from the jet exits [21–23]. As for square and triangular jets, studies have demonstrated that their ring-vortices will undergo rotations such that the vertices will be displaced by 45° and 60° respectively [20]. More complex jet exit geometries in terms of azimuthal variations have also been investigated and found to possess correspondingly complex vortex behaviour [44]. More likely than not, these highly three-dimensional vortex structures and behaviour improve mixing characteristics between the jet flows and their surroundings.

By now, it should not escape the attention of the reader that one common but important characteristic across the jet flow scenarios discussed above is that accentuation of three-dimensional vortex dynamics or generation of additional three-dimensional streamwise flow structures goes a long way towards improving overall jet-mixing and/or control characteristics. By promoting breaking-down of coherent, large-scale vortex structures, mixing between jet and surrounding fluids is effectively encouraged. In particular, deliberately producing and making use of streamwise vortices to accomplish that is sometimes more easily achievable than manipulating large-scale vortex structures directly. Furthermore, it may be necessary to satisfy certain jet exit geometry requirements for some applications, such that it is neither practical nor feasible to vary the jet exit azimuthal shape. In that case, persistent production of streamwise vortices at optimal locations around the jet exit may then be the best way forward. There are several popular ways to do that —Firstly, through active means such as injecting comparatively smaller positive or zero mass-flux jets radially into the primary jet such that the radial jets will be deflected to form counter-rotating vortex-pairs similar to a jet-in-cross-flow phenomenon [49]. However, note that while this technique offers greater versatility in terms of optimizing jet control (especially if appropriate feedback control mechanisms are in place to vary the radial jet flows accordingly), it requires additional energy input, more complex system design and control. Secondly, through passive means such as implementing physical tabs (often delta-shaped) around the azimuthal of and protruding slightly into the jet exit, where flow separations around the tabs will produce streamwise counter-rotating vortex-pairs. However, this technique is often associated with jet thrust loss due to the blockage effects generated by the tabs. Last but not least, by employing passive trailing-edge modifications to the jet nozzles such that their axial lengths vary azimuthally. In this case, large-scale vortex structures will undergo non-uniform distortions, accompanied by the formation of streamwise vortices at discontinuous or cusp points of the geometrically modified nozzles. As compared to the other two techniques, the main advantages associated with trailing-edge modified nozzles are that they do not lead to blockage effects (and hence thrust loss) and remain passive by nature. For some applications where complexity, robustness and cost-effectiveness are of significant concerns, trailing-edge modified nozzles may prove to be viable solutions.

* *In the interest of brevity, trailing-edge modified nozzles will simply be abbreviated as "modified nozzles", as opposed to "non-modified nozzles" with no physical alterations to their trailing-edges.*

# 2 Historical Understandings

## *2.1 Vortex Rings from Inclined Nozzles*

Discrete vortex-rings are generally considered and treated as basic "building-blocks" of free jet flows and extensive efforts had been expended to understand their formation, propagation, onset of wavy instabilities and their eventual breakdown behaviour [16, 19, 41–43, 64, 74, 100]. Typically, a vortex-ring is formed by a sudden ejection of a "cylindrical slug" of fluid through an orifice, usually generated through an impulsive, pre-programmed motion of a piston within a straight tube located upstream of the orifice. As the fluid slug exits from the orifice, a vortex sheet forms around the entire orifice circumference due to the presence of a significant velocity gradient between the fluid slug and the quiescent ambient fluid, which proceeds to roll up into a vortex ring [16]. Depending on the initial conditions, a single, discrete vortex-ring will be formed, or a vortex-ring followed by a trailing-jet [18, 45, 72]. Within the present scope, we will only focus upon discrete, laminar vortex-rings generated well within the optimal formation number, without any occurrence of a trailing-jet. Historical evidence has pointed out that the convective velocity of the vortex-ring decays with downstream distance and the vortex-ring core diameter increases as a result of ambient fluid entrainment. Eventually, at a sufficiently far distance downstream, short azimuthal bending wave instabilities will typically set in and produce regular sinusoidal distortions along the entire vortex-ring core. These wavy instabilities will proceed to grow before finally breaking up the coherent vortex-ring structure rapidly [74, 100, 101]. For in-depth literature reviews of research performed on discrete vortex-rings, the reader is referred to excellent reviews by Shariff and Leonard [77] and Lim and Nickels [36].

Some of the most interesting investigations on vortex-rings issuing from modified nozzles focused upon them produced by simple inclined nozzles. Considering the simplicity of such simple inclined nozzles, they produce surprisingly rich and three-dimensional flow fields, which attest to the notion that modified nozzles can be used as effective passive tools of manipulating vortex-ring and jet shear layer circulation. Recent studies have looked at the effects of the nozzle inclination and Reynolds number upon the resultant vortex-ring behaviour [32, 35, 84, 98]. Webster and Longmire [98] observed that inclined vortex-rings with non-uniform exit velocity, circulation and entrainment distributions are produced, where their maximum magnitudes are found along the shortest nozzle length locations. Using two very different vortex-ring Reynolds number of $Re$ = 2800 and 23,000 and different nozzle inclinations, they also found that the migration of inclined vortex-rings away from the nozzle centreline increases with the nozzle inclination. As a result, coherence level of the highly inclined vortex-rings reduces significantly with correspondingly lower convective velocities. More interestingly, flow interpretations from their time-sequenced dye visualizations and particle-image velocimetry (PIV) results reveal that the inclined vortex-ring is initially connected to the vorticity within the inclined nozzle via a pair of branched vortex-tubes located at the longest nozzle length location when it first exits from the inclined nozzle. They further postulated that as the flow develops further, the branched vortex-tubes will eventually detach away from the inclined nozzle, form a closed-loop vortex filament and propagate through and ahead of the inclined vortex-ring. Inherent limitations in dye visualizations and two-dimensional PIV measurements during this study meant that more intricate flow details were not available.

Much of the limitations faced by the preceding study were successfully overcome however, by a follow-up study by Troolin and Longmire [84], during which inclined vortex-rings at $Re$ = 2500 were investigated using three-dimensional volumetric PIV measurements. Other than successfully identifying the primary vortex-ring, their measurements also managed to capture the vortex loop associated with the nozzle asymmetry partially located within the inclined nozzle, as well as the stopping piston (i.e. "stopping" vortex-ring). Note that this stopping vortex-ring was postulated not to be the piston vortex emerging out of the inclined nozzle but instead, induced to form by the surge in ambient fluid into the inclined nozzle exit region when the primary vortex-ring begins to detach and move away from the inclined nozzle exit. As such, it has a rotational sense opposite to that of the primary vortex-ring. Interestingly, results suggest that part of the stopping vortex-ring filament originally at the longer nozzle length location then proceeds to be entrained by the primary vortex-ring and moves towards the shorter nozzle length location.

On the other hand, the opposite happens to the stopping vortex-ring filament portion originally at the shorter nozzle length location, where it travels towards the longer nozzle length location instead. Incidentally, it encounters the vortex loop along the way and through their mutual interactions, the latter will disconnect vortically into two separate vortex entities. On the other hand, the stopping vortex gets gradually stretched by the primary vortex-ring due to its entanglement and form the branch vortex-tubes postulated earlier on by Webster and Longmire [98], as shown in Fig. 1.

It should be noted that the flow dynamics occurring within the inclined nozzles as interpreted in the above study were based on the behaviour of the various vortex structures discerned *outside* of the inclined nozzle. However, this was addressed by a later investigation by Le et al. [32], when they performed a numerical study which modelled closely after Webster and Longmire [98] and Troolin and Longmire [84]. Their results confirmed many findings derived from the experimental studies, in particular the role of the stopping vortex-ring in producing the branched vortex-tubes. The ability of the numerical study to elucidate the vortex dynamics both *inside and outside* of the inclined nozzle demonstrates not only the surprisingly rich flow fields associated with the generation of inclined vortex-rings, but also provides further clarity upon the flow interactions between and evolutions of the dominant vortex structures. Equally important, the numerical investigation also confirmed the existence of a circumferential flow within the primary inclined vortex-ring core, travelling from the longest nozzle length location towards the shortest nozzle length location, as first noted by Lim [35] in his systematic and detailed flow visualization study. They attributed the source of this circumferential flow to the lower pressure incurred at the shortest nozzle length location, where it leads to a circumferential pressure gradient along the vortex-ring core and hence, induces the circumferential flow observed experimentally by Lim [35].

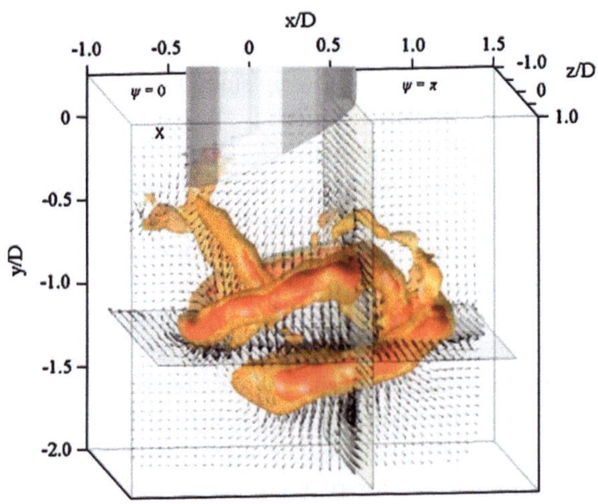


**Fig. 1** Three-dimensional flow field associated with a vortex-ring issuing from an inclined circular nozzle (With kind permission from Springer Science+Business Media: Troolin and Longmire [84], Fig. 9)

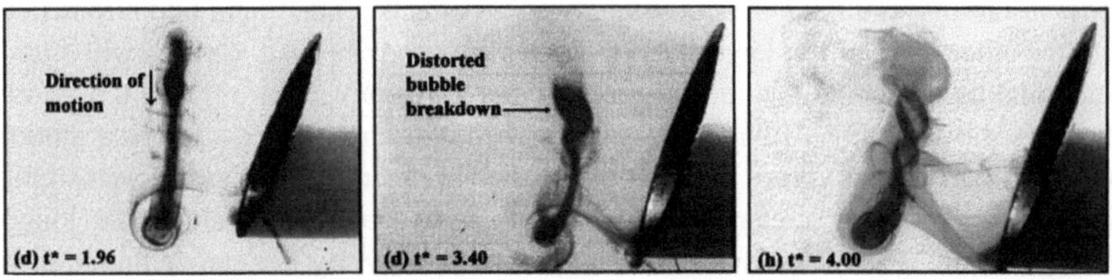


**Fig. 2** Vortex-core bulging, bubble-type and double-helix type vortex breakdowns in an inclined vortex-ring filament (Reprinted with permission from Lim [35]. Copyright 1998, AIP Publishing LLC)

It is worthwhile to point out at this point that the above studies focused upon the near-field dynamics and interactions of the primary inclined vortex-rings and other secondary vortex structures. In contrast, Lim [35] focused on the stability of the inclined vortex-ring produced under different Reynolds number, piston stroke-length and nozzle inclination combinations. Time-sequenced flow visualization results revealed that depending on the exact combination, the inclined vortex-ring may suffer either bulging or breakdown of its core. For the same piston stroke-length, this typically occurs at lower Reynolds number when the nozzle inclination increases and illustrates the effects of increased flow strain (and hence vortex-stretching) associated with a larger nozzle inclination. By carefully interpreting the motions of the coloured-dye infused inclined vortex-rings, Lim [35] deduced for the first time that a symmetrical circumferential flow from the longest nozzle length location towards the shortest nozzle length location exists. As it meets its counterpart at the shortest nozzle length location, it reflects and travels in the opposite direction. In doing so, wavy instabilities will begin to form before manifesting into a bubble-type vortex breakdown of the inclined vortex-ring core. However, this is only a temporary state, where it further evolves into a double helix-type vortex breakdown, as shown in Fig. 2. Occasionally, the bubble-type breakdown will lead to a spiral-type breakdown instead. Bulging of the vortex-ring core only occurs when the circumferential flow is too low to destabilize it, due to low Reynolds numbers and/or small nozzle inclinations used.

## 2.2 Circular Modified Nozzle Jets

One of the earliest investigations on incompressible single-stream jets issuing from modified circular nozzles were conducted by Kibens and Wlezien [24] and Wlezien and Kibens [103]. Their investigations on air jets exhausting from simple inclined circular nozzles led to important observations that shear layer ring-vortices possess both the outlines and inclinations as the inclined nozzles, so long the incline-angle remains small-to-moderate. Furthermore, these ring-vortices will retain these qualitative characteristics within the near-field regions of the nozzle exits before flow instability, mutual interactions and viscous effects render them into turbulence. On the other hand, when larger incline-angles are used, the ring-vortices will "turn" such that their inclinations reduce gradually as they convect away from the nozzle exits, as shown in Fig. 3. The vortex-turning behaviour resembles a pivoting motion whereby part of the vortex filament along the shorter nozzle length moves significantly faster in the streamwise direction than its counterpart along the longer nozzle length. This behaviour can be explained, firstly, by the earlier initiation of ring-vortices along the shorter nozzle length, where the rolling-up of the shear layer propagates from the shorter nozzle length towards the longer nozzle length. As such, the more developed part of the ring-vortex along the shorter nozzle length will convect along the jet flow directions even before its counterpart along the longer nozzle length has been fully formed. And secondly, non-uniform distribution of inclined ring-vortex circulation confers additional self-induced velocities upon the ring-vortex part along the shorter nozzle length, which reinforces the vortex-turning behaviour. Due to dissimilar behaviour along the non-uniform nozzle axial lengths, redistributions of the shear layer energy were observed in these studies and indicate that jet flows may be controlled appropriately by adjusting the nozzle lengths accordingly. Note that it was also during those studies that they first termed trailing-edge modified nozzles as "indeterminate-origin" nozzles, where non-uniform axial lengths around the azimuthal of these nozzles render their exact origins indeterminate. To resolve this issue, some studies define the origins of trailing-edge modified nozzles to be the average of the maximum and minimum nozzle axial lengths.

Investigations on more complex nozzle designs soon followed, such as crown-shaped circular nozzles studied by Longmire et al. [38]. In their study, the number and height of peaks/troughs were varied to look at the influences upon the vortex dynamics of a $Re = 19{,}000$ turbulent jet. The number of nozzle "tooth" ranged from two to eight, while tooth length varied from 0.25 to 2 times that of the jet diameter. All peaks and troughs were sharp in their configurations. Experimental cross-stream results taken for jets forced at St = 0.51 showed that for each peak in the crown

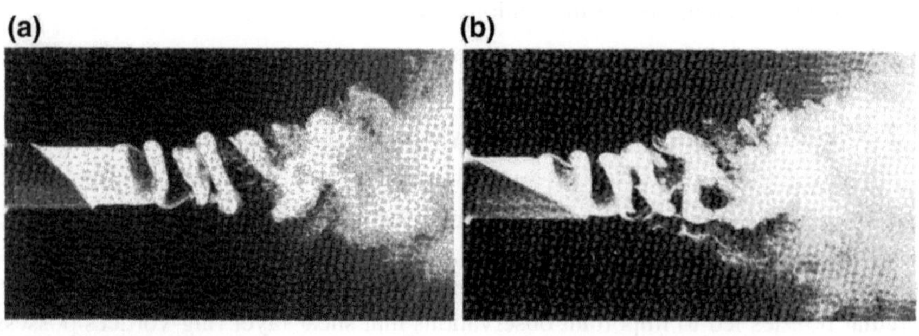


**Fig. 3** Differences in the vortex roll-up behaviour for a jet issuing from inclined circular nozzles of different inclination (From Wlezien and Kibens [103]. Reprinted by permission of the American Institute of Aeronautics and Astronautics, Inc.)

nozzle, outwards-spreading mushroom-shaped counter-rotating vortex pairs (i.e. abbreviated as CVPs hereafter) are formed along the jet periphery, as shown in Fig. 4. On the other hand, inwards entrainment of ambient fluid into the jet body occurs at the trough locations to produce relatively similar mushroom-shaped CVPs. Additionally, increasing the tooth length tends to accentuate the strength and coherence of the CVPs at both peak and trough locations, at least with the geometrical configurations used in this study. Since fluid ejection/entrainment is dependent upon whether the nozzle location consists of a peak or trough, it should not be surprising that the investigation further revealed differences in the jet-spreads along peak and trough locations. In this case, it should be quite clear that the jet-spread along the peak locations are wider than that along the trough locations. However, clarifications on the detailed vortex dynamics were not available until more recent studies by New et al. [63], Shu et al. [79–81] and New and Tsovolos [54–57] were conducted.

As for the investigations performed by Shu et al. [79–81], their approach to the crown-shaped nozzles was different from previous studies in two interesting ways: Firstly, the nozzles were designed such that their peaks were flat rather than sharp, such that stronger streamwise vortices may be produced at the peak locations. And secondly, the nozzles themselves were tapered close to their exits, so that the lower static pressure might increase subsequent jet entrainment. Due to these unique nozzle features, they were named as "tapered four-point crown" indeterminate-origin nozzles. And despite the differences in the physical nozzle geometry between these crown-shaped nozzles and those used in earlier studies, some of the basic flow behaviour observed in these studies are strikingly similar to those observed in earlier investigations. For example, outwards- and inwards-spreading streamwise vortices are formed at the peak and trough locations respectively (i.e. similar to observations made by Longmire et al. [38]), though in this case they are indeed accentuated by the

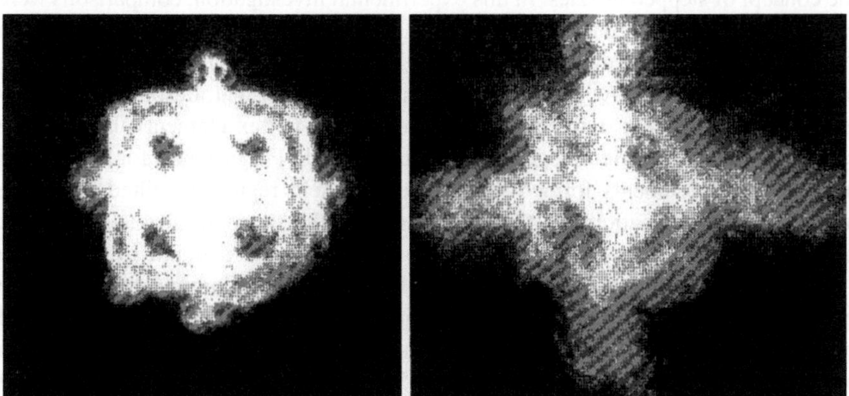

**Fig. 4** Mushroom-shaped streamwise-aligned counter-rotating vortex-pairs produced at each of the four peak locations of sharp crown-shaped nozzles (From Longmire et al. [38]. Reprinted by permission of Ellen Longmire)

atypical crown-nozzle designs used here. Particle image velocimetry measurements revealed that the half-jet width and momentum thickness along the peak locations typically exceed those along the trough locations within a short distance of the nozzle exit. Furthermore, significant differences in the distribution and magnitudes of streamwise Reynolds normal stresses were discerned as well. These quantitative differences were also observed in New and Tsovolos [54–57].

It should be mentioned that other than looking at inclined circular nozzle jets, Wlezien and Kibens [103] also compared their behaviour with those associated with stepped nozzles. Instead of having straight inclines connecting the shortest and longest length locations of the nozzles (i.e. as viewed from their sides), stepped nozzles literally possess 90° "steps" joining the shortest and longest nozzle length locations. From their side-views, their outline resemble backwards facing steps. In this study, the step junctions terminated at and aligned with the jet centerlines. Clearly, stepped nozzles are not defined by any nozzle incline-angles but rather, by the shortest and longest nozzle length differences. Due to the abrupt geometrical change brought about by the step, it was observed that unlike the case for inclined nozzles, the vortex roll-ups emerging from the shortest and longest nozzle length regions appear to be independent of each other initially, at least for the step sizes studied by Wlezien and Kibens [103]. As such, little self-excitation behaviour could be observed in stepped nozzles in that study, as that would have required shear layer instabilities to propagate azimuthally. However, the nozzle steps prevented that from happening, even though it has to be mentioned that interactions between the two different groups of vortex roll-ups would occur after they separated from the stepped nozzles entirely. Despite that, flow changes brought about by stepped nozzles were observed to be more muted as compared to inclined nozzles. Certainly, significant vortex-turning behaviour observed in inclined nozzles did not appear in stepped nozzles.

A more recent study by Longmire and Duong [37] offered an interesting take on the concept of stepped nozzles. In this experimental investigation, comparisons were conducted between a conventional stepped nozzle and one which was based on a "sawtooth" design, at Reynolds numbers of $Re = 7600$ and 22,000, as well as under unforced and forced flow conditions. The study revealed that, despite clear differences in the near-field dynamics of their respective vortex elements, both nozzle types possess some intriguing gross similarities as well. For instance, for the stepped nozzle, vortex filaments formed within the near-field region are tilted along the nozzle general inclination. On the other hand, for the sawtooth nozzle, the vortex filaments appear to loop continuously around the nozzle azimuth in an infinite "spiral-like" manner, which tilts the vortex filaments as well. Common to both nozzle types however, is the significant generation of streamwise vorticity by the tilted vortex filaments under forced flow conditions, which bifurcates the resultant jet into a primary and secondary jet, as shown in Fig. 5. Due to the overall tilt direction, the jet generally bifurcates along the plane intersecting the shortest and longest nozzle axial length locations. No such jet bifurcations were observed under unforced flow conditions in this investigation, even for high Reynolds number test condition. In many ways, this reinforces the importance of flow perturbations (i.e. inherent within the jet

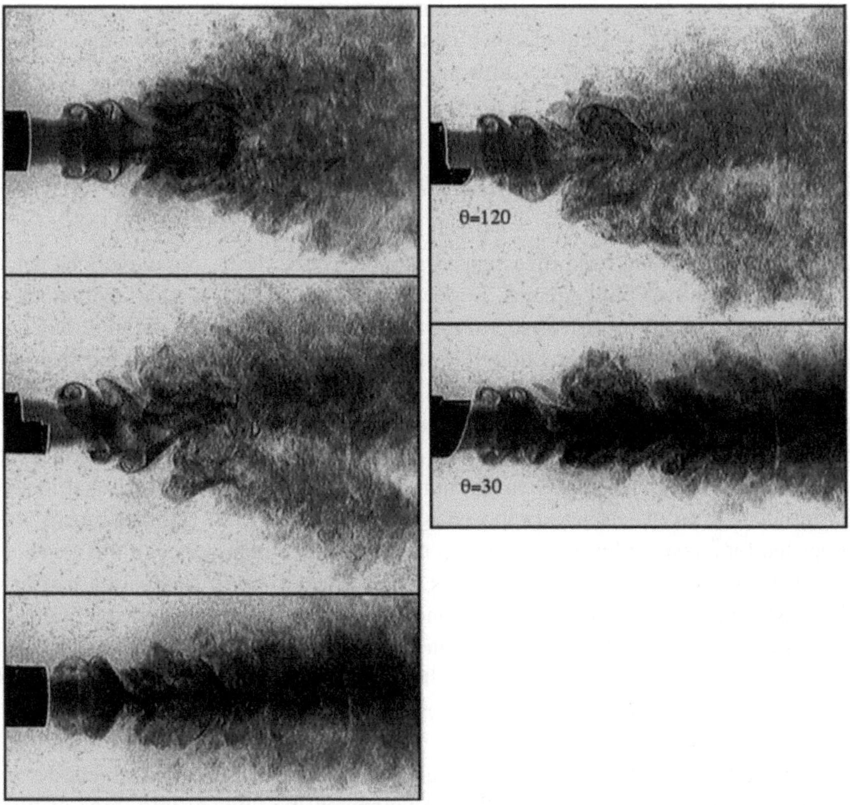


**Fig. 5** Ejections of jet fluid from the primary jet in the formation of secondary jet columns in step and sawtooth nozzles (Reprinted with permission from Longmire and Duong [37]. Copyright 1996, AIP Publishing LLC)

flows due to significant turbulent intensity levels or intentionally imparted through artificial means), as well as illustrating the interesting situation, in which these passive jet control nozzle performance can be optimized.

At the other end of the spectrum where high-speed jets are concerned, numerous experimental, numerical and theoretical studies on trailing-edge modified nozzles have been conducted by several research groups, such as Tam et al. [82], Raman [68, 69], Samimy et al. [76], Kim and Samimy [25, 26], Viswanathan [92], Knowles and Saddington [27], Alkislar et al. [2], Babucke et al. [3], Uzun and Hussaini [90], Zakharin et al. [107], Viswanathan and Czech [95], Powers and McLaughlin [66], Bridges [8] and Xia and Tucker [104], just to name a few. These studies mainly focused upon either high-subsonic or supersonic jets issued from serrated (i.e. chevron) or bevelled nozzles and the relationships between any alterations to the jet noise and exact nozzle designs. A good review on the historical evolution of high-speed jet noise control strategies based on such modified nozzles can be found in Zaman et al. [108] and readers are advised to refer to it for more information, as the present chapter concerns itself more with the fundamental vortex dynamics at much lower flow speeds.

## 2.3 Circular Coaxial Modified Nozzle Jets

Coaxial jets have long been investigated as an effective mean to passively enhance mixing between two fluid streams. Depending on the geometry, area- and velocity-ratio, nozzle wall thickness (on top of the usual Reynolds number, initial turbulence level), amongst others, interactions between the primary circular jet and the secondary annular jet will lead to distinctly different vortex structures and behaviour along the intermediate shear layers. Early studies were typically related to the enhancement of propulsion efficiency of double-spool bypass turbo-jet engines, as well as the reduction of their exhaust noise. An analytical two-dimensional model on turbulent coaxial jets was proposed by Morton [46], while another model was proposed for a coaxial burning jet by Beer et al. [7]. The importance of the velocity-ratio between the two coaxial jet streams with respect to noise attenuation was first demonstrated by Williams et al. [102] and their study was soon followed by those of Champagne and Wygnaski [14], Kwan and Ko [29–31] and Ko and Au [28], just to name a few. On the other hand, influences due to other flow parameters of interest, such as the initial turbulent intensity and the wall thickness, were investigated by Matsumoto et al. [40].

In an attempt to consolidate the typical behaviour of circular coaxial jets, Kwan and Ko [30] proposed that the jet flow development be divided into three distinct regions—the initial merging zone, intermediate merging zone and fully merged zone. Along the same vein, Ko and Au [28] postulated that the vortical behaviour associated with the coaxial jets could be treated as the super-positioning of two single-stream jets, where two separate trains of vortices and their mutual interactions (i.e. bifurcation and amalgamation) governing the final flow state. However, subsequent studies would show that such treatments might be too idealistic, considering the many physical and flow variables associated with coaxial jets. In reality, coaxial jets are proven to be far more complex. In particular, the study by Dahm et al. [15] demonstrated that coaxial jet behaviour might not be adequately understood by the simple super-positioning of two single-stream jets. Through a series of detailed laser-induced fluorescence visualization experiments, their investigation showed that the absolute jet velocities matter as much as the velocity-ratios, as the end-results were observed to be highly sensitive towards both these two parameters. Highly dynamic flow behaviour with strong coupling between the two jet streams were demonstrated to be highly non-linear with respect to the other initial flow conditions as well, such as shear layer thickness and wake defect. As a result, correspondingly complex jet entrainment and mixing behaviour were observed.

This prompted a series of follow-up studies to look more closely upon the impact of the initial flow conditions. For instance, the influence of the initial turbulent intensity on the jet developments was examined by Wicker and Eaton [99] and Buresti et al. [9], where detailed skewness and kurtosis data were presented. Further data and insights upon higher-order flow statistics were also presented in detail more recently by Sadr and Klewicki [73]. Interestingly, the formation of a recirculating bubble within the primary jet potential core when the velocity-ratio exceeds a critical value was observed by Rehab et al. [70] in their large velocity-ratio coaxial jets, where its low-frequency but large-amplitude oscillations conferred significant impact upon the overall jet-mixing characteristics. This was followed by a study by Villermaux and Rehab [91] where they looked into the mixing characteristics of such large velocity-ratio coaxial jets in terms of the jet shear layer behaviour, velocity differences and gap thickness, among others. The presence of the recirculation bubble was later confirmed and explored in greater detail by numerical studies performed by Reynier and Ha Minh [71] and Balarac and Metais [5]. In particular, the effects on the interactions between the two jet streams, as well as the recirculation bubble, due to variations in the nozzle lip thickness was investigated by the latter study as well. Influences due to the presence of flow forcing upon the secondary annular jet was also studied by Balarac et al. [6], where different excitation modes were used to explore their possible enhancements to the jet-mixing characteristics.

As compared to their single-stream counterparts, studies focusing upon the more fundamental aspects of indeterminate-origin coaxial jet flow dynamics are considerably more limited. Most of the interest stems from a desire to reduce jet noise associated with bypass turbofan engines, with the Boeing 787 Dreamliner chevron engine nacelle design as the most outstanding example at this point in time. Note that while extensive studies have been conducted in this case, only selected recent studies will be highlighted here for the sake of brevity. Other than incurring less thrust penalty, another important reason why chevrons are preferred over tabs in terms of engine noise control lies in their ability to better control the overall noise characteristics. For instance, Martens [39] noted that, while tabbed nozzles or mixers have the ability to shift the lower frequency noise to higher frequencies, chevrons are capable of actually reducing the low-frequency noise. Even though high frequency noise remains essentially the same as a conventional nozzle, audible engine "rumble" that propagates into the passenger cabin will be considerably reduced. A later study on the use of such chevrons on high bypass ratio nozzles by Saiyed et al. [75] confirmed this, when they compared between tabs and chevrons under both take-off and landing conditions as well. Chevrons were again observed to be more efficient towards jet noise reductions as compared to tabbed nozzles. Furthermore, their efficiency could be significantly improved by a moderate level of chevron penetration into the primary jet stream. On the other hand, the lower efficiency associated with tabs within their study could be attributed to an increase in the low-frequency jet noise during thrust increment.

In attempts to shed further light on the low- and high-frequency jet noise generation, a series of studies on coaxial jets and different modifications of chevron nozzles at the core jet were conducted by Callender et al. [11–13]. In their first investigation, chevron nozzles were observed to be more effective at lower acoustic frequencies and aft directivity angles. A high-penetration, eight-lobe nozzle was found to be the most optimal configuration among all the test nozzles examined, where it produces minimal acoustic noise at both low and high acoustics frequencies. Note that a combination of high-penetration nozzle and large velocity-ratio was observed to lead to aggressive mixing between the core and bypass jets, which in turn generates significant high-frequency jet noise. Therefore, intermediate velocity-ratios tend to be most beneficial for jet noise reduction using chevrons. These findings were subsequently supported with more experimental details using acoustic and PIV experiments in the two follow-up studies. A significant number of other experimental and numerical investigations on chevron nozzles have also been conducted by several research groups recently, such as Yan et al. [106, 105], Eschricht et al. [17], Page et al. [65], Tinney and Jordan [83], among many others.

In some ways, chevron nozzles can be seen as a conservative design, whereby minimal changes to the overall engine nacelle geometry are proposed. However, it has been recently demonstrated that imposing a significant inclination upon one or both nozzles in a turbofan engine could offer alternative jet noise control strategies. Take for instance, by inclining the primary jet nozzle only, Viswanathan [93, 94] was able to show that jet noise could be significantly reduced in the aft quadrant, while incurring only a slight increase at the lower polar angles. This idea was developed further and resulted in an extensive numerical investigation on the effects of inclining both the jet core and bypass nozzles [96], where the influences of using different inclination angles were examined. While significant information upon the resultant asymmetric flow behaviour was captured and analysed, a definitive conclusion on whether such nozzles would indeed bring tangible jet noise reductions (or not) could not be reached. This is due to the immense challenges associated in correlating the underlying flow phenomena and production of jet noise, which remains till this day. On the other hand, these developments also suggest that looking at inclined coaxial nozzles from a more fundamental point-of-view, in terms of the vortex dynamics, may serve as a valid starting point towards better identification of the flow-noise causality.

Over the next few sections, investigations on the vortex structures and dynamics associated with the use of trailing-edge modified nozzles by the authors will be covered. In particular, the effects of various trailing-edge modified nozzles on single-stream jets and dual-stream coaxial jets at low-to-moderate Reynolds numbers will be elaborated in detail. At these Reynolds numbers, the vortex dynamics can be discerned more clearly for less ambiguous flow interpretations.

# 3 Experimental Setup and Procedures

## *3.1 Recirculating Water Tank*

As the authors conducted the studies over a period of several years and significantly different nozzle geometries and types were investigated, several different experimental setups had been used, mostly in the University of Liverpool, UK. However,

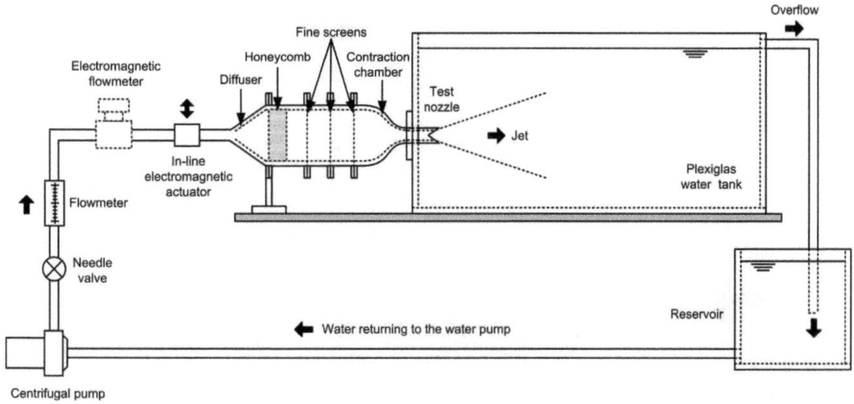


**Fig. 6** Schematics of a typical experimental setup used by the authors in their studies (Reprinted with permission from New and Tsovolos [54]. Copyright 2009, AIP Publishing LLC)

their designs were fundamentally similar and for the sake of simplicity, the most commonly used experimental setup is shown in Fig. 6. The setup essentially comprised of a small magnetic centrifugal pump channelling water from a small reservoir into a flow-conditioning jet apparatus, before the water exhausted from the apparatus into a quiescent 400 mm (H) × 400 mm (W) × 800 mm (L) Plexiglas water tank. Before the water from the pump entered the jet apparatus, its flow rate were controlled by a needle valve and metered using an electromagnetic flow meter, as well as a variable-area flow meter. In addition, an in-line electromagnetic actuator was located just before the jet apparatus to impart regular perturbations to the jet flows when flow forcings were required. Most the studies to be covered here made use of forced jet flows to better organize the fundamental vortex dynamics for ease in interpretations. Once the water entered the jet apparatus, it would pass through a honeycomb section, three layers of fine-grid screens and a contraction chamber, before exiting it via the test nozzle. To maintain the water level in the tank at a constant height, a pair of outlets located at the rear tank walls diverted any overflow back into the small reservoir and rendered the setup a recirculating one.

## 3.2 Dye-Visualization and Planar Laser-Induced Fluorescence

To facilitate clear visualizations of the near-field vortex dynamics, diluted blue-coloured liquid food dye was released circumferentially into the jet shear layers via a specially-designed cylindrical dye-port as shown in Fig. 7. While the exact design would vary according to the nozzle base geometry (i.e. circular or elliptic), the basic idea remains the same. Gravity-fed coloured dye was channelled into the cylindrical dye port via flexible PVC tubing and the four dye ports. The cylindrical dye port

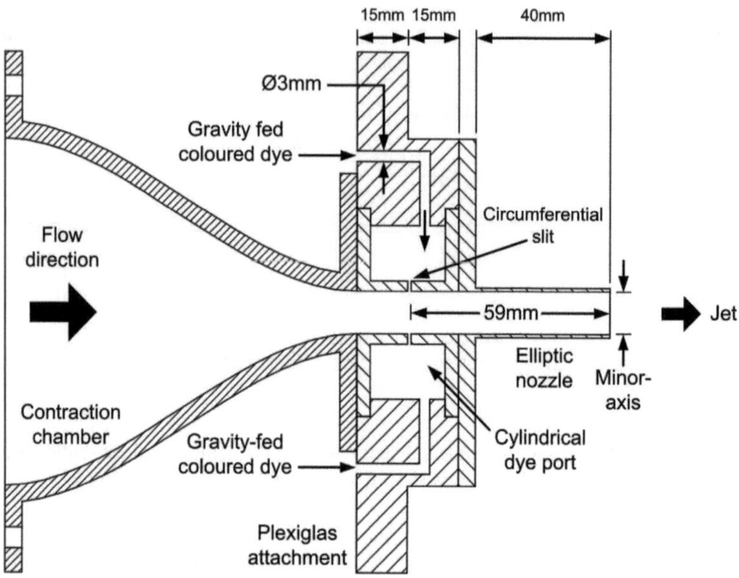


**Fig. 7** Schematics of the cylindrical dye port used for circumferential dye-visualizations (With kind permission from Springer Science+Business Media: New [47], Fig. 3)

would be completely filled with the coloured dye and exited from a 1 mm circumferential slit to trace out the shear layer as the jet exhausted from the test nozzle. Two 500 W halogen floodlights were used to illuminate the jet flows to reduce the presence of shadows and a digital 3CCD colour video camera with a TV zoom lens captured the flow visualizations directly into digital video files for subsequent analysis.

Planar laser-induced fluorescence (PLIF) technique was used to obtain cross-sections of the jet flows along both the streamwise and cross-stream directions to discern the vortex structures and their behaviour. For single-stream jets, fluorescein disodium salt was dissolved in and premixed with the water in the small reservoir before the experiments, such that during the experiments, the entire jet flow would fluoresce green when excited by laser sheet illumination. For dual-stream coaxial jets, the inner primary and outer annular jets would be seeded with premixed fluorescein disodium and rhodamine fluorescent salts respectively. When excited by the laser sheet, the secondary jet would fluoresce orange. To visualize the jet cross-sections along any desired plane of interest, laser beam from a 532 nm wavelength continuous-wave diode-pumped solid-state laser was steered into the appropriate location using periscopic beam-steering mirrors and formed into a thin laser sheet using a rotating mirror setup. The laser sheet thickness was limited to approximately 1.5 mm by using slits to restrict the laser sheet entering the water tank. Similar to the dye-visualization procedures described above, a colour video camera and TV zoom lens were used to capture the PLIF visualizations as digital videos.

### 3.3 Digital Particle-Image Velocimetry

To measure the jet flow fields quantitatively, a Dantec Dynamics two-dimensional digital PIV system was used. The system consisted of a 15 Hz-capable, 50 mJ double-pulse New-Wave Research Nd:YAG laser coupled with a 1600 px by 1200 px Flow Sense CCD camera with a Nikon 60 mm f2.8 lens. Timing and triggering of the laser and CCD camera, as well as the transferring of the captured particle image-pairs, were performed by National Instruments timing and image-grabber cards on-board a workstation respectively. To control the operations of these preceding hardware components, data-acquisition and subsequent post-processing, Dantec Dynamics Flow Manager™ software was used. To seed the water tank and jet flows, 1.03 g/cm$^3$ specific gravity polyamide seeding particles were used. For single-jet flows, 50 μm nominally-sized seeding particles were used. On the other hand, 20 μm seeding particles were used for dual-stream coaxial jet flows, so as to better resolve the multiple jet shear layers. The beam from the Nd:YAG laser was directed into the measurement regions by beam-steering optics and expanded into a laser sheet by sheet-forming optics. To restrict the laser sheet thickness such that it could be assumed to be a two-dimensional illumination source satisfactory, it passed through physical slits prior to its entry into the measurement regions such that its thickness was limited to approximately 1.5 mm.

Scattered light from the seeding particles were captured at either full system frequency of 15 Hz without any flow phase preference for time-averaged measurements, or triggered by jet flow forcing signals for phase-averaged measurements. Depending on the exact test jet configuration, a total of 500 or 1000 image-pairs were captured by the PIV system. To process the captured image-pairs, multi-pass, multi-grid cross-correlations were used. For single-jet flows, the initial and final interrogation window sizes used were 128 px × 128 px and 32 px × 32 px respectively. For dual-stream coaxial jet flows, the corresponding interrogation window sizes used were 64 px × 64 px and 16 px × 16 px respectively. After the raw velocity vector maps were determined, they were subjected to a series of global and local validation criteria to reject spurious vectors and subsequently smoothed by 3-point-by 3-point neighbour schemes to arrive at the final velocity vector maps. Derived flow quantities such as vorticity, Reynolds stresses and turbulent kinetic energy were then determined for further analysis.

## 4 Effects of Axial Length on Circular Notched and Hybrid Inclined Jets Vortex Dynamics

Much of the studies to be detailed here originated from the study conducted by the first author [63], where two-tooth V-notched nozzles were investigated. Note that these notched nozzles were formed by cutting the nozzle exit using a "V-shaped" notch, which produced two smooth peaks and two sharp troughs. In this case, the nozzle exit geometry is different from nozzles that possess sharp peaks and troughs [38]. Essentially, each half of the V-shaped notch was specifically designed to resemble a half of an ellipse with a desired aspect-ratio, as shown in Fig. 8. As such, these nozzles are distinguished from one another by the half-ellipse aspect-ratios upon which they are based on. Intriguingly, the study on these V-notched nozzles demonstrated that outwards-spreading mushroom-shaped CVPs are formed at both peak and trough locations, resulting in four of them been visualized along the jet periphery. This is in stark contrast with Longmire et al. [38], where outwards- and inwards-spreading CVPs were produced at the peak and trough locations respectively. As a result, the jet entrainment regions are also different between these two studies. Where they occurred along trough locations in Longmire et al. [38], results from New et al. [63] showed that they exist in-between peak and trough locations, as shown in Fig. 9.

At this point in time, it is believed that the jet initial conditions such as Reynolds number and azimuthal momentum thickness distribution are responsible for these discrepancies, as a follow-up study that made use of low Reynolds number jets (i.e. $Re = 2100$) on similar notched nozzles by New and Tsovolos [54] reproduced the earlier results observed earlier in New et al. [63]. However, this needs to be confirmed through further studies. Based on the visualization results obtained, they reconstructed and proposed a vortex flow model to explain the near-field vortex dynamics. The vortex flow model conjectured that vortex lines followed the

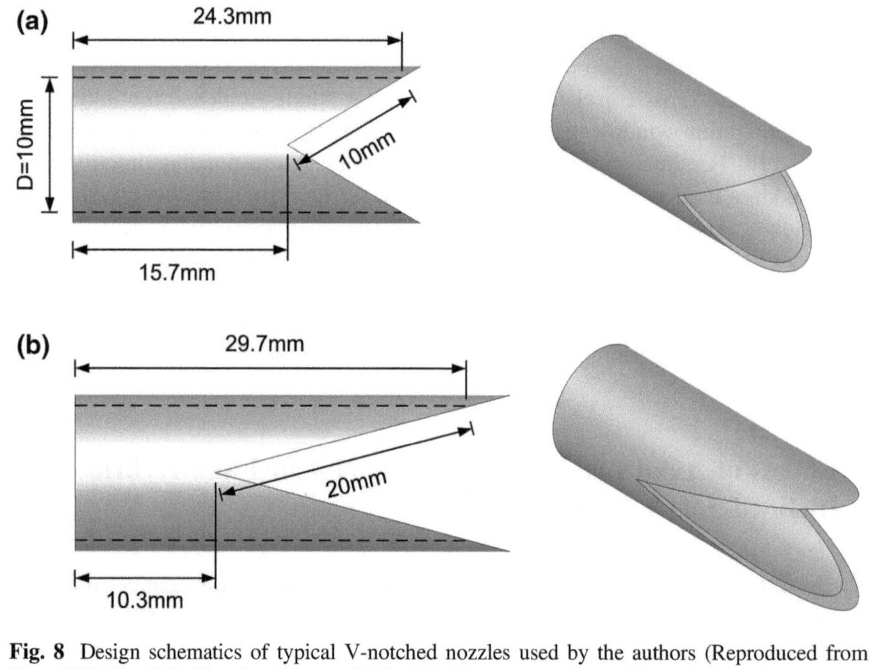


**Fig. 8** Design schematics of typical V-notched nozzles used by the authors (Reproduced from New [48] (© 2010 Elsevier Masson SAS. All rights reserved). **a** AR = 2 V-notched nozzle. **b** AR = 4 V-notched nozzle

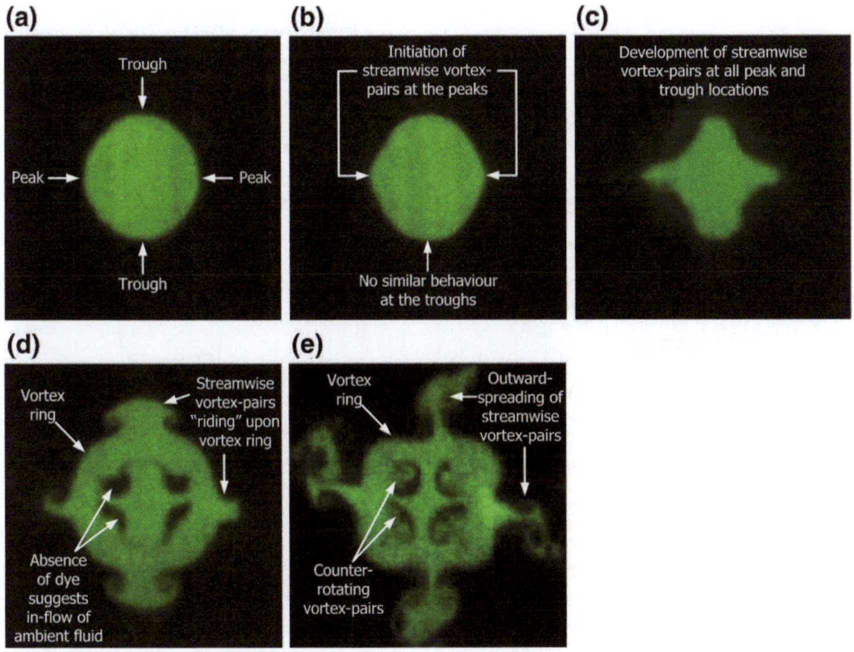


**Fig. 9** Jet cross-sections depicting the formation of outwards-spreading streamwise vortices at both peak and trough locations of a V-notched nozzle (Reprinted with permission from New et al. [63]. Copyright 2005, AIP Publishing LLC)

notched nozzle exit contour very closely to resemble a pre-bent elliptic vortex filament, before undergoing a series of vortex-bending processes that lead to the formation of outwards-spreading streamwise-aligned vortices at peak and trough locations, as shown in Fig. 10. Those at the peaks are postulated to be produced by the closely-spaced portions of the pre-bent vortex filament, while those at the troughs are believed to be formed by the vortex line bending towards the upstream direction. Hypothetical cross-sections of these streamwise vortices will reproduce the mushroom-shaped CVPs observed during the experiments. By shedding these streamwise vortices later on, the vortex lines will take after the outline of a bent vortex filament once again.

In a follow-up study by New and Tsovolos [54] where the effects of increasing the tooth length was carried out, it was discovered that the near-field vortex dynamics is sensitive towards the tooth length or peak/trough "sharpness". To be more specific, notched nozzles with relatively sharper peaks/troughs will lead to vortex filament bending directions opposite to that observed in nozzles with peaks/ troughs that are considerably less sharp, after the streamwise vortices have formed and shed, as shown in Fig. 11. Experimental evidence suggested that vortex disconnection and reconnection behaviour shown schematically in Fig. 12, due to differences in the proximity of the various vortex filament portions, accounts for these difference observations. For a "sharper" nozzle, vortex filament "arms" will

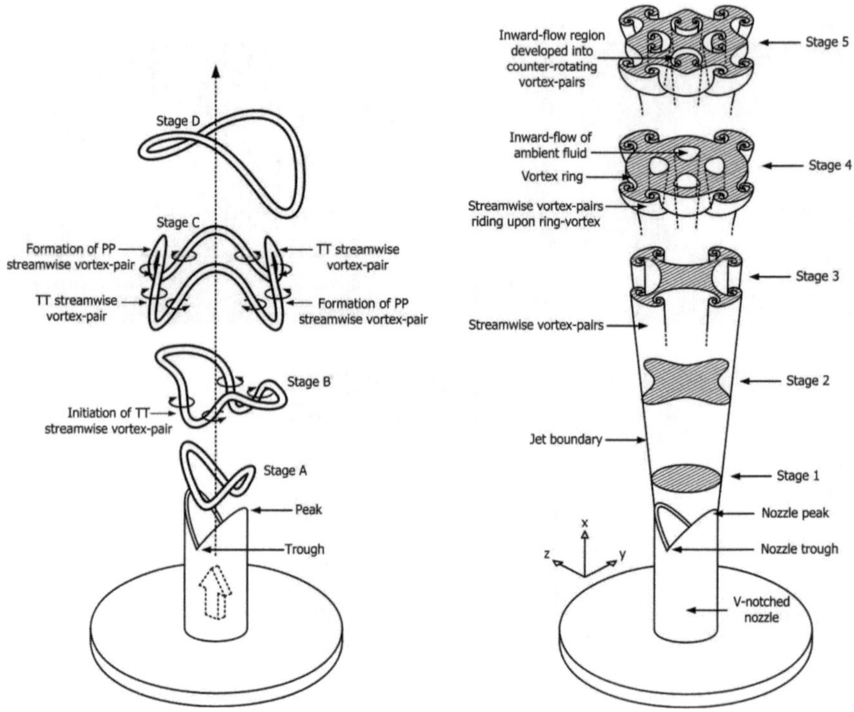


**Fig. 10** Vortex flow model postulated for a V-notched nozzle of moderated nozzle sharpness (Reprinted with permission from New et al. [63]. Copyright 2005, AIP Publishing LLC)

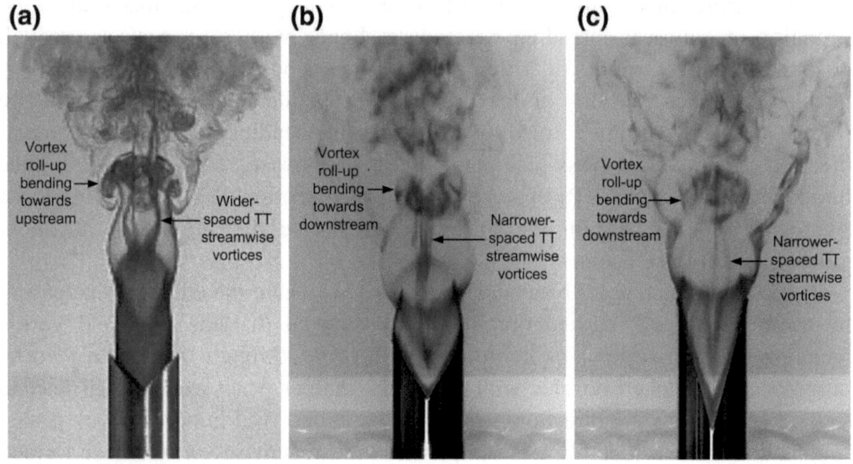


**Fig. 11** Changes to the vortex filament near-field bending direction as the relative sharpness of a V-notched nozzle varies (Reprinted with permission from New and Tsovolos [54]. Copyright 2009, AIP Publishing LLC)

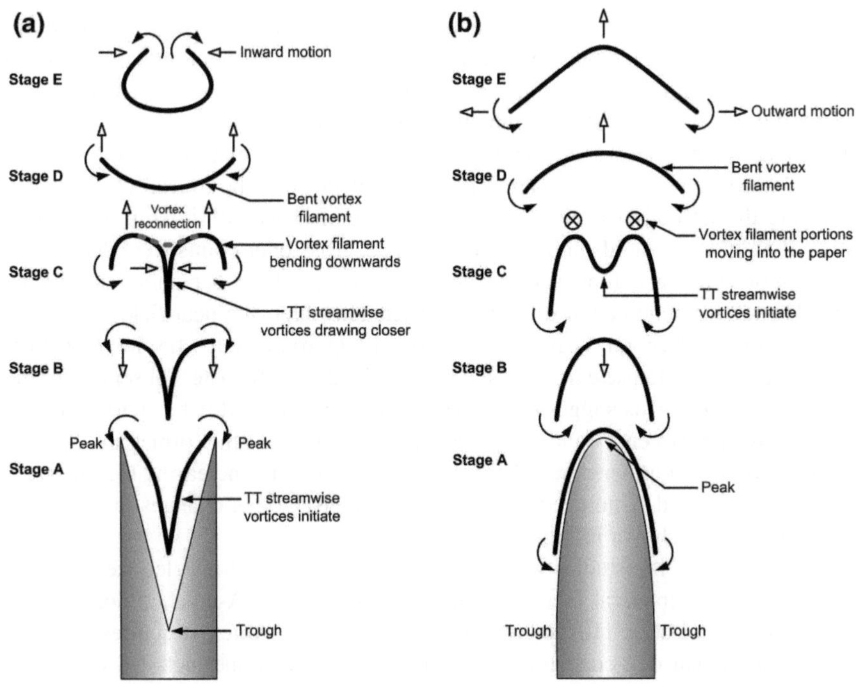


**Fig. 12** Vortex flow model proposed for a significantly "sharper" V-notched nozzle (Reprinted with permission from New and Tsovolos [54]. Copyright 2009, AIP Publishing LLC)

pair upmuch faster at the trough locations and produce conditions conducive towards formation of a vortex filament bending towards the downstream direction (i.e. when viewed along the peak-to-peak plane), after vortex disconnections and reconnections are completed. In contrast, more widely-spaced vortex filament "arms" in a less sharp nozzle will lead to a vortex filament bending towards the upstream location (i.e. when viewed along the peak-to-peak plane). It is insightful that such significant changes to the near-field vortex dynamics can be instigated simply by increasing the tooth length or the relative nozzle sharpness. More importantly, it infers that potentially impactful effects on the jet-mixing and control characteristics can be achieved through simple nozzle design changes.

At this point, the readers should have noted that imposing a "V-shaped" cut across the nozzle exit is not the only way to create a notched-nozzle—it is possible to use an "A-shaped" cut as well. In this case, rather than the two smooth peaks and two sharp troughs in V-notched nozzles, A-notched nozzles possess two sharp peaks and two smooth troughs instead. A study was conducted to see whether sharp or smooth configuration of the peaks/troughs has any appreciable influence upon the near-field vortex dynamics of notched nozzles and flow visualizations results revealed that they are essentially similar, at least at the low Reynolds number of $Re$ = 2000 [50], as shown in Fig. 13. Using the same V- and A-notched nozzles, hot-wire anemometry velocity-field measurements were taken at a much higher Reynolds number of

*Re* = 20,500. Results show that jet centreline velocity characteristics did not differ significantly between the two notched-nozzle types and a conventional non-notched nozzle. In-line with this observation,the exact configuration of the nozzle peak and trough did not have any strong influence upon the half-jet widths taken along the peak/trough planes. However, there is evidence to suggest that the use of V- or A-notched nozzle exerts more apparent flow influences upon the overall flow field by distorting the jet body (or cross-sections, to be more specific) differently at a significantly higher Reynolds number. For instance, cross-over points were detected in both nozzle types and hinted at presence of axis-switching behaviour due to the vortex-bending phenomenon described earlier. However, the near-field cross-over point occurred later for A-notched nozzle at approximately 4.5 jet diameters downstream of the nozzle exit, almost doubling the 2.5 jet diameters observed for the V-notched nozzle. This suggests that, while the near-field vortex behaviour may be largely similar between V- and A-notched nozzles, those for the former are likely to be more dynamic. Furthermore, a later study [48] found that increasing the relatively sharpness of V- and A-notched nozzles confers considerable changes in the cross-over locations, as shown in Fig. 14.

While much of the studies involving the use of modified nozzles were experimental-based, a numerical investigation on the use V- and A-notched nozzles was performed by Cai et al. [10] to compare their near-field vortex structures. Using a unique “numerical dye concentration visualization” technique, detailed cross-sections of jets issuing from these two nozzle types were obtained by solving the associated Navier-Stokes equations through finite-volume method. Similar to the observations made by New et al. [63] and New and Tsovolos [54], outwards-spreading streamwise vortices were produced at both peak and trough locations. In particular, changes to the jet cross-sections at progressively downstream locations matched up between these studies, as shown in Fig. 15, albeit some differences caused by the inherent difficulties in simulating the scalar fluorescent dye used in the two experimental studies. Interestingly though, the numerical failed to reproduce the vortex flow model proposed by New et al. [63], as no distinct vortex lines were observed to form immediately beyond the notched nozzle exits. However, note that

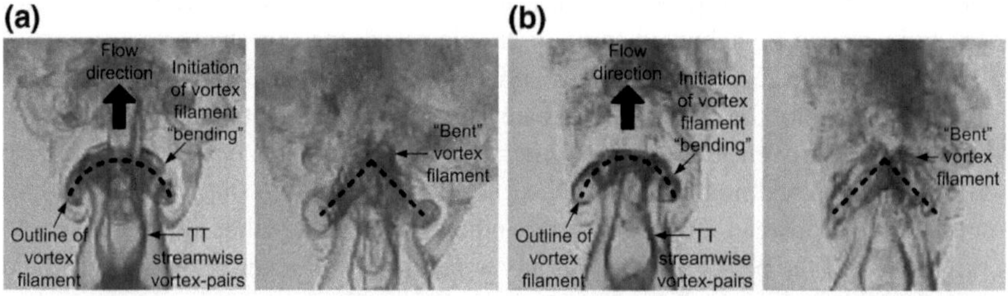


**Fig. 13** Relatively similar vortex structures and developments associated with a **a** V-notched and **b** A-notched nozzle jet at low Reynolds numbers (From New and Tsai [50]. Reprinted by permission of the American Institute of Aeronautics and Astronautics, Inc.)

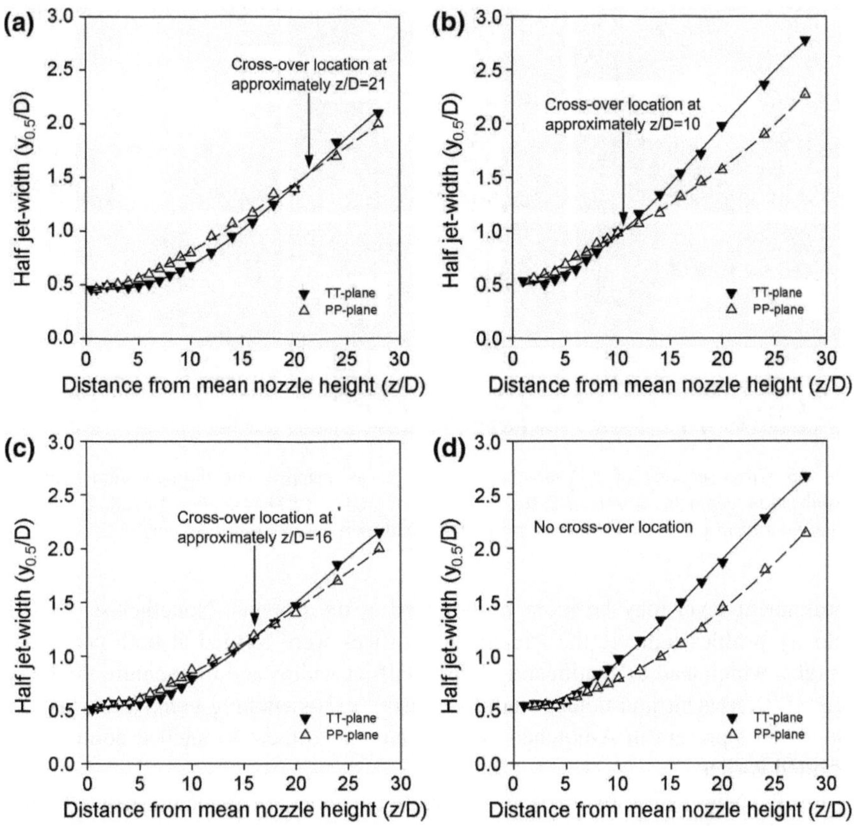


**Fig. 14** Variations to the cross-over locations associated with V- and A-notched nozzles of different relative sharpness (Reproduced from New [48] © 2010 Elsevier Masson SAS. All rights reserved)

no deliberate flow forcing was used in the numerical study and may explain this outstanding discrepancy. Flow forcing is often used to organize and accentuate the vortex formations, which in this case would allow the vortex lines to follow the nozzle exit contours faithfully. In its absence, the nozzle peaks and troughs would only serve as passive flow disturbances through their different initial boundary conditions, which was reflected in Cai et al. [10] as well. It will be interesting indeed to see if artificially perturbed V- and A-notched nozzle jet flow simulations will reproduce the experimental results more closely.

A-notched nozzles were explored further when their relative nozzle sharpness was varied systematically during an experimental study [57]. As expected, the use of a sharper A-notched nozzle serves to accentuate the underlying vortex dynamics. Intriguingly, the presence of the streamwise vortices at the troughs is not as obvious in the immediate near-field of the nozzle exit, compared to those at the peaks. Secondary vortices detected close to the jet centreline also suggest that

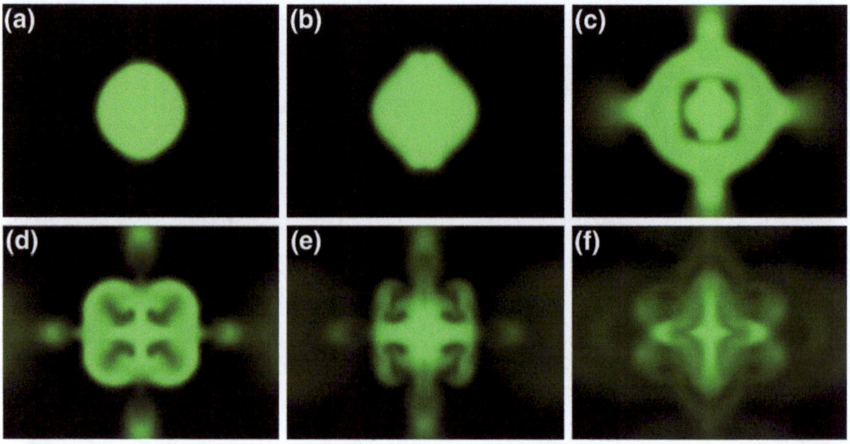


**Fig. 15** Cross-sections of a V-notched nozzle jet as mapped out through numerical dye visualization technique. **a** $x/D = 2.8$. **b** $x/D = 3.8$. **c** $x/D = 4.8$. **d** $x/D = 5.8$. **e** $x/D = 6.8$. **f** $x/D = 7.8$ (Reprinted from Cai et al. [10]. Copyright 2010, with permission from Elsevier)

entrainment level may be more pronounced at the troughs. Nonetheless, half-jet velocity profiles indicate the streamwise vortices were formed at both peaks and troughs, which lead to significantly larger half-jet widths and momentum thickness than an equivalent non-notched nozzle. Lastly, axis-switching behaviour was also found to be present in A-notched nozzles, similar to their V-notched counterparts reported earlier.

Lastly, New and Tsovolos [57, 58] explored other forms of non-conventional modified nozzles, named as hybrid inclined nozzles. These nozzles were designed such that each was a combination of one flat and one inclined half-nozzle sections, as shown in Fig. 16. In some ways, they resembled stepped and sawtooth nozzles, while possessing distinct inclined sections. Hence, it was of significant interest then to find out whether they behave more like simple inclined nozzles, or undergo jet bifurcations. Using $Re = 2100$ forced jets, the studies showed that they resembled more like the former, while exhibiting certain elements of the latter. For instance, hybrid inclined ring-vortices which veered towards the flat-section were produced when the relative sharpness of the inclined-section was moderate. This behaviour will be quite similar to that observed by Webster and Longmire [97] for their simple inclined jets. On the other hand, hybrid inclined ring-vortices undergo more convoluted flow distortions in the immediate near-field when the inclined-section was relatively sharp.

Interestingly though, regardless of the inclined-section relative sharpness (at least for these two studies by the authors), cross-sections of hybrid inclined jets deduced from PIV measurements appeared to correlate well with those of bifurcated jets, especially those of Longmire and Duong [37]. However, closer inspections of the cross-stream flow visualization results indicated that this is unlikely to be the case. Instead, strong formations of streamwise vortices by the inclined-section gave the impression that a weaker secondary jet is gradually detaching itself away from

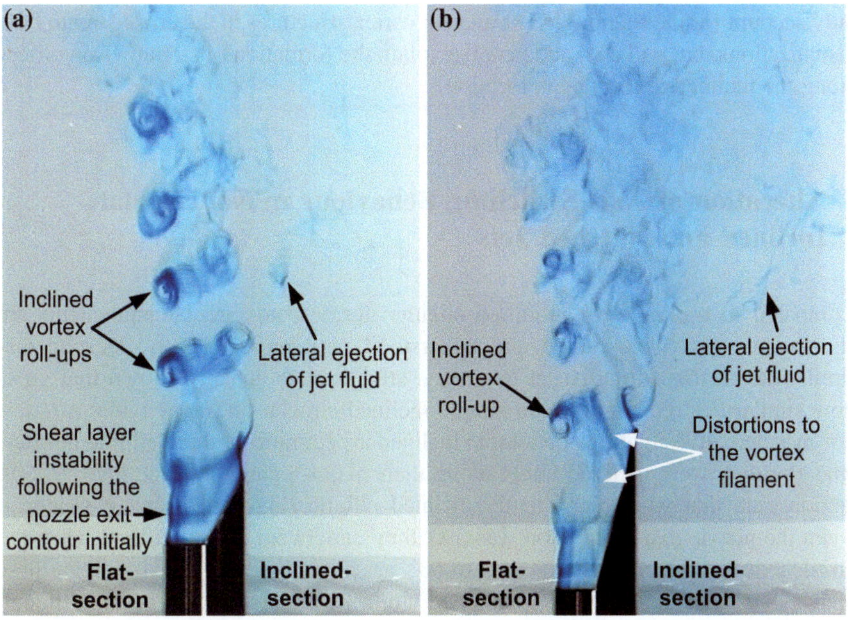


**Fig. 16** Near-field vortex structures associated with low and high aspect-ratio hybrid inclined nozzles. A "sharper" configuration produces stronger ejections of streamwise vortices. **a** AR2 hybrid inclined, $0 = \pi/2$ plane. **b** AR4 hybrid inclined, $0 = \pi/2$ plane (with kind permission from Springer Science+Business Media: New and Tsovolos [57], Fig. 3b, c)

the more dominant primary jet. Note that this is by no means conclusive, since Longmire and Duong [37] had earlier noted the sensitivity of jet bifurcation towards the forcing amplitude used. The forcing amplitude used for hybrid inclined jets was kept relatively low at approximately 2.2 %, just sufficiently large to induce regular large-scale azimuthal vortex roll-ups to be produced. Hence, it may be a topic of interest in future studies to clarify this issue.

Interestingly, the near-field dynamics of hybrid inclined jets share a few similar flow features with the flow mechanism put forward by Troolin and Longmire [84] to explain the behaviour of inclined vortex-rings. In particular, cross-stream flow visualization results indicated the presence of additional vortex structures that are not part of the large-scale ring-vortices produced by the forced jet flows. Instead, they are believed to be associated with vortex-loops induced by and trailing behind each of the large-scale ring-vortices. Each ring-vortex possesses an outline resembling closely the hybrid inclined nozzle exit contour and as it progresses to detach itself away from the nozzle exit, close proximity between the ring-vortex and the vortex-loop leads to their pairing along the flat-section. However, unlike the ring-vortex which leaves the nozzle completely in an inclined state, part of the vortex-loop remains mostly attached to the inclined-section while the pairing process accelerates the part along the flat-section. As a result, the vortex-loop is inclined in the opposite direction when it leaves the nozzle, in direct contrast to the ring-vortex. This explanation will not only account for the seemingly extraneous vortex structures in the cross-stream flow visualizations, but will also adequately explain the formation of streamwise vortices along the inclined-section peak location.

## 5 Alteration of Axis-Switching Behaviour in Noncircular Inclined and Notched Jets

Compared to trailing-edge modified circular nozzles, noncircular equivalents are significantly less well-studied and understood until more recently. Despite dissimilarities in the azimuthal jet geometry, studies have actually shown that gross flow similarities remain between simple inclined circular and noncircular nozzles. For instance, investigations on simple inclined aspect-ratio of three elliptic nozzles, whereby they were inclined either along their major- or minor-plane [47, 59, 60], demonstrate that while persistently inclined elliptic ring-vortices were produced when the nozzle exit inclination was 30°, they underwent vortex-turning when the nozzle exit inclination was increased to 60°, as shown in Fig. 17. This is regardless of whether the elliptic nozzle is inclined along its major- or minor-plane, though vortex-turning is more gradual in the latter case, as shown in Fig. 18.

Nevertheless, these observations attest to the critical role played by the initial boundary conditions imposed by trailing-edge modifications (To ease identification, nozzles inclined along their major- and minor-planes are known as major-plane and minor-plane inclined nozzles respectively hereafter). Experimental observations thus far suggest that simple inclined exits have significantly larger effects upon the initial jet circulation than those associated with the azimuthal jet geometry. Having said that though, influences coming off the azimuthal jet geometry and the exact incline-plane used do appear eventually as the simple inclined elliptic jets convect

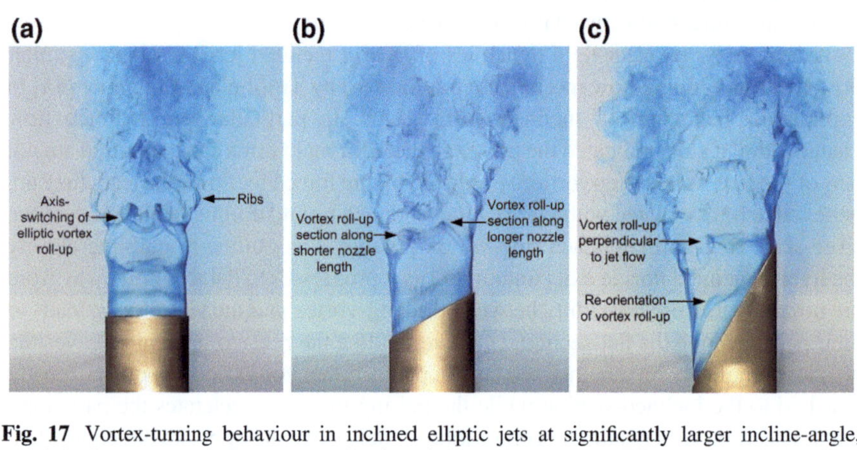


**Fig. 17** Vortex-turning behaviour in inclined elliptic jets at significantly larger incline-angle, which leads to progressively suppressed axis-switching phenomenon (With kind permission from Springer Science+Business Media: New [55], Fig. 4)

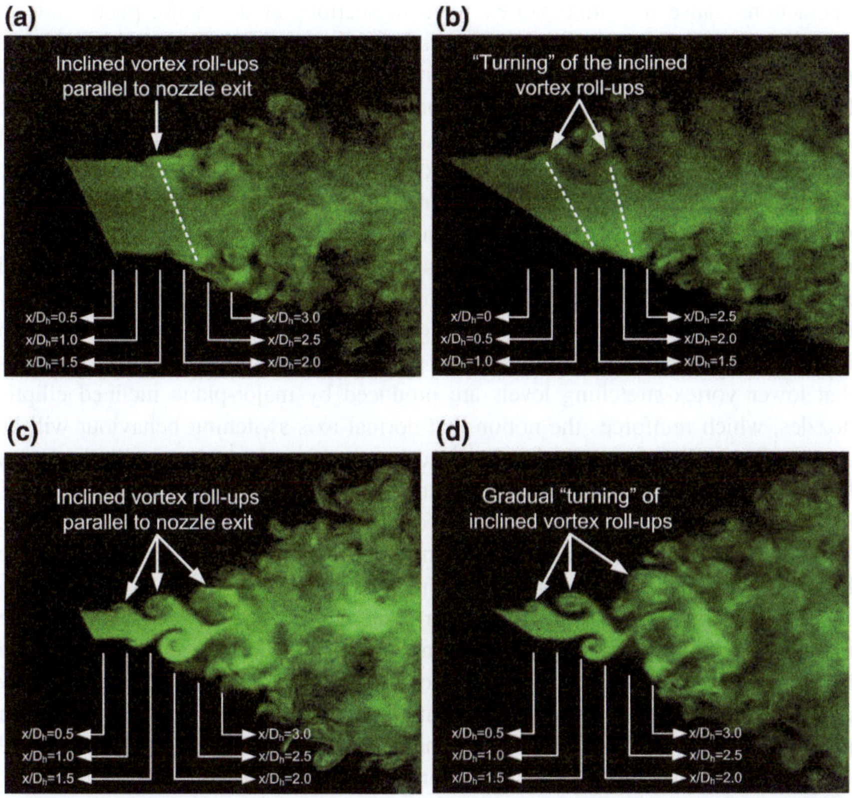


**Fig. 18** Significant vortex-turning observed in major-plane inclined elliptic jets, which is absent in minor-plane inclined elliptic jets at similar incline-angles. **a** 30° inclined-major plane. **b** 30° inclined-minor plane. **c** 60° inclined-major plane. **d** 60° inclined-minor plane (Reproduced from New and Tsovolos [59] © 2011 Elsevier Masson SAS. All rights reserved)

further away from the nozzle exits. As mentioned earlier, noncircular jets such as elliptic jets have a strong tendency to undergo axis-switching behaviour when they are non-inclined. Therefore, any physical changes should lead to alterations to the axis-switching behaviour.

Indeed, this was found to be the case for major-plane inclined elliptic nozzles, where no clear axis-switching phenomenon was observed in them, as compared to a non-inclined elliptic nozzle. The production of streamwise rib-structures along both ends of the nozzle major-planes was also found to become increasingly dissimilar as the nozzle incline was varied. For instance, formation of rib-structures along the shorter nozzle length reduces as the nozzle incline increases, while that along the longer nozzle lengths remains practically unaffected. According to earlier investigations on elliptic jets [21–23], rib-structures arise from the mutual interactions between the major-plane ends of the ring-vortex and the braid vortex filaments, caused by non-uniform self-induced velocities along both filaments. These interactions cause the braid vortex filament sections at the major-plane ends to realign in the streamwise direction to produce rib-structures. Correlating with the observations made for major-plane inclined elliptic nozzles, it would appear that their formation mechanism is only suppressed along shorter nozzle lengths. The exact reasons remain unclear at this point, though firstly, it is likely that the inclined nature of the ring-vortex and braid vortex filaments in major-plane inclined elliptic nozzles simply does not lead to initial circulation distributions favourable towards rib-structure formations at the shorter nozzle lengths. Phase-averaged measurements of the ring-vortices vorticity levels have indicated that major-plane inclined exits not only lead to an expected asymmetric distribution of ring-vortex vorticity between the shorter and longer nozzle lengths, but also produce an overall reduction in the vorticity level of the entire major-plane inclined ring-vortex. This suggests that lower vortex-stretching levels are produced by major-plane inclined elliptic nozzles, which reinforces the notion that normal axis-switching behaviour will be altered. Secondly, vortex-turning incurred by significantly inclined elliptic vortex filaments at the shorter nozzle lengths further likely disrupts any favourable self-induced velocity distributions necessary for successful production of rib-structures. This will offer some plausible explanations as to why rib-structures continued to be formed along the longer nozzle lengths instead.

Interestingly, in contrast to their major-plane inclined counterparts, minor-plane inclined elliptic nozzles produce increasingly higher vorticity levels in the inclined ring-vortices which arguably increases vortex-stretching levels, as demonstrate by phase-averaged PIV measurements. And despite some resemblance between inclined circular and minor-plane inclined ring-vortices, closer inspections will nevertheless reveal some interesting differences shortly after they are fully formed. Take for instance, Webster and Longmire [97] has shown that the inclination of the inclined circular ring-vortices tends to increase beyond that of the nozzle exit as they convect away from the nozzle exit, particularly when the nozzle exit inclination is smaller. In contrast, while minor-plane inclined elliptic ring-vortices exhibit grossly similar trends, vortex-turning in the 60° inclined nozzle will reduce their inclinations gradually till they are less inclined than the nozzle exit [60]. It should be noted that for the corresponding 30° nozzle, inclination of the ring-vortices kept increasing, similar to inclined circular ring-vortices as depicted in Fig. 19. This opposing behaviour raised an interesting question: is there a critical incline-angle for a minor-plane inclined elliptic nozzle where the inclined ring-vortices will remain the same as the nozzle exit inclination? Incidentally, the variations of the ring-vortex inclination angle at different locations downstream of the 30° and 60° minor-plane inclined elliptic nozzle exits exhibit relatively linear trends within the measurement range and therefore could be described by linear equations reasonably, except for locations very close to the nozzle exits. Subsequent analysis indicated that a critical angle possibly exists at approximately 41°—that is to say, inclined ring-vortices associated with a 41° minor-plane inclined elliptic jet are expected to retain their original inclinations in a persistent manner. Having said that, note that near-field vortex dynamics are ultimately highly dynamic and non-linear, and hence this critical incline-angle should be treated as a

**Fig. 19** Differences in the vortex roll-up inclinations for major- and minor-plane inclined elliptic jets (Reprinted from New and Tsovolos [60], with permission from Elsevier)

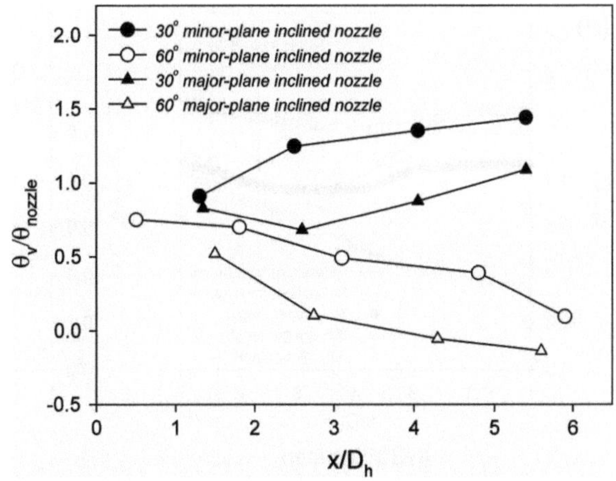


gross approximation based on the flow measurement data available at that point. Note that this exercise was not performed on major-plane inclined elliptic jets, as there were insufficient data points that corresponded to $\theta_v/\theta_{nozzle} = 1$, due to rapid dissipation of the ring-vortices through viscous effects.

Regardless of whether the elliptic jet is major- or minor-plane inclined, PIV measurements indicate that jet entrainment becomes more asymmetric along the incline-plane as the incline-angle increases. In this case, jet entrainment will be higher along the longer nozzle length regions, though it should be mentioned that the higher jet entrainment levels are only limited to the immediate vicinity of the nozzle exits for 60° inclined nozzle exits. Furthermore, all inclined elliptic jets achieved lower centreline velocities than a non-inclined elliptic jet from approximately three hydraulic diameters onwards. Visual and quantitative evidence pointed towards higher levels of near-field flow dynamics to be responsible for such observations of possibly better jet-mixing characteristics. Half-jet width (shown in Fig. 20) and momentum thickness information were also extracted from PIV measurements and they showed that axis-switching behaviour not only continues to be present in minor-plane inclined elliptic jets but appears to have been accentuated, albeit in an asymmetric manner due to the nozzle inclination. This is in contrast to the non-presence of strong axis-switching behaviour in major-plane inclined elliptic jets noted earlier. Thus, it would appear that the continuing presence of axis-switching behaviour is dependent upon whether significant vortex-turning behaviour is promoted by the inclined nozzle exits. The mixing layers of major- and minor-plane inclined elliptic jets were also significantly altered and rather complex to analyse, as the nozzle lengths varied continuously around the exit circumferences. Nevertheless, for minor-plane inclined elliptic jets, the mixing layers along the shorter and longer nozzle length regions are considerably thicker than that along their major-planes, particularly for the 60° inclined test case. This agrees well with the flow visualization results that showed progressive narrowing of the jet column, as well as the earlier occurrences of axis-switching behaviour, in minor-plane

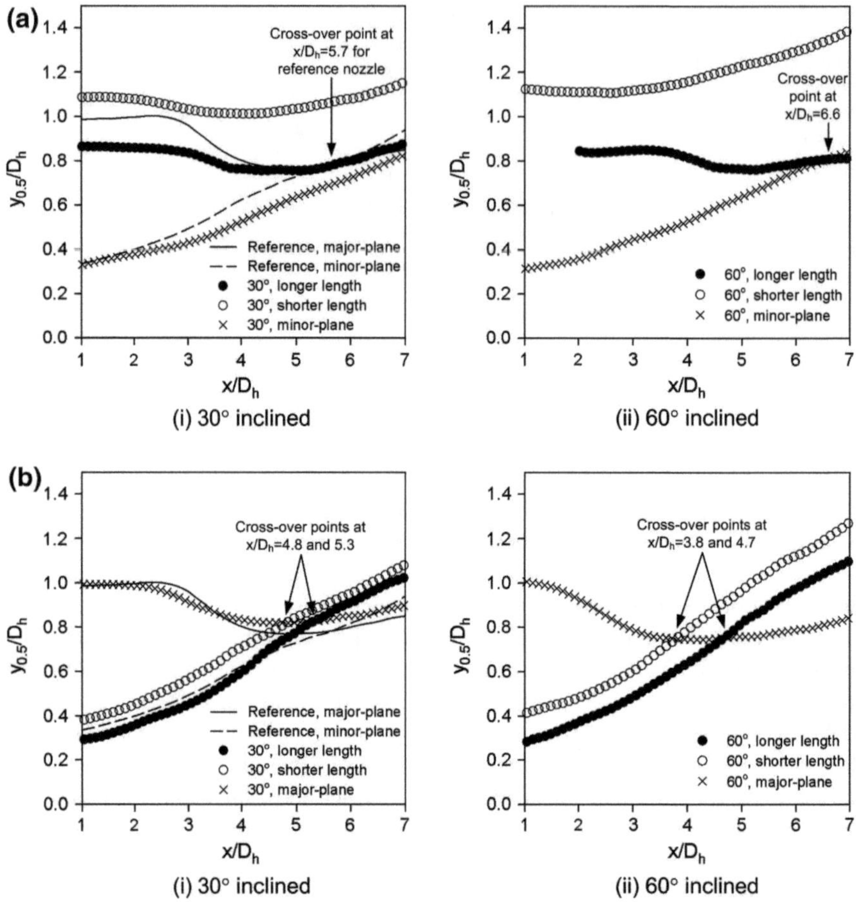


**Fig. 20** Half-jet width developments of major- and minor-plane inclined elliptic jets show sensitivity towards the exact incline-plane and incline-angle used. **a** Major-plane inclined nozzles. **b** Minor-plane inclined nozzles (Reproduced from New and Tsovolos [59] © 2011 Elsevier Masson SAS. All rights reserved)

inclined elliptic jets as the incline-angle increases. Interestingly though, the mixing layer along the shorter nozzle length region tends to be thicker for both major- and minor-plane inclined elliptic jets within four to five hydraulic diameters downstream of the nozzle exits, before being overtaken by that along the longer nozzle length region. However, this does not imply that the use of inclined nozzles here merely displaces the jet shear layer developments further downstream along the longer nozzle length region, as both the half-jet width and momentum thickness growth rates and trends are quite different along these two distinct locations. Instead, it should be seen as the source of asymmetric flow strain distributions, which in turn lead to the behaviour observed in these inclined elliptic jets so far.

Similar to modified circular jets discussed earlier, efforts were made to shed light upon elliptic jets issuing from more complex nozzle configurations. One of them focuses upon the use of hybrid inclined elliptic nozzles, where the general flow behaviour was relatively similar to that exhibited by hybrid inclined circular nozzles, except for some differences arising from the use of an elliptic geometry [61]. For the sake of brevity, their detailed findings will not be covered here. Instead, the authors would like to shift the attention to V-notched elliptic jets under forced flow conditions, which can be seen as a natural extension to V-notched circular jets studied in their earlier investigations. While it is tempting to assume that the flow fields would be very similar between the two nozzle types, experimental findings indicate that they are actually very different [62, 88, 89], due to the non-uniform circumferential distribution of the initial jet shear layer characteristics of the conventional unaltered elliptic nozzle to begin with. In this particular study, 60° V-shaped notches were imposed such that they were in-plane with either major- or minor-plane. Clearly, this further varies the initial jet shear layer characteristics and results demonstrated that very different flow states can indeed be achieved.

As shown in Fig. 21(a), for the elliptic nozzle with the V-shaped notch aligned such that it crosses the major-plane (i.e. Nozzle A), the vortex filaments at the troughs paired up very shortly after they manifested from the flow forcing and leads

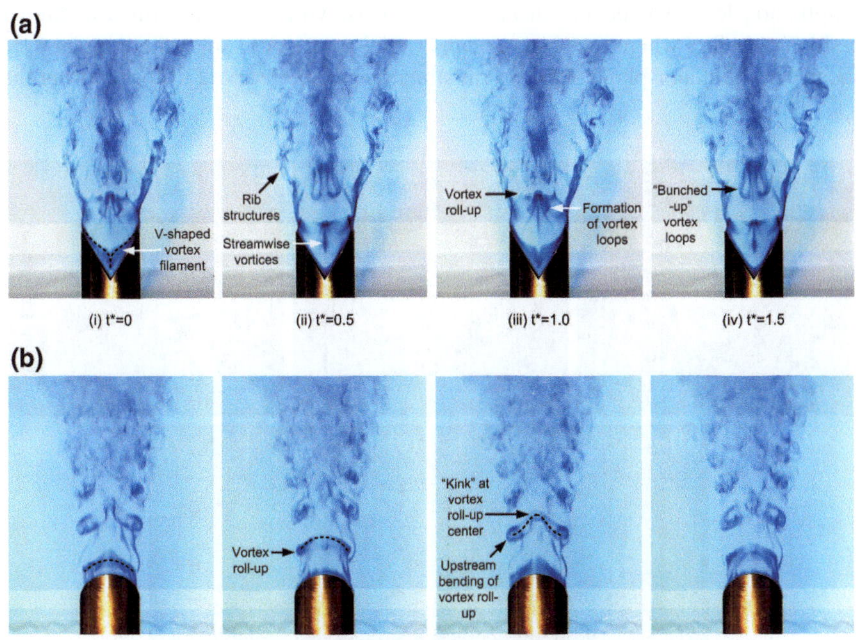


**Fig. 21** Differences in the near-field vortex structures between two different V-notched elliptic nozzles (Reprinted from New and Tsovolos [62], with permission from Elsevier). **a** V-notch aligned along minor-plane (Nozzle A). **b** V-notched aligned along major-plane (Nozzle B)

to the immediate formation of streamwise vortices. In particular, as the vortex filament convect downstream and rolled up, it congregated upon itself to form repeating loops that appeared to "bunch up" together along the jet centreline. Note that for conventional non-notched elliptic jets, the vortex filament would have underwent more straightforward bending behaviour associated with axis-switching phenomenon. Despite possessing vortex dynamics that depart significantly from those associated with non-notched elliptic jets, rib structures continued to form at both major-plane ends (where the smooth peaks are located), though closer to the nozzle origin. Cross-sections of the V-notched elliptic jet indicate that the vortex dynamics are also significantly more complex than its non-notched equivalent. For instance, braid vortex structures induced by the V-shaped vortex roll-ups are observed in the cross-section flow images to travel along the latter peripheral, before proceeding to form part of the bunched-up vortex loop with the vortex roll-up. The vortex-loop cross-section can also be clearly seen to consist of four pairs of mushroom-shaped CVPs, as shown in Fig. 22.

Jet vortex dynamics associated with the elliptic nozzle with the V-shaped notch aligned across the minor-plane (i.e. Nozzle B) reveal the formation of elliptic vortex roll-ups which followed closely the nozzle exit contour, as shown in Fig. 21(b). In contrast to the preceding modified elliptic nozzles however, the elliptic vortex roll-ups maintained their gross outlines far further. A closer look at the flow images will nonetheless uncover some flow similarities between the two different modified elliptic nozzles however. For instance, streamwise vortices were formed in similar fashion as before where the sharp troughs were located. And since the troughs were located at the major-plane ends, no rib structures were observed to form during the

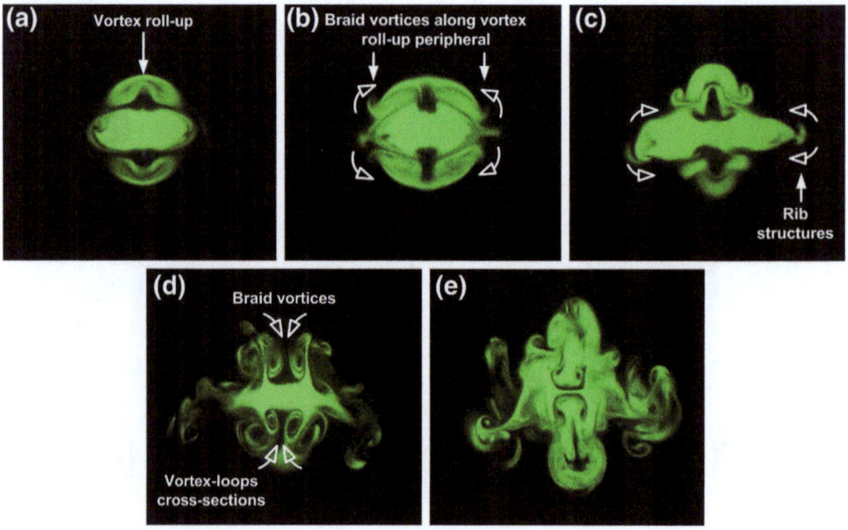


**Fig. 22** Near-field vortical formations and interactions associated with Nozzle A (Reprinted from New and Tsovolos [62], with permission from Elsevier)

study. Along the peaks however, due to the nozzle exit contour, the vortex roll-ups were visualized as bending towards the upstream direction. While they largely retained such outlines as they convect further downstream, a small kink was observed to manifest at their centres which led to rapid localized distortions. Nevertheless, the distorted vortex roll-ups would continue to preserve their general upstream-bending outlines. In several ways, the overall flow dynamics were more muted than the preceding modified elliptic jet and the jet cross-sections in the near-field did not show too much discrepancy from those of the non-notched elliptic jet until further downstream at about three hydraulic diameters away from the mean jet origin. Instead of the expected outwards-spreading streamwise vortices at major-plane ends detected further upstream, inwards-spreading vortex structures were formed. This has not been observed in non-notched or the preceding modified elliptic jets before, though it was speculated that they were associated with braid vortices.

Qualitative vortical differences were found to translate into tangible quantitative flow differences as well. And it is interesting to note that both modified elliptic nozzles led to an increase and decrease in the erosion of the jet centreline velocities respectively, as compared to the non-notched elliptic jet. This observation is in good agreement with more intense vortex dynamics associated with Nozzle A, as well as with the more subdued dynamics ones observed for Nozzle B. These are likely to be the reasons behind why no axis-switching behaviour was found for the first modified nozzle but only merely delayed further downstream for the second modified nozzle, as shown in Fig. 23. While the exact vortex dynamics of the elliptic modified jets covered in this section continue to be examined in detail by the authors, experimental data gathered so far provide further evidences that the axis-switching behaviour of an elliptic jet can be drastically altered through appropriate nozzle exit shaping.

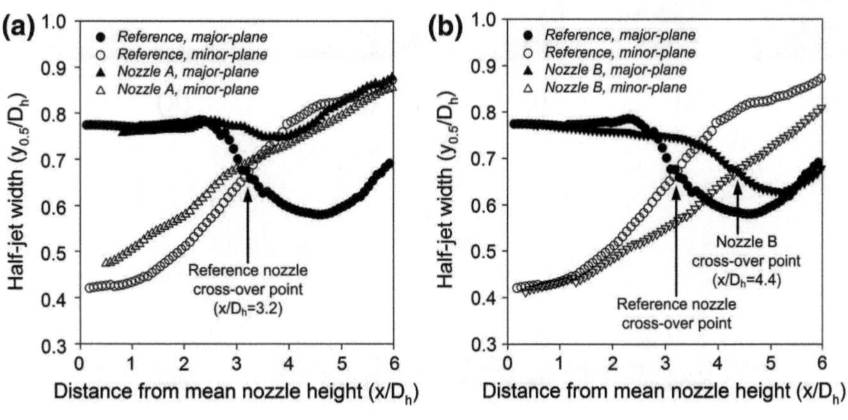


**Fig. 23** Cross-over points in half-jet width profiles for V-notched elliptic jets (Reprinted from New and Tsovolos [62], with permission from Elsevier)

Most recently, an attempt to compare and distinguish between freely-exhausting inclined rectangular and elliptic jets was made by Shi and New [78]. The study revealed that if rectangular and elliptic nozzles are inclined along their minor-planes, they produced persistently inclined vortex roll-ups similar to what have been reported previously. Intriguingly, if they are inclined along their major-planes, the resultant vortex dynamics will be sensitive towards the exact nozzle base geometry, particularly the vortex-turning behaviour. While both major-plane inclined rectangular and elliptic jets undergo vortex-turning shortly after they exhaust from the nozzles, the inclined rectangular jet undergoes over-turning whereby the inclined vortex roll-ups not only gradually reduce their inclination but increase their inclination beyond the cross-stream axis. From the flow visualization results shown in Figs. 24 and 25, inclined elliptic vortex roll-ups undergo significant distortions during the vortex-turning stage, which is reminiscent of axis-switching behaviour. On the other hand, inclined rectangular vortex roll-ups do not suffer such flow distortions. The detailed flow behaviour was extracted and presented in Fig. 26. Hence, it would appear that the lack of axis-switching in the inclined rectangular jet within that study allowed vortex over-turning to occur.

This study essentially demonstrates that axis-switching behaviour typically observed for noncircular jets adds another layer of complexity when investigating inclined noncircular jets.

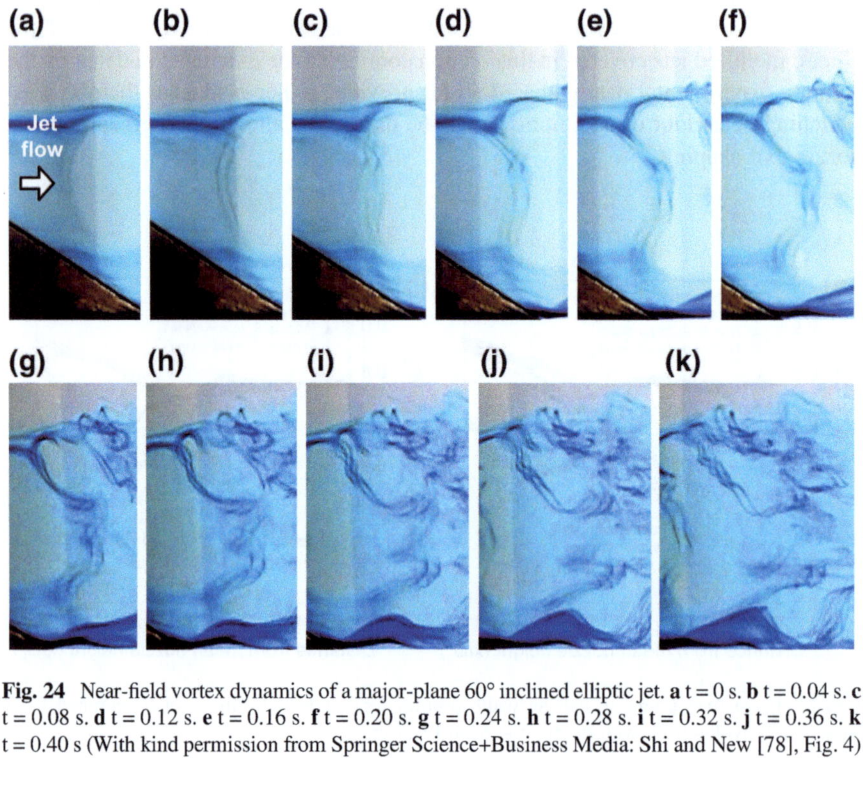


**Fig. 24** Near-field vortex dynamics of a major-plane 60° inclined elliptic jet. **a** t = 0 s. **b** t = 0.04 s. **c** t = 0.08 s. **d** t = 0.12 s. **e** t = 0.16 s. **f** t = 0.20 s. **g** t = 0.24 s. **h** t = 0.28 s. **i** t = 0.32 s. **j** t = 0.36 s. **k** t = 0.40 s (With kind permission from Springer Science+Business Media: Shi and New [78], Fig. 4)

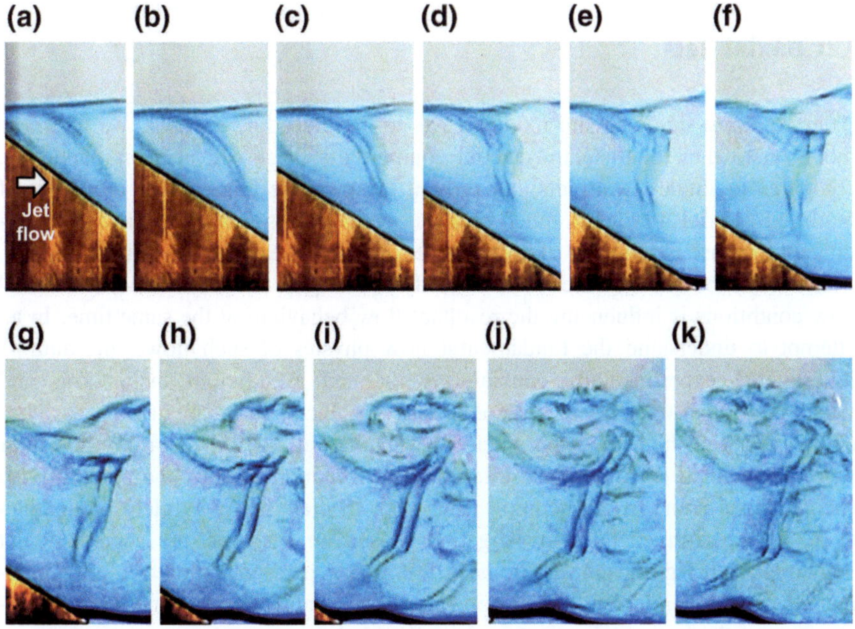


**Fig. 25** Near-field vortex dynamics of a major-plane 60° inclined rectangular jet. **a** t = 0 s. **b** t = 0.04 s. **c** t = 0.08 s. **d** t = 0.12 s. **e** t = 0.16 s. **f** t = 0.20 s. **g** t = 0.24 s. **h** t = 0.28 s. **i** t = 0.32 s. **j** t = 0.36 s. **k** t = 0.40 s (With kind permission from Springer Science+Business Media: Shi and New [78], Fig. 5)

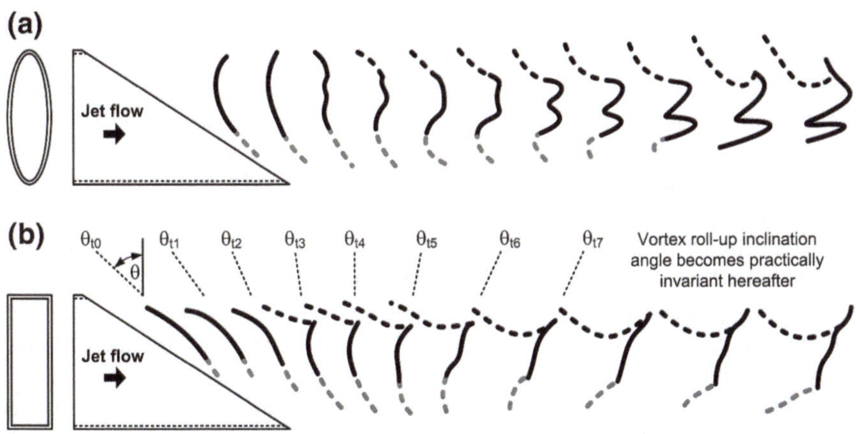


**Fig. 26** Extracted outlines of the jet ring-vortices produced by major-plane inclined elliptic and rectangular jets (With kind permission from Springer Science+Business Media: Shi and New [78], Fig. 6)

## 6 Unsteady "Serpentine" Motions in Inclined Circular Coaxial Jets

After covering single-stream modified jets in the preceding sections, we will now move on to their dual-stream coaxial counterparts. It should be mentioned that, compared to single-stream jets, our present grasp on the fundamental dynamics of modified coaxial nozzles is much more limited. This is primarily due to the significant increase in the level of flow complexity when dealing with coaxial nozzles, where a combination of area- and velocity-ratios, gap thickness and other initial flow conditions is influencing the resultant flow behaviour at the same time. In an attempt to understand the fundamental flow physics of such flows, the authors investigated experimentally coaxial jets inclined at different inclinations and velocity-ratios at transitional Reynolds numbers. Incidentally, this study was conducted at about the same time when Viswanathan et al. [96] reported upon their work on inclined coaxial nozzles, and thus further motivated by the possibility to shed light on some of flow complexities seen in their study. While Viswanathan et al. [96] performed their investigations based on subsonic and supersonic jet velocities with significant variations in the nozzle pressure ratios, a more fundamental flow dynamics study can nonetheless be useful to get a first-hand understanding of the basic inclined coaxial jet flow behaviour.

Early dye-visualization and laser-induced fluorescence experiments by Tsioli and New [85] suggested that their near-field formation and vortex-pairing behaviour are just as sensitive towards the incline-angle as the velocity-ratio. Analysis of the qualitative flow visualization results for 45° and 60° inclined coaxial jets indicated that these events occur significantly more rapidly along the shorter nozzle region, compared to those along the longer nozzle region. Furthermore, the discrepancy increases with the incline-angle and velocity-ratio (i.e. 0.5–2.0, in this case). Tsioli and New [86] and New and Tsioli [51] demonstrated that increasing the area-ratio of inclined coaxial jets from 2.25 to 4.0 reproduces relative similar flow behaviour. However, it was also observed that the vortex formation and pairing behaviour becomes less intense, presumably because the larger annular gap between the primary and secondary jets leads to a large physical separation distance between them and a reduction between their mutual interactions. Intriguingly, more detailed follow-up studies by New and Tsioli [51, 52] revealed that, due to asymmetric vortex-pairing and merging processes between the shorter and longer nozzle regions, the primary jet column will take on an unsteady "serpentine"-shaped outline when the velocity-ratio reaches 1.0 and beyond, as shown in Fig. 27. This meandering jet column has actually been observed prior to this study by Webster and Longmire [97] in inclined circular jets, as well as by New and Tsovolos [59] in minor-plane inclined elliptic jets. However, the flow mechanisms leading up to such a phenomenon appear to be quite different between these two flow scenarios. In inclined circular jets, the serpentine outline of the jet can be attributed to the increasingly asymmetric displacement of the jet shear layer along the shorter and longer nozzle regions as the incline-angle increases. This in turn leads to an increase in the initial vortex roll-up incline-angle

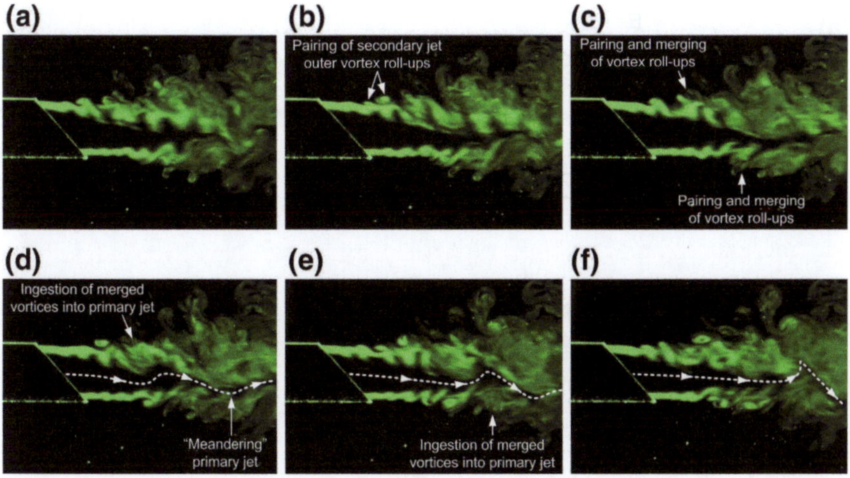


**Fig. 27** Unsteady "serpentine"-shaped primary jet column produced by a 45° inclined coaxial jet at a velocity-ratio of 1.0. **a** t = 0 s. **b** t = 0.16 s. **c** t = 0.32 s. **d** t = 0.48 s. **e** t = 0.64 s. **f** t = 0.80 s (With kind permission from Springer Science+Business Media: New and Tsioli [52], Fig. 5)

and causes jet entrainment to become progressively asymmetric, thus producing a serpentine jet outline. On the other hand, for inclined coaxial jets at a velocity-ratio of 1.0, it is mainly the interactions between the primary and secondary jet ring-vortices that result in a similar serpentine primary jet outline. In this case however, the process is typically associated by pairing of the secondary jet outer shear layer vortices which subsequently merged with primary jet shear layer vortices. The coalesced vortices subsequently move towards the jet centreline, thus producing significant physical indentation and entrainment of ambient fluid into the primary jet. The detailed flow mechanism is illustrated in Fig. 28.

Interestingly, while a velocity-ratio of 2.0 also produces a serpentine primary jet outline, the flow mechanism again differs from that at a velocity-ratio of 1.0, as shown in Fig. 29. In this case, due to the relatively faster secondary jet, the rotation sense of the primary jet shear layer vortices will be reversed and their interactions with secondary jet outer shear layer vortices will be much more limited. Incidentally, this was reflected in the significantly higher turbulence levels along the primary and secondary jet interface at a velocity-ratio of 1.0, rather than at a velocity-ratio of 2.0. As such, the serpentine primary jet outline will be a result of the asymmetric primary jet shear layer vortices along the shorter and longer nozzle length regions.

It was also observed that inclined coaxial jets incurred vectoring of the primary jet towards the longer nozzle length regions, where either a larger incline-angle or higher velocity-ratio serves to accentuate this behaviour, as shown in Fig. 30. It remains unclear the extent to which the serpentine jet behaviour described above contributes towards the jet vectoring, though it should be rather significant. Nevertheless, some of inclined coaxial flow features were validated by numerical

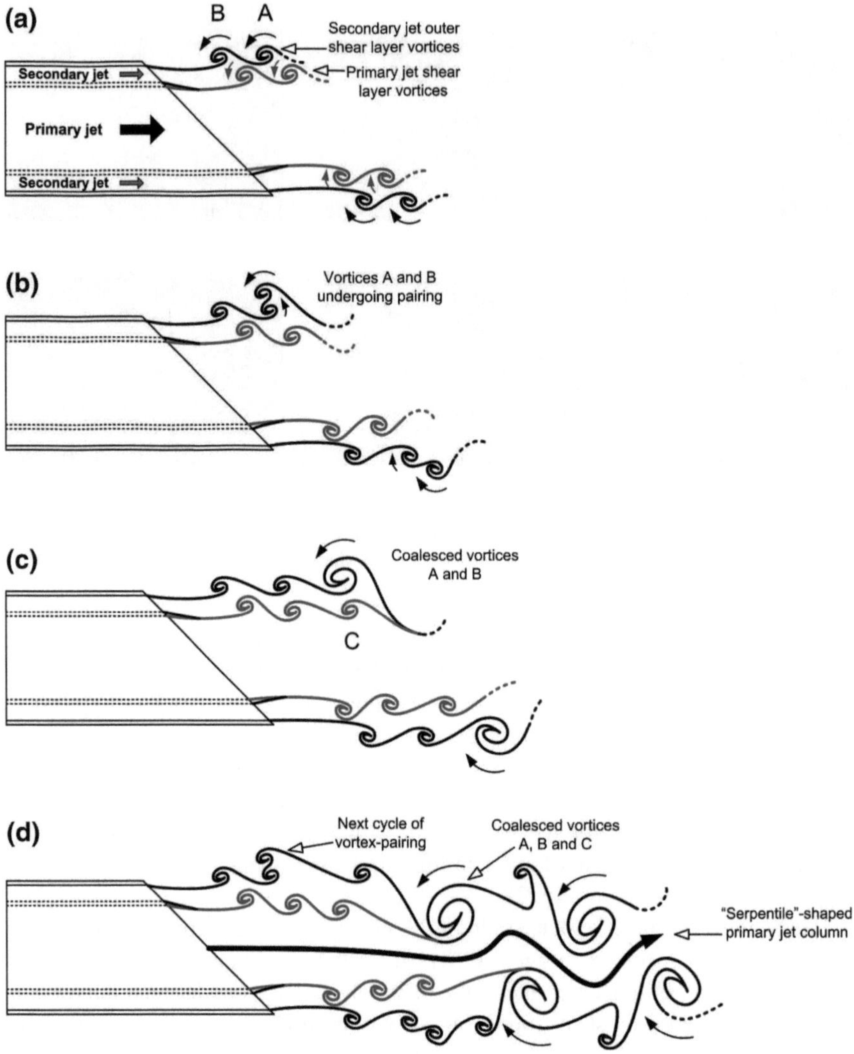


**Fig. 28** Flow mechanism leading up to serpentine-shaped primary jet column in an inclined coaxial jet from velocity-ratio of 1.0 (With kind permission from Springer Science+Business Media: New and Tsioli [59], Fig. 6)

simulations conducted by Tsioli et al. [87]. In particular, inclined ring vortices along the secondary jet shear layers could be clearly discerned in their numerical results, which would agree well with the experimental results reported by New and Tsioli [52] earlier, as shown in Fig. 31.

Lastly, a more recent study conducted by New and Tsioli [53] further investigated the effects of increasing the coaxial nozzle area-ratio on their earlier work, particularly with respect to the serpentine jet behaviour. With the aid of PIV

**Fig. 29** Flow mechanism leading up to serpentine-shaped primary jet column in an inclined coaxial jet at a velocity-ratio of 2.0 (With kind permission from Springer Science+Business Media: New and Tsioli [59], Fig. 7)

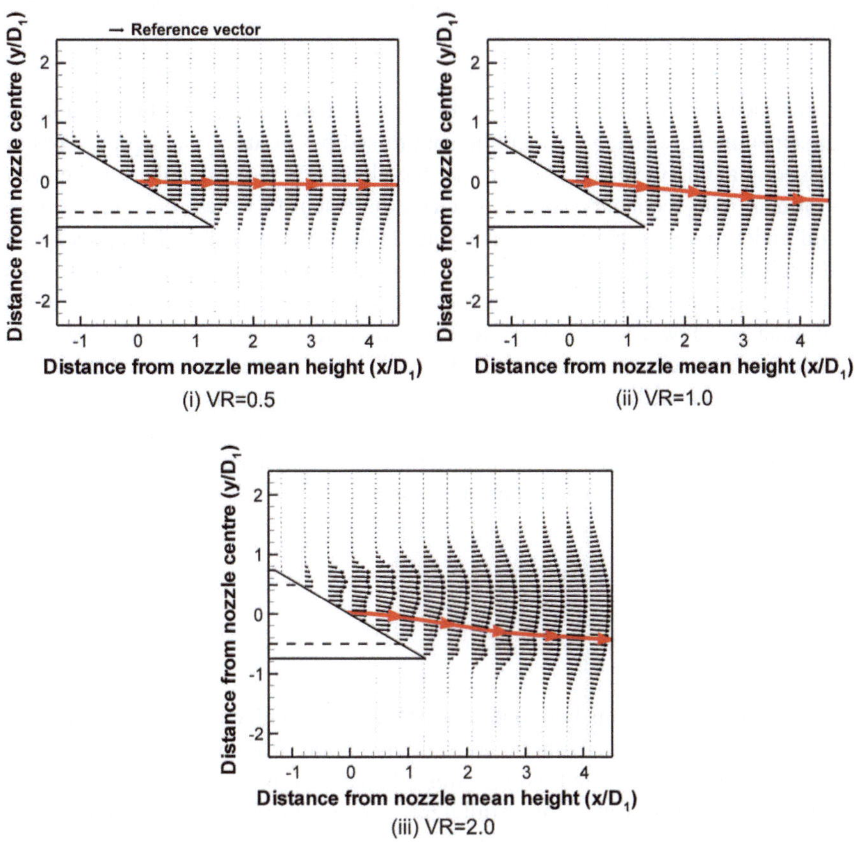


**Fig. 30** Increased vectoring of primary jet momentum towards the longer nozzle length regions in 60° inclined coaxial jet as the velocity-ratio increases (With kind permission from Springer Science+Business Media: New and Tsioli [52], Fig. 10a)

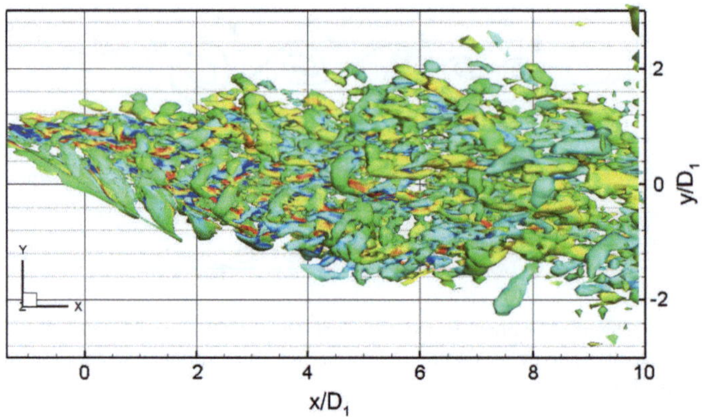


**Fig. 31** Large-eddy simulation result of an inclined coaxial jet shows good agreement with earlier experimental results (Reproduced from Tsioli et al. [87])

measurements to complement their earlier flow visualizations, this study confirmed that serpentine jet behaviour could be suppressed by the use of a larger area-ratio. Detailed quantitative results showed that the physically wider annular gap between the primary and secondary nozzles reduced the extent to which their ring-vortices could undergo vortex-pairing and merging. However, the observed suppression of self-excited jet oscillations was only restricted to lower velocity-ratios—they resumed once the velocity-ratio reached VR = 2 in the study. Examination of the centreline jet flow characteristics revealed that, as the area-ratio increases, the penetration and subsequent mixing between the two jet streams reduced as a result of the lack of jet oscillations. Once the velocity-ratio reached VR = 2 however, they increased drastically instead. As such, this and earlier study showed that coaxial nozzles could be controlled by the appropriate use of inclined exits, velocity-ratios, as well as area-ratios, upon which a fascinating range of jet flow behaviour is dependent.

## 7 Summary

Research highlights from key experimental studies conducted by the authors on single-stream jets and dual-stream coaxial jets issuing from trailing-edge modified nozzles have been covered within this chapter. Significant and wide-ranging influences arising from the use of different geometrical modifications and base jet geometry in single-stream jets, as well as velocity- and area-ratio in dual-stream coaxial jets, have been reported and compared to earlier studies for a more holistic and balanced approach. More often than not, imparting geometrical modifications to these nozzles as simple as an inclination can produce drastically dissimilar and intriguing flow behaviour. Depending on the exact modification, both large- and small-scale vortex structures may be altered or even formed, such that the resultant flow phenomenon may be suitable to enhance mixing or control. While there are infinite number of possible modified nozzle configurations, studies have demonstrated that certain physical modifications and methodologies are more effective than some others. As more engineering applications demand passive jet-mixing and control techniques that achieve a balance between design simplicity, cost-effectiveness, robustness and a more targeted solution, a strong grasp of the fundamental underlying vortex dynamics and their impact upon high-speed flow applications is important and necessary. It is the wish of the authors that their work reported here will motivate other researchers to explore more trailing-edge modified nozzle designs and continue to advance this particular research topic in the future.

**Acknowledgments** The authors gratefully acknowledge the encouragement and support provided by UK Engineering and Physical Science Research Council, The Royal Society, University of Liverpool, Defence Science and Technology Laboratory (Dstl), S & C Thermofluids Ltd, Temasek Laboratories at National University of Singapore, as well as Nanyang Technological University for their fundamental research investigations on trailing-edge modified nozzles.

## References

1. Abramovich, G. N. (1963). *The theory of turbulent jets*. Cambridge: MIT Press.
2. Alkislar, M. B., Krothapalli, A., & Butler, G. W. (2007). The effect of streamwise vortices on the aeroacoustics of a Mach 0.9 jet. *Journal of Fluid Mechanics, 578*, 139–169.
3. Babucke, A., Kloker, M., & Rist, U. (2008). Direct numerical simulation of a serrated nozzle end for jet-noise reduction. *High performance computing in science and engineering 07* (pp. 319–337). Berlin, Heidelberg: Springer.
4. Bai, S. (1954). *Fluid dynamics of jets*. New York: Van Nostrand.
5. Balarac, G., & Métais, O. (2005). The near field of coaxial jets: A numerical study. *Physics of Fluids, 17*, 065102.
6. Balarac, G., Métais, O., & Lesieur, M. (2007). Mixing enhancement in coaxial jets through inflow forcing: A numerical study. *Physics of Fluids, 19*, 075102.
7. Beer, J. M., Chigier, N. A., & Lee, K. B. (1963). Modeling of double concentric burning jets. In *Symposium (International) on Combustion* (Vol. 9(1), pp. 892–900). Amsterdam: Elsevier.
8. Bridges, J. E. (2012). Acoustic measurements of rectangular nozzles with bevel. *National Aeronautics and Space Administration, Glenn Research Center*. NASA/TM-2012-217674.
9. Buresti, G., Talamelli, A., & Petagna, P. (1994). Experimental characterization of the velocity field of a coaxial jet configuration. *Experimental Thermal and Fluid Science, 9*(2), 135–146.
10. Cai, J., Tsai, H. M., & Liu, F. (2010). Numerical simulation of vortical flows in the near field of jets from notched circular nozzles. *Computers & Fluids, 39*(3), 539–552.
11. Callender, B., Gutmark, E. J., & Martens, S. (2005). Far-field acoustic investigation into chevron nozzle mechanisms and trends. *AIAA Journal, 43*(1), 87–95.
12. Callender, B., Gutmark, E., & Martens, S. (2008). Near-field investigation of chevron nozzle mechanisms. *AIAA Journal, 46*(1), 36–45.
13. Callender, B., Gutmark, E. J., & Martens, S. (2010). Flow field characterization of coaxial conical and serrated (chevron) nozzles. *Experiments in Fluids, 48*(4), 637–649.

14. Champagne, F. H., & Wygnanski, I. J. (1971). An experimental investigation of coaxial turbulent jets. *International Journal of Heat and Mass Transfer, 14*(9), 1445–1464.
15. Dahm, W. J., Frieler, C. E., & Tryggvason, G. (1992). Vortex structure and dynamics in the near field of a coaxial jet. *Journal of Fluid Mechanics, 241*, 371–402.
16. Didden, N. (1979). On the formation of vortex rings: Rolling-up and production of circulation. *ZeitschriftfürangewandteMathematik und Physik ZAMP, 30*(1), 101–116.
17. Eschricht, D., Panek, L. J. Y., Michel, U., & Thiele, F. (2008). Jet noise prediction of a serrated nozzle. *AIAA Paper, 2971*.
18. Gharib, M., Rambod, E., & Shariff, K. (1998). A universal time scale for vortex ring formation. *Journal of Fluid Mechanics, 360*, 121–140.
19. Glezer, A. (1988). The formation of vortex rings. *Physics of Fluids, 31*(12), 3532–3542.
20. Grinstein, F. F., & DeVore, C. R. (1996). Dynamics of coherent structures and transition to turbulence in free square jets. *Physics of Fluids, 8*(5), 1237–1251.
21. Husain, H. S., & Hussain, F. (1991). Elliptic jets, part 2. Dynamics of coherent structures: pairing. *Journal of Fluid Mechanics, 233*, 439–482.
22. Husain, H. S., & Hussain, F. (1993). Elliptic jets, part 3. Dynamics of preferred mode coherent structure. *Journal of Fluid Mechanics, 248*, 315–361.
23. Hussain, F., & Husain, H. S. (1989). Elliptic jets, part 1. Characteristics of unexcited and excited jets. *Journal of Fluid Mechanics, 208*, 257–320.
24. Kibens, V., & Wlezien, R. W. (1985) Active control of jets from indeterminate-origin nozzles. In *AIAA Shear Flow Control Conference*, AIAA-1985-0542.
25. Kim, J. H., & Samimy, M. (1999). Mixing enhancement via nozzle trailing edge modifications in a high speed rectangular jet. *Physics of Fluids, 11*(9), 2731–2742.
26. Kim, J. H., & Samimy, M. (2000). On mixing enhancement via nozzle trailing-edge modifications in high-speed jets. *AIAA Journal, 38*(5), 935–937.
27. Knowles, K., & Saddington, A. J. (2006). A review of jet mixing enhancement for aircraft propulsion applications. *Proceedings of the Institution of Mechanical Engineers, Part G: Journal of Aerospace Engineering, 220*(2), 103–127.
28. Ko, N. W. M., & Au, H. (1985). Coaxial jets of different mean velocity ratios. *Journal of Sound and Vibration, 100*(2), 211–232.
29. Ko, N. W. M., & Kwan, A. S. H. (1976a). The initial region of subsonic coaxial jets. *Journal of Fluid Mechanics, 73*, 305–332.
30. Kwan, A. S. H., & Ko, N. W. M. (1976b). Coherent structures in subsonic coaxial jets. *Journal of Sound and Vibration, 48*(2), 203–219.
31. Kwan, A. S. H., & Ko, N. W. M. (1977). The initial region of subsonic coaxial jets, part 2. *Journal of Fluid Mechanics, 82*, 273–287.
32. Le, T., Borazjani, I., Kang, S., & Sotiropoulos, F. (2011). On the structure of vortex rings from inclined nozzles. *Journal of Fluid Mechanics, 686*, 451–483.
33. Lee, J. H. W., & Chu, V. H. (2003). *Turbulent jets and plumes: A Lagrangian approach*. Dordrecht: Kluwer Academic Publishers.
34. Liepmann, D., & Gharib, M. (1992). The role of streamwise vorticity in the near-field entrainment of round jets. *Journal of Fluid Mechanics, 245*, 643–668.
35. Lim, T. T. (1998). On the breakdown of vortex rings from inclined nozzles. *Physics of Fluids, 10*(7), 1–6.
36. Lim, T. T., & Nickels, T. B. (1995). Vortex rings. *Fluid vortices* (pp. 95–153). Netherlands: Springer.
37. Longmire, E. K., & Duong, L. H. (1996). Bifurcating jets generated with stepped and sawtooth nozzles. *Physics of Fluids, 8*(4), 978–992.
38. Longmire, E. K., Eaton, J. K., & Elkins, C. J. (1992). Control of jet structure by crown-shaped nozzles. *AIAA Journal, 30*(2), 505–512.
39. Martens, S. (2002). Jet noise reduction technology development at GE aircraft engines. *ICAS Paper, 842*.
40. Matsumoto, R., Kimoto, K., & Tsuchimoto, N. (1973). A study on double concentric jets (1st report, experimental results of air-air flow). *Bulletin of JSME, 16*(93), 529–540.

41. Maxworthy, T. (1972). The structure and stability of vortex rings. *Journal of Fluid Mechanics, 51*, 15–32.
42. Maxworthy, T. (1974). Turbulent vortex rings. *Journal of Fluid Mechanics, 64*, 227–240.
43. Maxworthy, T. (1977). Some experimental studies of vortex rings. *Journal of Fluid Mechanics, 81*, 465–495.
44. Mi, J., Nathan, G. J., & Luxton, R. E. (2000). Centreline mixing characteristics of jets from nine differently shaped nozzles. *Experiments in Fluids, 28*(1), 93–94.
45. Mohseni, K., & Gharib, M. (1998). A model for universal time scale of vortex ring formation. *Physics of Fluids, 10*(10), 2436–2438.
46. Morton, B. R. (1962). Coaxial turbulent jets. *International Journal of Heat and Mass Transfer, 5*(10), 955–965.
47. New, T. H. (2009). An experimental study on jets issuing from elliptic inclined nozzles. *Experiments in Fluids, 46*(6), 1139–1157.
48. New, T. H. (2010). Effects of notch geometry and sharpness on turbulent jets issuing from indeterminate-origin notched nozzles. *European Journal of Mechanics-B/Fluids, 29*(4), 309–320.
49. New, T. H., & Tay, W. L. (2006). Effects of cross-stream radial injections on a round jet. *Journal of Turbulence, 7*(N57), 1–20.
50. New, T. H., & Tsai, H. M. (2007). Experimental investigations on indeterminate-origin V- and A-notched jets. *AIAA Journal, 45*(4), 828–839.
51. New, T. H., & Tsioli, E. (2010). On the flow behaviour of jets issuing from issuing from inclined coaxial nozzles. In *14th International Symposium on Flow Visualization*, pp. 1–8.
52. New, T. H., & Tsioli, E. (2011). An experimental study on the vortical structures and behaviour of jets issuing from inclined coaxial nozzles. *Experiments in Fluids, 51*(4), 917–932.
53. New, T. H., & Tsioli, E. (2014). Effects of area-ratio on the near-field flow characteristics and deflection of circular inclined coaxial jets. *Experimental Thermal and Fluid Science, 54*, 225–236.
54. New, T. H., & Tsovolos, D. (2009a). Influence of nozzle sharpness on the flowfields of V-notched nozzle jets. *Physics of Fluids, 21*(8), 084107, 1–18.
55. New, T. H., & Tsovolos, D. (2009b). Characterization of jets issuing from circular nozzles with A-shaped notch lip-modifications. *Journal of Turbulence, 10*(24), 1–20.
56. New, T. H., & Tsovolos, D. (2009c). Flow characteristics of inclined elliptic nozzles. In *6th International Symposium on Turbulence and Shear Flow Phenomena* (Vol. 2, pp. 941–946).
57. New, T. H., & Tsovolos, D. (2009d). A digital particle image velocimetry study on jets issuing from hybrid inclined nozzles. *Flow, Turbulence and Combustion, 83*(4), 485–509.
58. New, T. H., & Tsovolos, D. (2010). Cross-stream behaviour and flow characteristics of hybrid inclined nozzle jets. *Journal of Turbulence, 11*(N18), 1–23.
59. New, T. H., & Tsovolos, D. (2011). On the vortical structures and behaviour of inclined elliptic jets. *European Journal of Mechanics-B/Fluids, 30*(4), 437–450.
60. New, T. H., & Tsovolos, D. (2012a). On the flow characteristics of minor-plane inclined elliptic jets. *Experimental Thermal and Fluid Science, 38*, 94–106.
61. New, T. H., & Tsovolos, D. (2012b). Vortex behaviour and velocity characteristics of jets issuing from hybrid inclined elliptic nozzles. *Flow, Turbulence and Combustion, 89*(4), 601–625.
62. New, T. H., & Tsovolos, D. (2013). On the vortex structures and behaviour of notched elliptic jets. *Experimental Thermal and Fluid Science, 49*, 51–66.
63. New, T. H., Lim, K. M. K., & Tsai, H. M. (2005). Vortical structures in a laminar V-notched indeterminate-origin jet. *Physics of Fluids, 17*(5), 054108, 1–14.
64. Norbury, J. (1973). A family of steady vortex rings. *Journal Fluid Mechanics, 57*, 417–431.
65. Page, G. J., McGuirk, J. J., Hossain, M., Hughes, N.J., & Trumper, M. T. (2002). A computational and experimental investigation of serrated coaxial nozzles. *AIAA paper, 2554.*

66. Powers, R. W., & McLaughlin, D. K. (2012). Acoustic measurements of scale models of military style supersonic beveled nozzle jets with interior corrugations. In *18th AIAA/CEAS Aeroacoustics Conference, AIAA-2012-2116.*
67. Rajaratnam, N. (1976). *Turbulent jets. Developments in water science*. Amsterdam: Elsevier.
68. Raman, G. (1997). Screech tones from rectangular jets with spanwise oblique shock-cell structures. *Journal of Fluid Mechanics, 330*, 141–168.
69. Raman, G. (1999). Shock-induced flow resonance in supersonic jets of complex geometry. *Physics of Fluids, 11*, 692–709.
70. Rehab, H., Villermaux, E., & Hopfinger, E. J. (1997). Flow regimes of large-velocity-ratio coaxial jets. *Journal of Fluid Mechanics, 345*, 357–381.
71. Reynier, P., & Ha Minh, H. (1998). Numerical prediction of unsteady compressible turbulent coaxial jets. *Computers & Fluids, 27*(2), 239–254.
72. Rosenfeld, M., Rambod, E., & Gharib, M. (1998). Circulation and formation number of laminar vortex rings. *Journal of Fluid Mechanics, 376*, 297–318.
73. Sadr, R., & Klewicki, J. C. (2003). An experimental investigation of the near-field flow development in coaxial jets. *Physics of Fluids, 15*(5), 1233–1246.
74. Saffman, P. G. (1978). The number of waves on unstable vortex rings. *Journal of Fluid Mechanics, 84*, 625–639.
75. Saiyed, N. H., Mikkelsen, K. L., & Bridges, J. E. (2003). Acoustics and thrust of quiet separate-flow high-bypass-ratio nozzles. *AIAA Journal, 41*(3), 372–378.
76. Samimy, M., Kim, J. H., Clancy, P. S., & Martens, S. (1998). Passive control of supersonic rectangular jets via nozzle trailing-edge modifications. *AIAA Journal, 36*(7), 1230–1239.
77. Shariff, K., & Leonard, A. (1992). Vortex rings. *Annual Review of Fluid Mechanics, 24*(1), 235–279.
78. Shi, S., & New, T. H. (2013). Some observations in the vortex-turning behaviour of noncircular inclined jets. *Experiments in Fluids, 54*(11), 1614.
79. Shu, F., Plesniak, M., & Sojka, P. (2005). Indeterminate-origin nozzles to control jet structure and evolution. *Journal of Turbulence, 6*(26), 1–18.
80. Shu, F., Plesniak, M., & Sojka, P. (2005). Visualization of streamwise vortex pairs in an indeterminate origin (IO) nozzle jet. *Journal of Visualization, 8*(3), 195.
81. Shu, F., Plesniak, M., & Sojka, P. (2007). Evolution of vortical structures in indeterminate-origin nozzle jets. *Journal of Flow Visualization and Image Processing, 14*(2), 143–154.
82. Tam, C., Shen, H., & Raman, G. (1997). Screech tones of supersonic jets from bevelled rectangular nozzles. *AIAA Journal, 35*(7), 1119–1125.
83. Tinney, C. E., & Jordan, P. (2008). The near pressure field of co-axial subsonic jets. *Journal of Fluid Mechanics, 611*, 175–204.
84. Troolin, D. R., & Longmire, E. K. (2010). Volumetric velocity measurements of vortex ringsfrom inclined exits. *Experiments in Fluids, 48*(3), 409–420.
85. Tsioli, E., & New, T. H. (2009). Near field vortex structures of inclined coaxial jets. In *6th International Symposium on Turbulence and Shear Flow Phenomena* (Vol. 2, pp. 499–504).
86. Tsioli, E., & New, T. H. (2009). Vortex structures in the near-field of inclined coaxial jets. In *39th AIAA Fluid Dynamics Conference AIAA-2009-4316*, pp. 1–8.
87. Tsioli, E., Albets-Chico, X., & Kassinos, S. (2012). A numerical investigation of coaxial jets with complex nozzle geometries. In *9th International ERCOFTAC Symposium on Engineering Turbulence Modelling and Measurements.*
88. Tsovolos, D., & New, T. H. (2009). Alteration of axis-switching behaviour in indeterminate-origin elliptic jets. In *39th AIAA Fluid Dynamics Conference AIAA-2009-4008*, pp. 1–10.
89. Tsovolos, D., & New, T. H. (2010). An experimental study on the flow behaviour of V-notched elliptic jets. In *15th International Symposium on Applications of Laser Techniques to Fluid Mechanics.*
90. Uzun, A., & Hussaini, M. Y. (2009). Simulation of noise generation in the near-nozzle region of a chevron nozzle jet. *AIAA Journal, 47*(8), 1793–1810.
91. Villermaux, E., & Rehab, H. (2000). Mixing in coaxial jets. *Journal of Fluid Mechanics, 425*, 161–185.

92. Viswanathan, K. (2005). Nozzle shaping for reduction of jet noise from single jets. *AIAA Journal, 43*(5), 1008–1022.
93. Viswanathan, K. (2006). Elegant concept for reduction of jet noise from turbofan engines. *Journal of Aircraft, 43*(3), 616–626.
94. Viswanathan, K. (2006). Noise of dual-stream beveled nozzles at supercritical pressure ratios. *Journal of Aircraft, 43*(3), 627–638.
95. Viswanathan, K., & Czech, M. (2011). Adaptation of the beveled nozzle for high-speed jet noise reduction. *AIAA Journal, 49*(5), 932–944.
96. Viswanathan, K., Shur, M., Spalart, P. R., & Strelets, M. (2008). Flow and noise predictions for single and dual-stream beveled nozzles. *AIAA Journal, 46*(3), 601–626.
97. Webster, D. R., & Longmire, E. K. (1997). Vortex dynamics in jets from inclined nozzles. *Physics of Fluids, 9*(3), 655–666.
98. Webster, D. R., & Longmire, E. K. (1998). Vortex rings from cylinders with inclined exits. *Physics of Fluids, 10*(2), 400–416.
99. Wicker, R. B., & Eaton, J. K. (1994). Near field of a coaxial jet with and without axial excitation. *AIAA Journal, 32*(3), 542–546.
100. Widnall, S. E., Bliss, D. B., & Tsai, C. Y. (1974). The instability of short waves on a vortex ring. *Journal of Fluid Mechanics, 66*, 35–47.
101. Widnall, S. E., & Sullivan, J. P. (1973). On the stability of vortex rings. *Proceedings of the Royal Society of London. A. Mathematical and Physical Sciences, 332*(1590), 335–353.
102. Williams, T. J., Ali, M. R. M. H., & Anderson, J. S. (1969). Noise and flow characteristics of coaxial jets. *Journal of Mechanical Engineering Science, 11*(2), 133–142.
103. Wlezien, R. W., & Kibens, V. (1986). Passive control of jets with indeterminate origins. *AIAA Journal, 24*(8), 1263–1270.
104. Xia, H., & Tucker, P. G. (2012). Numerical simulation of single-stream jets from a serrated nozzle. *Flow, Turbulence and Combustion, 88*(1–2), 3–18.
105. Yan, J., Eschricht, D., & Thiele, F. (2009). Numerical investigation of the noise emission from serratednozzles in coaxial jets. *High Performance Computing in Science and Engineering, Garching/Munich 2007* (pp. 325–334). Berlin: Springer.
106. Yan, J., Panek, L., & Thiele, F. (2007). Simulation of jet noise from a long-cowl nozzle with serrations. *AIAA Paper-3635.*
107. Zakharin, B., Kit, E., & Wygnanski, I. (2009). On a turbulent mixing layer created downstream of a "Λ" notch simulating one wavelength of a chevronnozzle. *Flow, Turbulence and Combustion, 83*(3), 371–388.
108. Zaman, K. B., Bridges, J. E., & Huff, D. L. (2011). Evolution from 'tabs' to 'chevron technology'-a review. *International Journal of Aeroacoustics, 10*(5), 685–710.

[illegible]

# Impinging Jets

V. Tesař

**Abstract** This Chapter summarises existing knowledge about near field of fluid jets impinging upon a solid wall. In particular, the chapter focuses attention on submerged jets (those having the same density as the surrounding more or less stationary outside fluid) in which the impingement is complicated by vortical structures formed due to hydrodynamic instability on the jet outer edge. Because the most important engineering application of such flows is in transfer of heat and mass into/from fluid to the wall, the chapter pays considerable attention not only to the fluid mechanics of the impingement flow but also to basic problems of the convective transfer. Particular space is also devoted to interesting cases of impinging jets issuing from annular nozzles, the flowfield of which exhibits some special features of importance for applications. The last part of the chapter discusses, due to their recently increasing importance, impinging jets excited so as to get a pulsating and hybrid-synthetic character.

**Keywords** Impinging jets · Heat/mass transfer · Thermal boundary layer · Varying structure of annular impinging jets · Jet core · Synthetic jets · Hybrid-synthetic jets

## 1 Impinging Jets and Their Importance for Applications

Fluid jets impinging upon a solid wall are an important class of fluid flow configurations, possessing a number of features interesting from the fluid mechanics point of view. They are also of high practical importance in many engineering applications, particularly those concentrating on heating, cooling, and drying. Extensive classical surveys of the subject exist [1, 2], but they become gradually

V. Tesař (✉)
Academy of Sciences of the Czech Republic, Institute of Thermomechanics
AS CR v.v.i., Dolejškova 1402/5, 182 00 Praha 8, Czech Republic
e-mail: tesar@it.cas.cz



D.T.H. New and S.C.M. Yu (eds.), *Vortex Rings and Jets*,
Fluid Mechanics and Its Applications 111, DOI 10.1007/978-981-287-396-5_6

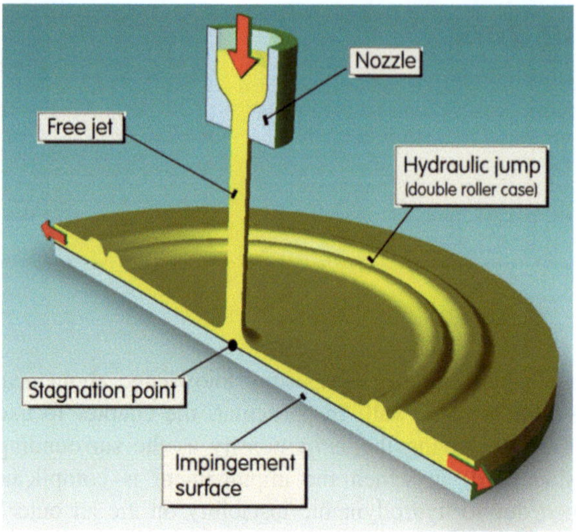


**Fig. 1** Diameter of a free jet (e.g. water jet in air) does not practically vary along its path from the nozzle up to the impingement surface (as long as it is not thinned by gravitation). Thickness of the liquid sheet that forms on the surface initially decreases because of radial divergence, but further downstream begins to increase since it is decelerated by friction. The hydraulic jump [25] is made from gravity waves that cannot propagate upstream because the liquid there flows too fast

outdated because of steady flow of new information produced by continuing intensive research activities in this field.

Very basic distinction has to be drawn between free and submerged impinging jets, as shown in Figs. 1 and 2 respectively. Free jets—of liquid issuing into gaseous atmosphere—have generally less intricate internal structures. Since their jet liquid does not mix with its surroundings and momentum transport from the jet by surface friction is rather weak, the shear stresses inside the free jet are mostly negligible. More important is effect of surface tension. It may cause instability waves to form and grow on the free jet surface, finally leading to jet breaking-up into a spray of droplets. This, however, is rarely encountered in impinging jet flows, unless the jet velocity is extremely high, because the distance to the impingement surface is usually much shorter than the streamwise distance from the nozzle to the location at which such instabilities become important.

On the other hand, the structure of a submerged jet (i.e., a liquid jet issuing into a liquid environment or a gas jet into a gaseous surrounding) is generally much more complex. A region devoid of shear stresses, called jet core, may also exist but only for a short streamwise distance. It may not be present at all if there are strong disturbances already in the flow supplied to the nozzle. The core is surrounded by mixing layer that in streamwise direction gradually grows in thickness. On the outer side of this layer, instability waves form (different from those in free jets as there are no surface tension effects), usually developing into vortices (Fig. 2). The distances at which the vortices form may be comparable with the distance from the nozzle to

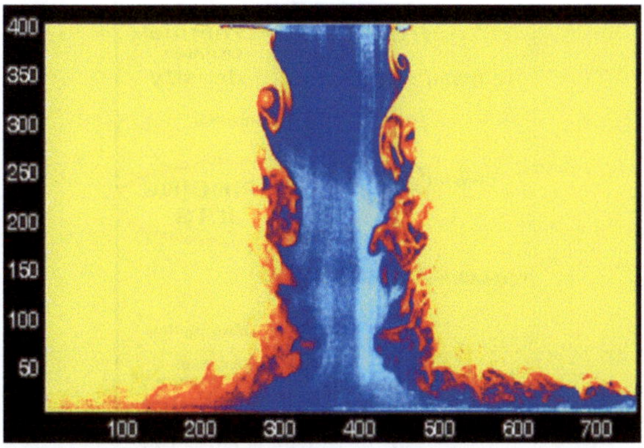


**Fig. 2** An example of submerged impinging jet. Air flow from the nozzle (on *top* of the picture, not shown) in this case contains condensing humidity. False colouring indicates droplet concentration in meridian section visualised by laser light sheet. Reynolds number is here high enough for turbulent chaotic eddies already forming in the mixing layer on the boundary between the jet and the outer air (the latter containing no droplets)

the impingement surface and hence can significantly influence the characteristics of the jet impingement. Vortices do not remain organised and unless the Reynolds number is low (i.e. such that the jet flow remains laminar), spectral transport of eddies that transfer energy to smaller length scales will lead to transition into chaotic turbulence.

Significant contributions towards understanding impinging jets had been made already in early 20th century, mainly driven by their practical importance in applications. Historically, heating or cooling by impinging jets has been an old method, known and practised since the pre-industrial ages (even if we neglect the much older direct heating by flames—which are impinging gaseous jets with exothermic chemical reaction). An advantage of this form of heating is easiness of creating a hot jet and aiming it at a particular location on the heated object that requires high intensity of thermal transfer. Equally important and widely used is the mass transfer, for instance drying by impinging jets, a common process in manufacturing paper, textile, ceramics and other products made using a liquid process. In drying, the heat and mass transfer are frequently combined.

The main advantage of impinging jets is their ability to achieve the highest realisable values of thermal power transfer density, $Q$ [W/m$^2$], per unit area of the heated surface, among all other convective transport processes. The transfer density is critically dependent upon the temperature difference, $\Delta T$ [K], between the heat-carrying fluid and the surface, and can be described by

$$Q = \alpha_T \Delta T \tag{1}$$

The coefficient $\alpha_T$ [W/m$^2$K] was introduced to cover all effects, mostly difficult to describe mathematically, other than the temperature. Even though situation

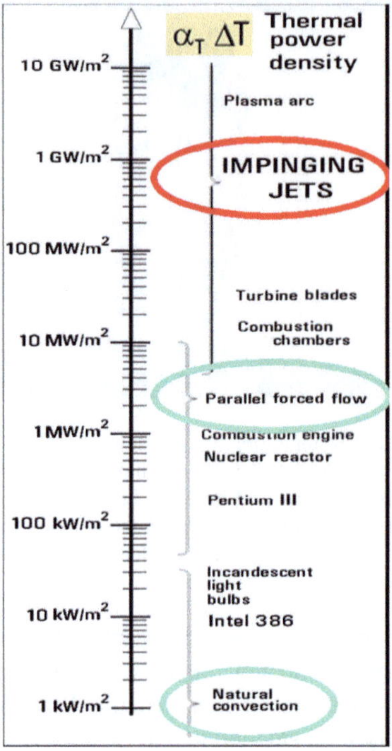


**Fig. 3** Magnitudes of surface density of thermal power transfer rate, achievable with the three basic methods of convective transfer (i.e. transfer by flow of heat-carrying fluid). Flows in direction parallel to the heated or cooled surface are divided into two classes. The natural convection is caused by the density difference created by heating. It is inexpensive to set up, but thermal transfer intensity is very low. More efficient are the forced heat-transferring flows, produced by a blower or similar device. These flows parallel to the transfer surface are the standard approach used in most engineering applications like radiators and coolers. Obviously, even at their most efficient installations at high temperature differences, they cannot compete in efficiency with impinging jets, which are perpendicular to the surface and can reach to the smallest normal distance, leaving the shortest remaining distance to be crossed by the ineffective conduction

gradually changes, values of $\alpha_T$ are still mostly evaluated experimentally, since computations still provide comparatively less accurate results. For basic geometric configurations the values are readily available in literature. As can be seen in Fig. 3, impinging jets can bring to a cold body (or remove from a hot one) thermal loads as high as 500 MW/m$^2$. Even higher thermal load levels at the top of the scale in Fig. 3 do exist, but are mostly exceptional cases, obtained with very high temperature differences $\Delta T$.

The cooling or heating task becomes particularly demanding if there is, as is often the case in cooling of electronic devices and components, a certain maximum surface temperature that must not be exceeded. For example, the junction temperature in some semiconductor lasers must be kept below 150 °C. As a result, the necessary thermal loads may reach 10 $MW/m^2$. Considering the typical not very low working temperature of the cooling fluid, these situations call for very high $\alpha_T$ values obtainable with impinging jets.

The main hindrance in convective heat transfer, which limits the highest achievable $\alpha_T$, is the thermal conduction sub-layer that develops on the heated or cooled surface. This layer of almost stationary fluid (its motion is restricted by viscous friction forces [3]) acts as a thermal insulation. Heat is transported across this layer through highly ineffective heat conduction mechanism. With gases, characterised by their typically very low thermal conductivities (for example, thermal conductivity of air is only $\lambda$ = 0.026 W/m.K) this layer so thin that its presence may be hardly noticeable. Typical thicknesses, $\delta_T$, of the thermal conduction sublayer are of the order of mere tenths of a millimetre. Despite this, however, this layer can dominate the overall resistance to heat transfer. In conductive heat transport the thermal power density $Q$ is governed by expression

$$Q = \lambda \Delta T / \delta_T \tag{2}$$

In heat transfer problems, the thickness $\delta_T$ cannot be directly evaluated as there is no identifiable outer limit. The insulation effect decreases asymptotically with distance from the wall, theoretically reaching infinity. It may, however, be evaluated indirectly, from computed or measured $\alpha_T$ through Eqs. (1) and (2):

$$\delta_T \approx \frac{\lambda}{\alpha_T} \tag{3}$$

Impinging jets are so effective transfer mechanisms because their flow direction (mostly perpendicular to the impingement surface) can bring the hot fluid to the smallest distance $\delta_T$ from the heated wall. Therefore, only a short distance remains to transfer the heat by conduction. The presence of thermal conduction sublayer can be discerned in the flow visualisation presented in Fig. 4, where it was obtained under extreme conditions making the layer exceptionally thick.

The importance of impinging jets for cooling applications in electronics may be demonstrated by the thermal power density $\alpha_T \Delta T$ that must be extracted from computer processors to keep their temperatures within acceptable limits. In the year 1984, Intel processor i386 generated $\alpha_T \Delta T$ = 0.01 $MW/m^2$ on its outer surface, which could be easily managed through natural convection. This was still the preferred cooling method for Intel Pentium in 1990, although the processor cases had to be provided with ribs to obtain the necessary surface area allowing the heat extraction rate to reach $\alpha_T \Delta T$ = 0.05 $MW/m^2$. Later processors necessitated cooling by a blower-generated forced flow. Interestingly, Moore's law of exponential growth of computer processor density and speed as a function of time [4] is equally valid for the processor cooling intensities. They currently reach levels [5] where extreme $\alpha_T \Delta T$ magnitudes became one of the basic considerations in processor design. It is worthwhile to note that processor heat transfer rates are actually

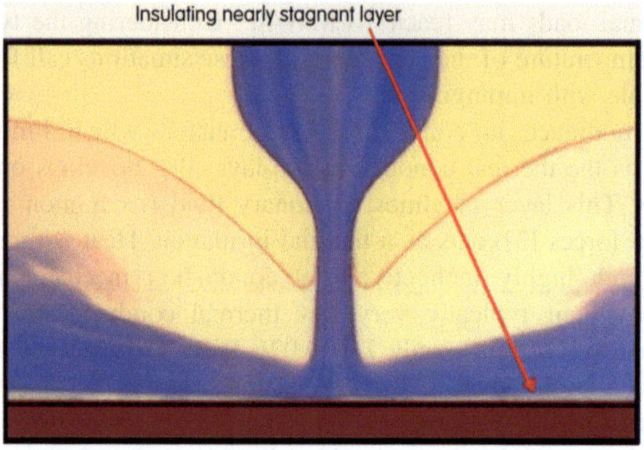


**Fig. 4** By a special flow visualisation method (sudden input of coloured water into the nozzle) it was possible to demonstrate in an impinging jet flow the presence of the insulating conduction layer, which retained the transparency of previously flowing clear water and thus became distinguishable

comparable to values in nuclear reactors, typically at $\alpha_T \Delta T = 2\ \mathrm{MW/m^2}$. For another comparison, values $\alpha_T \Delta T = 10\ \mathrm{MW/m^2}$ are reached in rocket engine nozzles.

While there are still many unknowns associated with the basic axisymmetric impinging jet configuration (typically characterised by the jet issuing from a single round nozzle), an increasing number of investigations has been nowadays reported in the literature aimed at exploring less understood more complex geometric variants. It is also not uncommon for the jet to impinge not perpendicularly but at an oblique angle to the surface. More often encountered complicating factor are industrial applications with a number of jet nozzles arranged in an array, as shown in Fig. 5. Similar problem of impinging jet arrays at large scale was studied in the context of vertical take-off aeroplanes where the impingement can remove and accelerate objects (like stones) from the landing surface. In the nozzle array, the parallel jets interact mutually so that their respective flowfields differ considerably from the simple case of a single nozzle. A complicating factor may also be the presence of the wall surrounding the nozzle exits, parallel with the impingement plane. This wall substantially influences the re-circulatory flow surrounding the actual impinging jets. The lack of available space in some applications, such as hollow gas turbine blades cooled from inside, have also led to some very special nozzle configurations. In this case blades are usually cooled by a multi-hole nozzle unit positioned inside the blade and shaped so as to fit into available restricted space. Similar multi-nozzle configurations for heating from inside are applied to the leading-edge of airplane wings for de-icing. In both cases, the analysis is complicated by the non-planar, complex shape of the impingement surface.

The row of round nozzles in Fig. 5 may be replaced by a single slit nozzle (Fig. 7). While cross-sectional geometry of the slit nozzle is two-dimensional, its current analyses in increasing measure need taking into the consideration

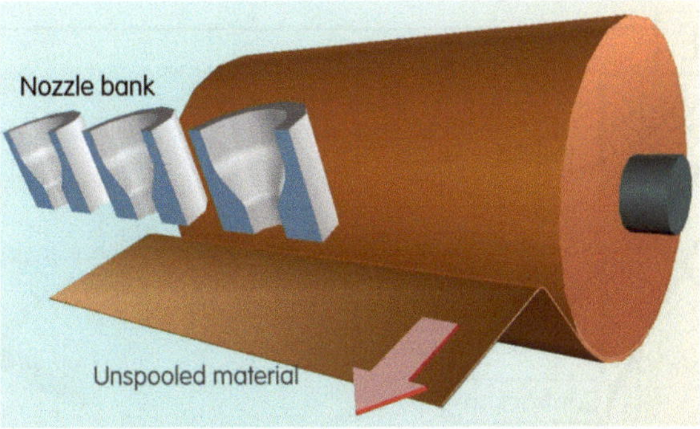


**Fig. 5** A configuration often used in applications is a linear row of nozzles perpendicular to the direction of motion of manufactured planar product such as dried paper or textile fabric made on hydraulic loom. Analysis of the flowfield is, of course, complicated by the mutual interaction of the jets

three-dimensional effects—especially at and near the slit ends. Recently becoming particularly important are annular nozzles. Their original purpose was to cover a larger surface area on the impingement wall with lower fluid flow rate than that of a circular jet. More importantly, their flowfields exhibit a range of interesting flow regimes. Particularly important effects of theoretical interest are found if the flow in an annular nozzle is periodically excited.

## 2 Structure of Impinging Jet Flow-Fields

Apart from a few extraordinary jet generation methods (and cases where cost minimization is of prime importance so that the jet simply issues from the end of a supply pipe), impinging jets are typically generated by an axially symmetric nozzle similar to that shown in Fig. 6 or with slit exit as shown in Fig. 7. Distances to the heat transfer surface are in the range $h/D$ (or $h/b$) $\approx$ 1–10, typical magnitude being $h/D = 5$. Rounding of internal edges as shown in Fig. 6 is usually recommended. It eliminates flow separation from the walls, which causes energy losses and unsteadiness. A recommended feature is a short constant cross-section exit channel of length approximately $2D$ to $3D$ to stabilise and direct the issuing jet.

The axisymmetric jet and planar impingement surface configuration is the subject of most investigations discussed in literature. Initially, the advantage was the plausibility of two-dimensional analysis (i.e. at least as the first approximation) —nowadays less important with availability of more powerful numerical computation facilities. Most jets in engineering applications are turbulent and turbulence is, of course, a three-dimensional phenomenon. Its effects are included into the analysis and design of impinging jets through a mathematical model. Most models currently used in practical engineering applications treat turbulence as if it were

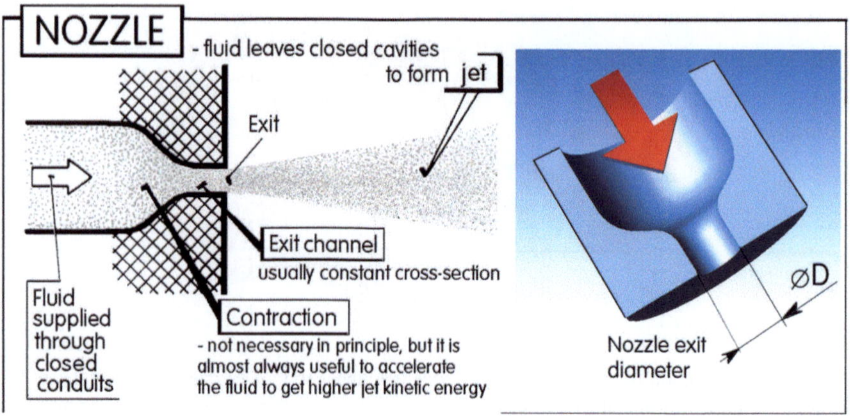


**Fig. 6** *Nozzles* Basic fact and nomenclature. Jets are generated by allowing the fluid to issue from the nozzle exit. Useful (though not necessary) is directing the outflow by a short constant-cross-section exit channel. Most common are axisymmetric nozzles, at *right*, characterised by diameter D

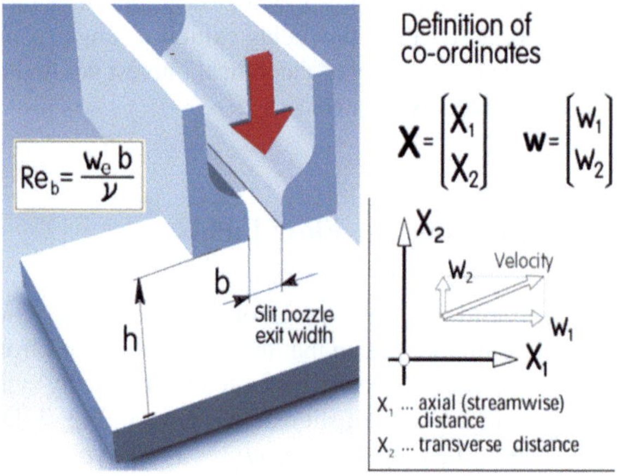


**Fig. 7** Slit nozzles are characterised by their exit width *b*. Character of generated impinging flow depends upon the relative height *h*/*b* and on Reynolds number $Re_b$—the parameter of flowfield similarity ($w_e$ [m/s] is the nozzle exit velocity, $\nu$ [$m^2$/s] is fluid viscosity). Many investigated slit nozzle impingement flows may be in the first approximation analysed as two-dimensional. The two spatial coordinates $X_1$ and $X_2$ form the position by vector **X**. The same two-dimensional approach may be applied to velocity vector **w**

isotropic (i.e. with the same properties in all three spatial directions) and hence fully described by a scalar variable. This makes the assumption of two-dimensionality reasonable. Clearly, the isotropy assumption is not acceptable for a swirling jet, but this is a rather exceptional case [6]. The mixing layer around the jet core may contain helical or annular vortical structures of obviously anisotropic character.

In the meridian plane sections through the jets (e.g., in Figs. 2 or 8) the presence of these structures is seen as alternatively positioned excrescences on both sides of the jet. Taking into account these structures in analyses—especially the helical ones—has been so far rarely attempted [6]. After all, within a short distance downstream, they decompose into the chaotic small-scale turbulence, usually successfully modelled by isotropic models. Approaches to identification of such motions exist—such as the application of proper orthogonal decomposition or POD [7]—but at this point of their development they are exceptionally demanding (and their connection with the practically important properties of the velocity and temperature aspects not is yet fully established).

After all, what an engineer needs as a result of the analysis for his design activities is a dependence, formulated as simply as possible, of distributions of time-mean velocity and temperature, on the parameters that may be varied in the design—for instance, the relative height *h*/*D* or Reynolds number *Re*. The two-dimensional approach with simple modelling of turbulence is therefore nowadays just what meets the demands. In fact, the usual practice is to go on further with the simplifications. One of the outcomes of this simplification trend is the separation of velocity field from the temperature field. The temperature may be quite successfully handled as a passive "contaminant", transported by the fluid flow. Another common treatment is the division of the flowfield into parts that are analysed separately, namely as A, B, and C shown in Fig. 8. This is, after all, in line with one of the basic axioms of science: complex phenomena are studied by decomposing them into simpler and more manageable separate parts. It may be noted that this procedure goes as far as using different local systems of spatial co-ordinates $X_1$ and $X_2$ in the parts A and C, both based on the usual convention of axis $X_1$ parallel with the main direction of the fluid flow.

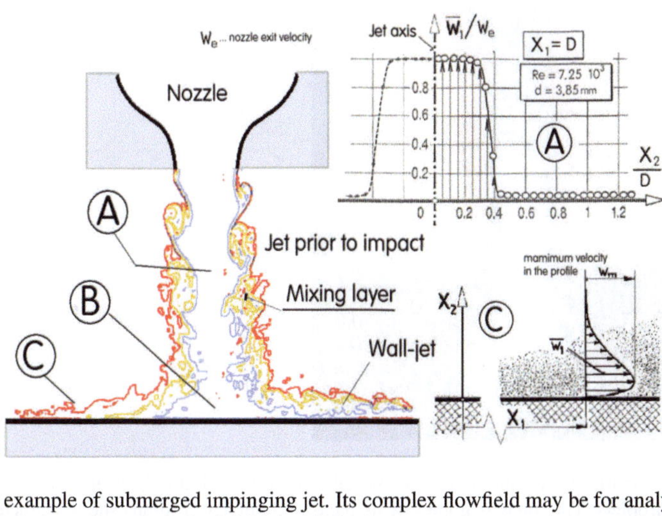


**Fig. 8** An example of submerged impinging jet. Its complex flowfield may be for analysis divided into separately investigated: *A* the initial jet, *B* the impact region near the stagnation point, and *C* the wall-jet flow. The cases *A* and *C* may use much of the knowledge already accumulated with non-impacting flows for which are shown two typical velocity profiles

With reference to Fig. 8, the part A contains the jet prior to the impingement. It may be surprising that the impact has usually less influence on the part A than might be expected. While there is no doubt that the fluid flow velocity along the nozzle axis has to decrease ultimately to zero when it reaches the stagnation point on the surface, this decrease takes place only over the final segment of the flow path to the wall. Typically, the axial velocity distributions deviate from their non-impinging jet counterparts only at distances from the impingement wall comparable to the nozzle diameter—i.e., only in the subsequent part B. Thus, significantly useful information about part A of the flowfield may be obtained from the analyses earlier developed for the non-impinging jets (i.e. like the one presented in Fig. 9). Examples of measurement results in this part of the flowfield are presented in the velocity profiles in Figs. 8 and 10, where the latter shows a distribution of the axial velocity along the radial co-ordinate $X_2$. In turbulent flow velocity measurements, the symbol $\overline{w_1}$ (i.e. Eq. 4) emphasises the fact that the plotted quantity is based on

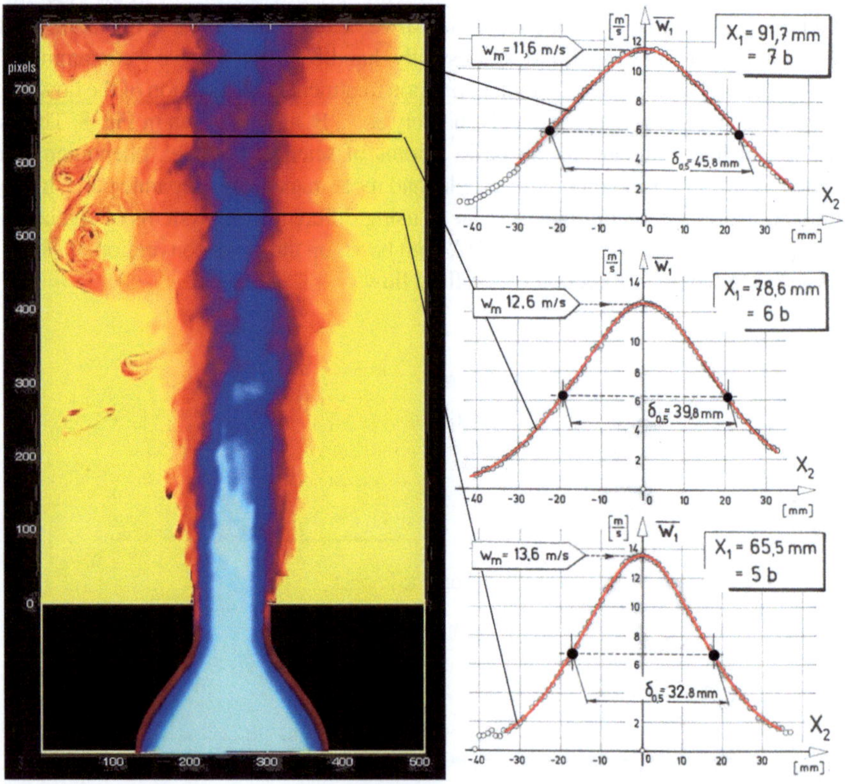


**Fig. 9** False colouring of submerged jet visualised by laser light sheet and scattering on sub-millimetre droplets. The three examples of its measured velocity profiles show characterisation of the jet width by the convention-defined diameter $\delta_{0.5}$. Although the similarity of profile shapes is not exact, it is an acceptable approximation on which is based the similarity transformation approach to analysis of jet flows [8, 9]

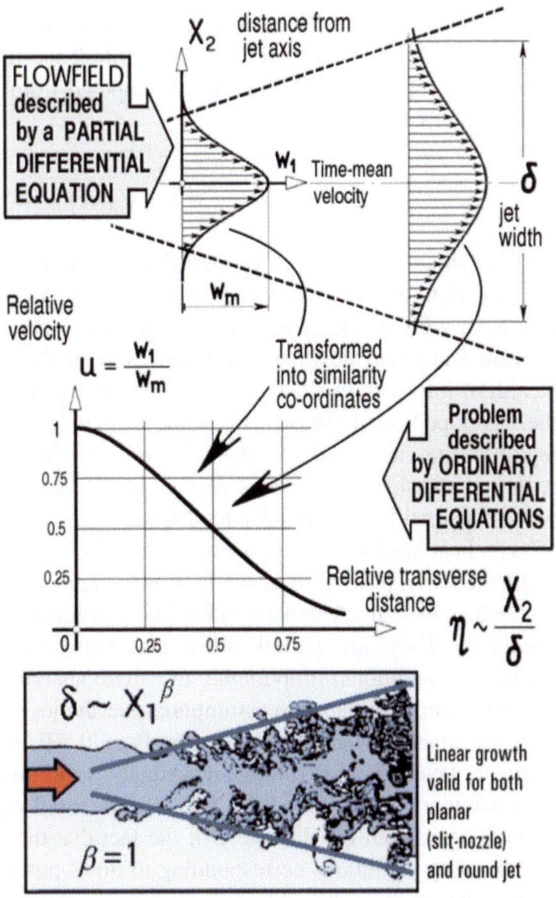


**Fig. 10** The similarity of velocity profiles is in turbulent regime, strictly speaking, only approximate. Nevertheless it simplifies substantially the governing equations (partial differential equations are converted into ordinary ones) and leads to useful results. A worse fact is the similarity-based solution is not valid at short axial distances $X_1$ from the nozzle —i.e. just the locations of main interest in studies of impinging flows. In the example of turbulent jet photograph below, the superposed straight lines demonstrate the linear growth of the characteristic width (or diameter) $\delta_{0.5}$ with increasing streamwise distance $X_1$. Adapted from [14]. *Courtesy* of Artech House Inc

the time-mean value. Velocity profiles are usually plotted (i.e. as depicted in part A of Fig. 8) non-dimensionalized by the velocity $w_e$ in the nozzle exit. It is important to note the other, temporal decomposition of velocities—different from their decomposition into the spatial co-ordinate directions—into the time-mean, organised coherent, and chaotic turbulent components. In the case of the streamwise-directed velocity component, $w_1$, these components are

$$w_1 = \overline{w_1} + w_1' + w_1'', \tag{4}$$

(the decomposition of $w_2$ components is, of course, analogous). Apart from the time-mean value, which is of highest importance in engineering design of a cooling or drying facility, the two additional components in Eq. (4) vary in time. As such, the values in Eq. (4) represent their instantaneous magnitudes. The coherent component $w_1'$ is of particular importance in periodically excited impinging flows. Some important measurement techniques, like hot-wire anemometry, do not discriminate it from the completely chaotic, stochastic time-mean fluctuation component $w_1''$—typically present due to turbulence. To characterise the intensity of unsteady components, RMS value is computed from values integrated over a certain length of time as

$$\widetilde{w}_1 = \sqrt{\overline{(w_1'')^2}} \tag{5}$$

The characteristic shapes of the velocity profile shown schematically in Fig. 10 correspond to the conditions in fully developed jets—while the profile in part A of Fig. 8 is, with its characteristic flat-top central part, typical for velocity profiles measured at small downstream distances from the nozzle. While a considerable wealth of information exists (obtained mostly in studies of non-impinging jets) for the developed jet flowfields, the same cannot be said for part B (i.e. the impact region in the vicinity of the stagnation point), where available information for this region is considerably less extensive. Until recently, this region was known to possess several less well-understood features, in particular the enigma of secondary off-axis heat and/or mass transfer maxima [26].

In the early history of shear flow investigations, the problem faced by many researchers were mathematical difficulties surrounding the governing Naviér-Stokes equations. They are partial differential equations, practically (i.e. for useful boundary conditions) impossible to solve analytically. Certain simplifications resultant from the plausible assumption that the jet is narrow (if planar) or slender (if axisymmetric) were put forward by Prandtl. His result, known generally as the boundary layer equation (though it is equally valid for a wide number of cases other than boundary layers), is still a partial differential equation. A substantial breakthrough came with the discovery of the fact that the solution of Prandtl equations for boundary conditions corresponding to flows possessing the character of thin or narrow layers may be obtained after transformation into ordinary differential equation (i.e. with dependences on only a single variable). The ordinary equations are much easier to solve. The condition under which this approach becomes applicable is that profiles of various quantities (velocities, of course, being the most important among them) at different streamwise distances should be mutually similar. For some of the shear flows, with developed jet among them, analytical solutions in the form of a formula describing the relation between individual variables (in their similarity-transformed forms) were actually found. This was, however, possible in only a small number of flows, the majority of them being of laminar character. Even if the derived ordinary differential equation for the transformed variables does not admit an analytical solution, it may be easily integrated numerically. Into the results of this integration it is then possible to insert various values of parameters and see what are the consequences. Compared with the problems posed by the governing partial differential equations, it is obvious how such a solution is valuable for the needs of an engineer and his design tasks. Among the shear flows that were solved this way was the laminar submerged jet (i.e. both planar and axisymmetric: Schlichting's solution, 1933).

For the similarity transformation, it is necessary to know the laws governing the streamwise variation of velocity and of the characteristic transversal dimension of the mixing layer, characterised by the conventional jet width $\delta_{0.5}$, as shown in Fig. 9. These were mostly derived already by the earliest investigators. As an example, the bottom part of Fig. 10 shows the extremely simple linear law of the streamwise growth of turbulent jets. Compared with the success of laminar jet solutions, the case of turbulent jets progressed slowly—despite a number of published attempts—mainly because of the inadequacies and shortcomings of the turbulence models used. An important attempt was undertaken already in 1926 by Tollmien (which preceded the Schlichting's laminar jet similarity solution). Some of the results derived by Tollmien, such as the characteristics of the spatial distribution of mixing length, which characterises the size of turbulent eddies, in a jet cross section, are still valuable today. Nevertheless, the solutions as a whole have encountered drawbacks because spatial distributions of turbulence parameters have properties that have to be expressed as convective transport and also modelled as gradient-type diffusion (which the then used algebraic turbulence models cannot do). The similarity solutions of jet flows based on more sophisticated turbulence transport equations models were not available until relatively recently [8, 9]. They are quite successful despite these models' simplification of assuming isotropy of turbulent motions. In fact, they may compete with numerical solutions of the flowfields described by the partial differential equations. Although these numerical solutions are currently more favoured over similarity transformations approach, it should be reminded that the numerical computations produce a solution valid only for a single particular set of boundary conditions. They simply do not lead to the simple and universally valid engineering formulas. In fact, they must be conducted anew for each case, thus failing to provide an adequate general view.

In the part C of the flowfield shown in Fig. 8, the fluid flows along the impingement surface radially away from the stagnation region. This flowfield is known as a wall jet. Its typical property is the velocity maximum near the wall and wider external flow very similar to half that of a jet. Again, under the assumption that the wall-jet is relatively thin, the flowfield can be assumed to be governed by the Prandtl equation. If the wall-jet is laminar, then there is an exact similarity solution for this case due to Glauert [10]. In the same reference [10], there is also a similarity-transformation solution for the turbulent flow It uses a turbulence model too simple—so that it may be acceptable only as a starting point for better approximations. The problem complicating the turbulent wall-jet flow is the presence of the viscous sub-layer on the solid surface. It varies along the radial streamwise co-ordinate differently than the flow above the sub-layer. This different behaviour has been always the main hindrance in attempts at a more exact solution of turbulent wall jets. Heat or mass transfer in this wall-jet region is less intensive and this is probably the reason why the part C has been so far receiving less attention in impinging jet studies, compared to the region nearer the stagnation point. There are some solutions based on assumptions corresponding to the flow near the stagnation point (i.e. part B in Fig. 8). Unfortunately again, none of them demonstrate the desirable universal validity.

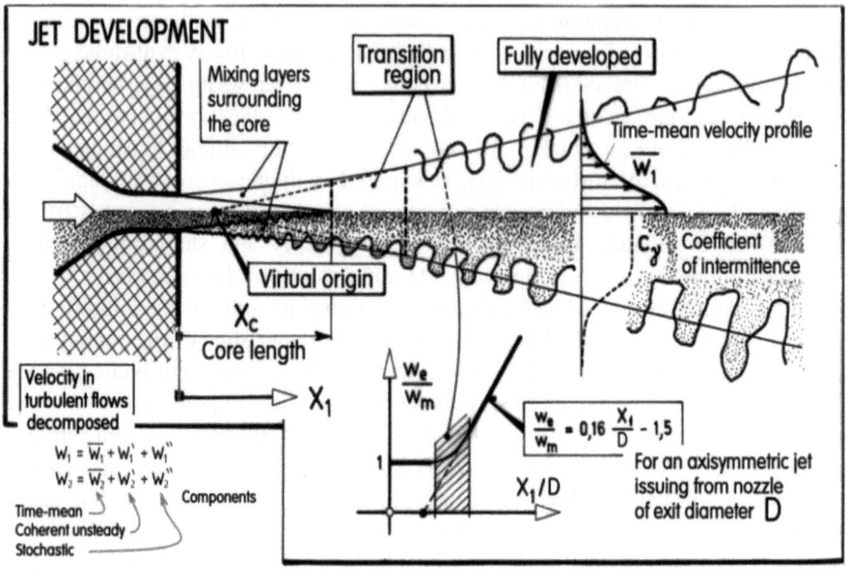


**Fig. 11** The basic facts about non-impinging jets. Note that the fully developed similarity region is even farther from the nozzle exit that the core length, because the flow has to pass first through the transition region of profile development. Another complicating factor is the intermittent presence of non-turbulent outer fluid carried along with the large vortices. *Courtesy* of Artech House Inc

Axial distribution of the maximum $w_m$ in velocity profiles should follow for a developed turbulent jet the $w_m \sim 1/X_1$ relationship, as shown in the bottom central part of Fig. 11. Strictly speaking, the similarity solutions are valid only for fully developed flows. Most shear flows, however, do not possess the proper full development characteristics at the location where they start. A certain streamwise development length at the beginning is needed for smoothing out the flow differences. In jet flows, the typical problem is that the character of velocity profiles at the nozzle exit does not agree with the developed shape shown in Fig. 12. It still does not agree at the distance equal to exit diameter $D$ as shown in part A of Fig. 8. In an analogous manner, the initial stages of the radial flow leaving the stagnation region, part B in Fig. 8, differ from the profile shape of a developed wall-jet shown as part C in the same figure. The development is of particular importance for impinging jet studies, because the impingement surface is usually placed at distances, $h$, comparable with the flow development lengths. The impingement is thus very often influenced by the presence of what is described as the jet core or, sometimes, potential core so called because, just like in non-viscous potential flows, there are no or very weak (and therefore negligible) shear stresses. The character of the core region depends on the wall contours of the nozzle. For instance, in a nozzle with very long exit channel there may be a trend towards generating velocity profiles approaching those known in pipe or tube flows, where velocity decreases in radial

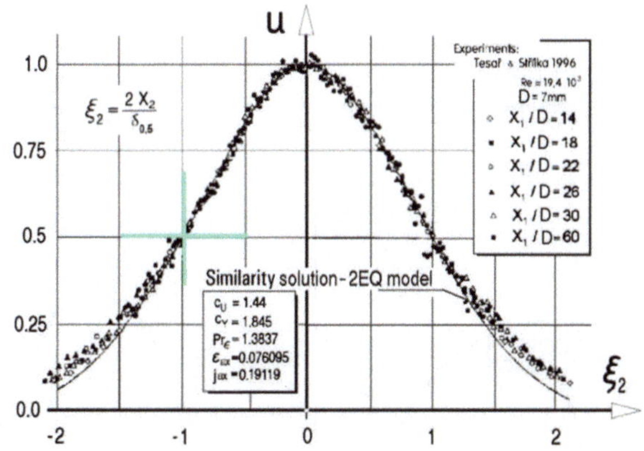


**Fig. 12** An example of compared computed and measured velocity profiles of developed turbulent plane jet. Similarity solution with two-equation model of turbulence does not agree well with data at the jet outer edge—where the differences are fortunately not of much importance, because the conditions there (intermittency and non-zero radial velocity) differ anyway from the assumptions of the computation model

direction. On the contrary, if the nozzle contraction is abrupt, the velocity profile in the exit may have peaks on the periphery and a shallow minimum in the centre. In most nozzles however, neither of these extremes dominates so that typical first velocity profiles downstream from the exit are mostly flat. The velocity in the core therefore is more or less constant and equal to the nozzle exit velocity $w_e$. It is the presence of this core that is revealed in measured profiles by the characteristic flat central part of velocity profiles, as seen in the example in the upper right-hand corner of Fig. 8. It is obvious, from the profiles plotted in Fig. 13, that in this region there can be no hope for velocity profile similar to that observed in Fig. 12. The similarity approach, nevertheless, may be successfully applied to the mixing layer at the outer edge of the developing jet [9]. The core length $X_C$ may be very large for Reynolds number values just below those associated with transition to turbulence, as shown in Fig. 14. This however, is rarely encountered in industrial impinging flows, as the Reynolds number is usually much higher and typical core lengths are $X_C \approx 4D$–$5D$. Immediately downstream from the core there is the transition region so that the full development is farther. On the contrary, Fig. 15 shows that core lengths $X_C$ are much shorter in flows with disturbances at the nozzle inlet or when the coefficient of turbulence intensity, as defined below, is high:

$$c_{tu} = \tilde{w}_1 / \overline{w_1}. \tag{6}$$

Note also that very short core lengths may be found if the nozzle has a very long exit channel and also in the cases with absent nozzle contraction.

Complete development is usually demonstrated by comparisons of computed velocity profiles with experimental data, as presented in the example in Fig. 11.

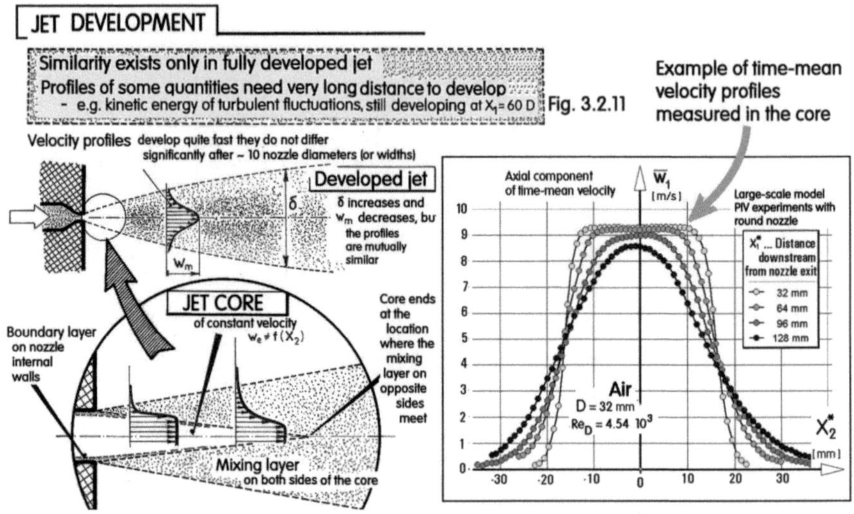


**Fig. 13** Similarity solution fails at short streamwise distances from nozzle, important in impinging jet flow studies. The jet is there in its development regime with the core, the length of which is often comparable with the distance h of the impingement surface. Adapted from [14]. *Courtesy* of Artech House Inc

**Fig. 14** With increasing Reynolds numbers in the laminar regime the jet core lengths $X_c$ increase—and may be much larger than generally expected—until its growth is reversed by turbulence. Adapted from [14]. *Courtesy* of Artech House Inc

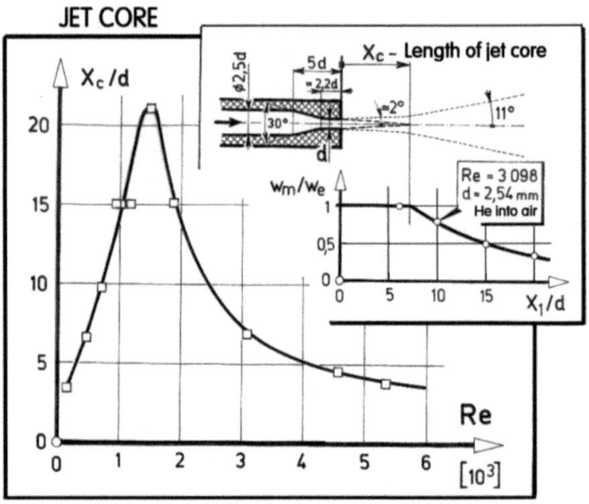


This is however, not a reliable comparison, because velocity is a quantity for which the attainment of reasonably good similarity (though not perfect—but the deviations are usually hidden in the inevitable scatter of experimental data) is exceptionally rapid. Other quantities often develop much more slowly with the axial distance from the nozzle, as may be seen in Fig. 16 for the case for specific energy of turbulent fluctuations, $e_f$ [J/kg = m$^2$/s$^2$], defined as

$$e_f = \frac{1}{2}\left[\overline{(w_1'')^2} + \overline{(w_2'')^2} + \overline{(w_3'')^2}\right]. \tag{7}$$

This is an important quantity used to characterise the intensity of fluctuations in almost all turbulence models. Fluctuations of not measured velocity component, $w_3$,

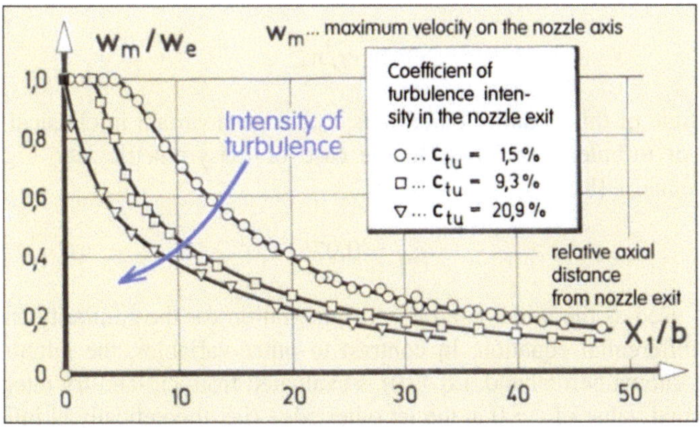


**Fig. 15** Potential core length decreases—up to complete disappearance—if a pulsation is introduced in the nozzle (or upstream). The pulsation intensity is here characterised by turbulence intensity coefficient defined by Eq. (6)

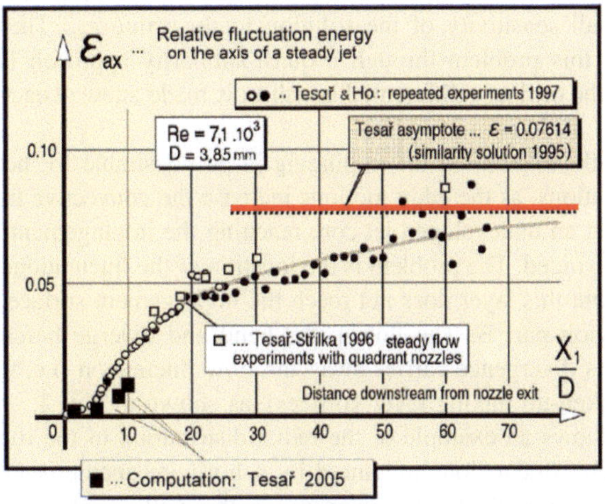


**Fig. 16** Axial distribution of non-dimensionalized specific energy of turbulent fluctuation measured by hot-wire anemometer. The constant value, on the existence of which rests the applicability of similarity solution, is only asymptotically approached at very long axial distances outside the region of interest in engineering applications. *Courtesy* of Elsevier B.V

are in currently usual two-dimensional analyses included into Eq. (7) on the strength of the assumption of isotropy. This permits simplification of Eq. (7) to

$$e_f \cong \frac{3}{4}\tilde{w}^2. \tag{8}$$

Similarity solutions, e.g. [8], operate with relative value (related to the maximum velocity $w_m$ in time-mean velocity profile at the same axial distance):

$$\varepsilon = e_f / w_m^2 \tag{9}$$

Magnitude of this quantity on jet axis, $\varepsilon_{ax}$, is an invariant in classical similarity solutions of turbulent jet flows. In the case of axisymmetric jets, Fig. 16, this constant value is [8]:

$$\varepsilon_{ax} = 0.07614 \tag{10}$$

This is used as the initial condition of integration for the similarity-transformed ordinary differential equation. In contrast to other variables, the initial values of which are known beforehand, Eq. (10) is evaluated from cut-and-try integrations to meet the final value of $\varepsilon = 0$ at the jet outer edge (i.e. theoretically at infinity). The value in Eq. (10), however, obviously cannot hold everywhere along the jet. Within the non-turbulent core $\varepsilon$ must be zero. From there, the value increases with increasing $X_1$, as shown by the experimental data presented in Fig. 16. The problem is that the increase is too slow. Note that the value in Eq. (10) is in Fig. 16 not yet fully attained at the distance as large as $X_1 = 60D$, i.e. far beyond the region of interest for practical impinging jets. That the similarity solutions are useful at all is due to the small sensitivity of the solution to the value $\varepsilon_{ax}$. There is a way of circumventing this problem through a quasi-similarity approach [11], where the integration of the ordinary differential equation is made anew at each axial distance of interest.

In general, fluctuations in the impinging jet are desirable for heat and/or mass transfer applications, as the eddy motions increase the convective transport effects. This means that an undeveloped jet core reaching the impingement surface should preferably be avoided. The problem is the location of the fluctuations, present in the mixing layer, but this layer does not reach the impingement surface, at least not in the impact region part B. The flow paths bend and diverge before reaching the surface and this divergence carries away the flow fluctuation (i.e. remnants of the paired and broken-up mixing layer vortices), as shown in Fig. 2.

Figure 17 shows an example of the radial distribution of the root-mean-square value of velocity fluctuations measured by a hot-wire anemometer. The diagram shows the typical low fluctuation intensity on the wall near the jet axis as a consequence of the jet core reaching up to this location. On the other hand, there are strong fluctuations at radial distances from the jet axis equal to $D/2$ and $2D$. This may be explained by the convective transport of turbulence generated in the mixing

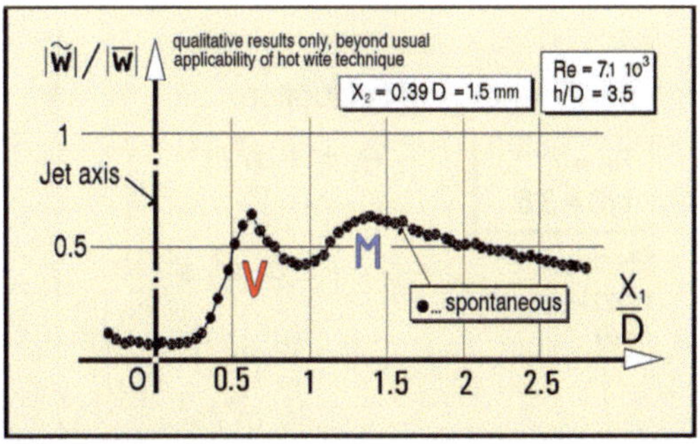


**Fig. 17** Radial distribution of the root-mean-square value of the fluctuating component of absolute velocity measured as a small height above the impingement surface. A typical feature are the two maxima *V* and *M* [26]

layer. Quite surprisingly, there is not only the expected single maximum. Instead, there are two peaks, labelled **V** and **M** in Fig. 17. This effect is associated with the enigmatic "secondary maxima" of heat and/or mass transfer, a phenomenon known for a long time by researchers working on impinging jets but one that remained unexplained until recently. The maxima are found in flows with relatively small nozzle height *h* above the impingement wall. The data in Fig. 17 were measured with the nozzle positioned at quite small $h/D \approx 3.5$. The currently accepted explanation for the off-axis secondary maxima is based on the notion of coherent vortical motions, present in the impinging jet in the form of vortex rings within the mixing layer. This explanation is only qualitative and has the drawback of not clarifying why such a localised stationary extreme may be a consequence of the vortex rings carried with the time-mean flow. Therefore, it is also plausible to base the explanation upon the influence of small-scale turbulence, into which the vortex rings rapidly disintegrate, as depicted in Fig. 2. They obtain energy from the large-scale motions by spectral cascade, transferring energy into smaller scale motions in the spectrum of turbulence [12]. At the location corresponding to the position of the peaks in Fig. 17, the vortices have already decomposed into stochastic turbulence. Moreover, the secondary maxima are also seen in results of numerical flowfield computations, which handle the turbulent fluctuations as being fully described by a scalar value—and cannot therefore include the effect of organised vortex rings. The influence of small-scale eddies is characterised by the fluctuation energy. Standard isotropic modelling, of course, is incapable of capturing the large eddies and their motions. Yet, the example results in Fig. 17 clearly demonstrate the presence of a region at the initial stages of the wall-jet part C (i.e. see Fig. 8) into which the fluctuation energy as described by Eq. (8), generated upstream in the mixing layer, is transported by convection as depicted in Fig. 19. Note that the local kinetic energy production and convective transport are two processes which are taken account of in present-day engineering modelling of turbulence—and were missing in early algebraic models.

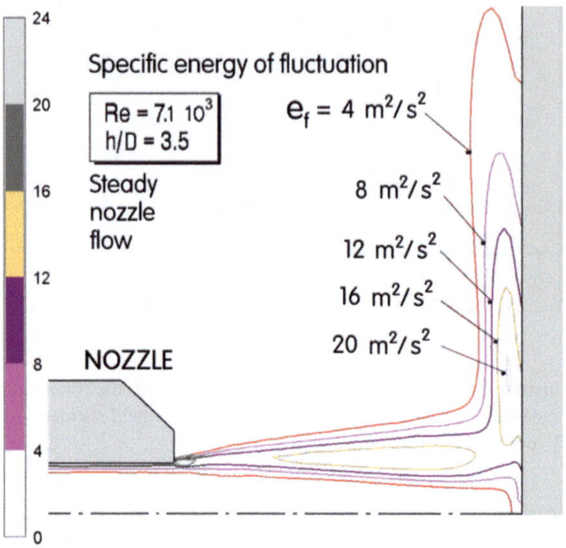


**Fig. 18** Spatial distributions in meridian plane of the fluctuation energy computed using an isotropic (two-equation) model of turbulence. This model is not capable of capturing the coherent eddies and yet it shows development of the off-axis local maximum

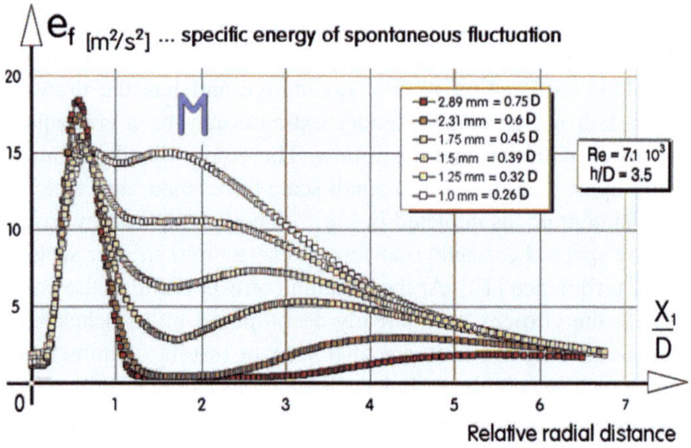


**Fig. 19** This computed radial distributions of specific energy of turbulent fluctuation plotted along straight lines at different heights $X_2$ above the impingement surface is a useful complement to Fig. 18. Convected turbulent fluctuations exhibit the double peaks, the second (downstream) one capable of providing an explanation for the maxima *V* and *M* in Fig. 17—adapted from Ref. [26]

Since eddies carried with the flow enhance convective transport, it is not surprising that the presence of two peaks (similar to what was observed earlier) is also observed in measurements of heat or mass transfer at small nozzle distances, *h*.

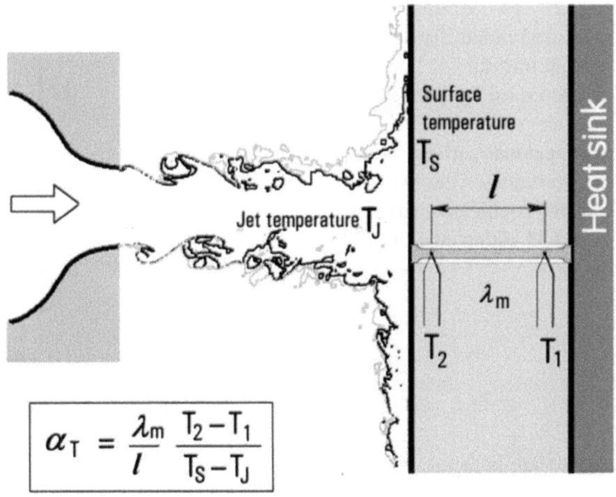


**Fig. 20** The principle of measuring local density of heat transfer into a heated surface. The coefficient $\alpha_T$ at a location of interest is evaluated from measured conduction through a thermally isolated small element of the surface

It may be useful to at least briefly describe here how the thermal transport is measured. The aim is to evaluate heat transfer coefficient $\alpha_T$, usually in the non-dimensionalized form of Nusselt number:

$$Nu = \frac{\alpha_T D}{\lambda} \tag{11}$$

Figure 20 shows schematically hot fluid at a temperature $T_J$ impinging upon a flat plate, upon which the thermal power from the fluid is transferred. There are alternative variants of the scenario but let us choose the case of constant surface temperature $T_S$ obtained by high thermal conductivity of the impingement surface material in the radial direction. The conductive solid-material layer carries the transferred heat into the sink. Its thermal-flux resistance (characterised by conductivity $\lambda_m$) generates a measurable temperature difference $T_2$–$T_1$ from which the axial heat flux is evaluated by means of an adapted variant of Eq. (2), applied to a thin cylinder of solid material of known thermal conductivity, $\lambda_m$, the same as that of the surrounding thick plate. Note that the element is separated from its surrounding by very narrow gas-filled gaps, which practically eliminates any transversal heat transport away from the element. The expression for $\alpha_T$ is given at the bottom of the picture. Typical radial distributions of Nusselt number Eq. (11) obtained by this or similar method in heat transporting impinging hot air jets issuing from a nozzle positioned at a small distance, $h$, are presented in Fig. 21. The analogy with Fig. 17 is obvious. The discrepancy in $Nu$ numerical values between the two data sources in Fig. 21 is typical—measurements of heat transfer exhibit generally accuracy lower by a decimal order of magnitude than measurements in fluid mechanics.

As was already mentioned in Sect. 1, one of the reasons why the insulating fluid conduction layer is difficult to investigate directly is its usually very small thickness. It may be estimated, using the value $Nu = 150$ presented in Fig. 21 and from Eq. (11), that the coefficient $\alpha_T$ defined in Eq. (1) has a value of 1300, so that the

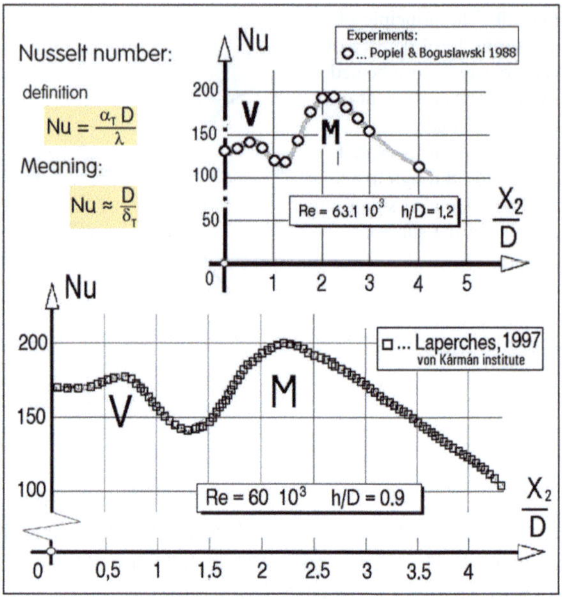


**Fig. 21** Two examples of measured radial distribution of heat transfer—characterised by Nusselt number—over the impingement surface. Note the presence of the two transfer coefficient maxima *V* and *M*. Adapted from [27] and [6]

effective thickness of the thermally insulating air layer on the surface under a typical $D = 3$ mm diameter nozzle is only 0.02 mm!

Because of the relatively small improvements achieved with the axial (varicose) pulsation of the supply flow, this idea is currently lacking wider enthusiasm. It should be, however, not forgotten that the secondary maxima M (Fig. 21) are indeed an increase in local heat transfer rate—and also that they are a consequence of the ring vortices carried in the mixing layer or in the synthetic jets.

Recent research activities concentrate on generation of longitudinal vortices carried with the jet. The potential advantage of this approach is the possibility of generating the corresponding maxima nearer to the jet axis. There are two approaches to the generation of longitudinal vortices: (a) passive one, using low past tabs [39] protruding into the jet at the nozzle exit, and (b) small auxiliary synthetic jets, generating the vortices in a manner analogous to [40]. Both approaches have already demonstrated an increase in heat transfer. Recent auxiliary small synthetic jets are applied to disturb the main jet to which they are inclined at a small angle (30 ° in [41]).

## 3 Annular Impinging Jets

Identification of the thermal conduction sublayer as the key factor that limits the achievable thermal power transfer rate gave rise to the obvious effort to increase the transfer efficiency by somehow "destroying" the layer or at least decreasing its influence. This target is clear but achieving it is by no means easy. The viscous

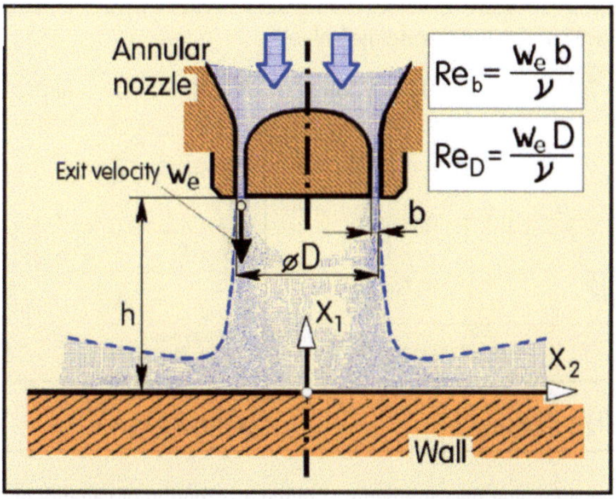


**Fig. 22** Layout and basic dimensions of an annular nozzle for generation of the investigated annular impinging jet. This particular geometry was characterised by D = 40 mm and b = 1 mm. The quantity $\nu$ in Reynolds number definition is the viscosity of the fluid. *Courtesy* of Elsevier B.V

effects that actually cause the layer to exist and to stay on the heat transfer surface are, of course, inescapable. Another aspect is the requisite cost-effectiveness and simple configuration of the impingement heating (cooling) solution. It would be an exceptional situation indeed, if it would be acceptable to increase the efficiency of impinging jet heating at the price of adding some complex and electric driving energy consuming mechanism performing the task of conduction layer removal. Ideally, the role of layer removing would be taken over by some fluid mechanical effect in the jet flow—perhaps generated in the nozzle as the place with highest concentration of fluid flow energy. It would be then necessary for this effect to reach to the impingement location on the wall.

Partly as an example of a scenario whereby it was indeed possible to demonstrate an increase in the thermal transport efficiency by simple means, and partly to demonstrate how complex impinging jet flows can become when topological changes in the flowfield characteristics are involved, it will be useful to show here in closer detail one particularly intriguing configuration [13]. It is the case of an impinging jet generated by outflow from an annular nozzle, as presented schematically in Fig. 22. It is not a configuration that may be described as wholly extraordinary. To design and manufacture an annular nozzle is by no means difficult, the main practical task being to arrange the struts that hold ned centre the inner body. The annular configuration has already been used for a long time in situations where there was a requirement to have only a single nozzle and yet to cover a large impingement surface area. Covering the large area may also be obtained with the usual round nozzle of a very large exit diameter *D*, but such a choice would be highly uneconomical because of the large flow rate of the heat-carrying fluid required.

Annular nozzle configurations were first used as fluidic no-moving-part detectors for sensing a presence of an object at a distance of the order ~10 mm [14].

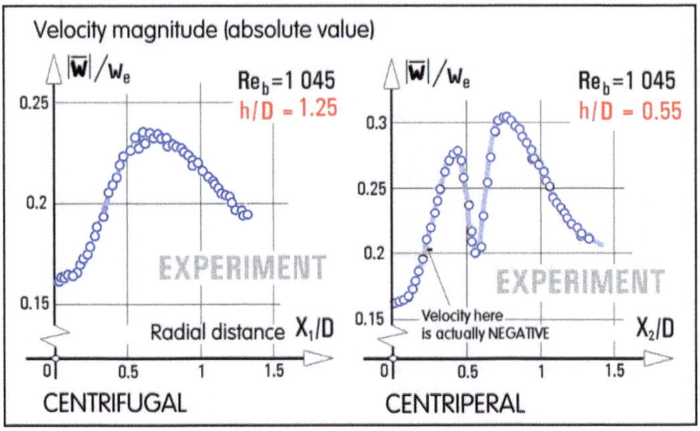


**Fig. 23** Hot wire anemometer measurements of radial distribution of velocity magnitude (absolute velocity) at a very small height $X_2$ above the impingement wall (as small as acceptable without mechanically endangering the probe)

Researchers who studied them were surprised by non-monotonous responses to variation of the distance $h$ between the nozzle exit and the impingement surface. Increasing $h$ leads to qualitative, topological changes of the flowfield (note that in a slightly different context, the changes are described in [15] as metamorphoses). Two examples of qualitatively different radial distributions of time-mean velocity, measured at a very small distance $X_2 = 0.01D$ above the impingement surface, are presented in Fig. 23. The nozzle was in both cases the same. Anemometric probe used cannot discriminate between the two components $w_1$ and $w_2$ in this two-dimensional flow and the plotted quantity is local magnitude of the velocity vector, $\mathbf{w}$. The values are non-dimensionalized by the nozzle exit velocity $w_e$. Explanation for the different characteristics of the two distributions appearing in Fig. 23 is obtained by comparisons with the corresponding flow-lines of the surface presented in Fig. 24. Ideally, the velocity in the local minimum of the centripetal case should be zero—with sign discrimination, the curve would have to pass through zero from positive to negative values. The non-zero value may be an effect of flow unsteadiness (Fig. 25).

Detailed studies of the annular jet impingement show that there are actually more than two topologically different possible flowfield configurations. This is demonstrated through computational results obtained for a particular nozzle (i.e. outer diameter $D = 40$ mm, slit width $b = 1$ mm, such as the device shown in Fig. 35) at gradually increased distance $h$, presented in meridian plane sections in Fig. 26. Topology is defined by the path-lines passing through the stagnation points. In all cases, there is a stationary vortex ring attached to the bottom of the central body of the nozzle. The configuration C with extremely extraordinary double stagnation point is the most interesting one from theoretical point of view. Practical importance for application of this configuration is small, however because of the difficulty of adjusting and keeping it. On the contrary, the configuration B is important from the

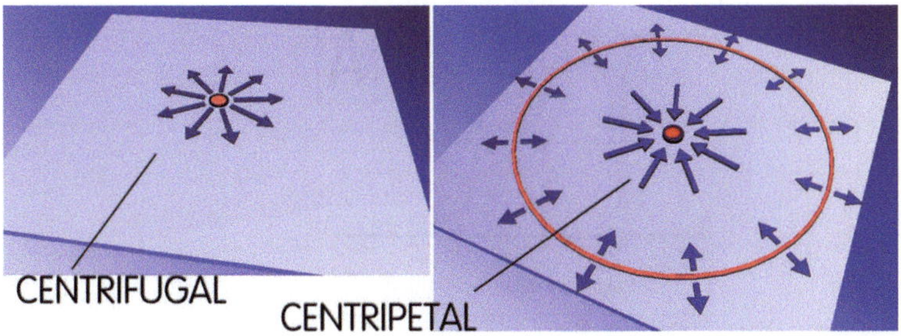


**Fig. 24** Flow directions on the impingement surface correspond to the two situations in previous Fig. 23. The centrifugal flow [15] (i.e. running away from the central stagnation point) is encountered at large nozzle heights $h$ above the surface while the centripetal flow (towards the central stagnation point) results from the recirculation bubble reaching up to the impingement surface. Radius of the stagnation circle in the centripetal case is $X_{1R}$ (the case B in Fig. 26). *Courtesy* of Elsevier B.V

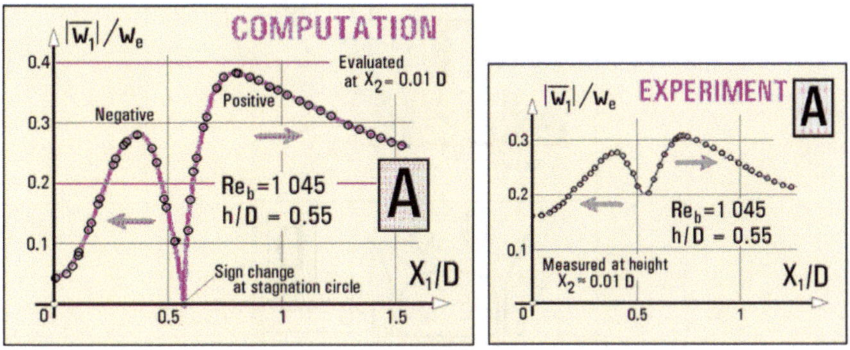


**Fig. 25** Comparison of computed (*left*) and measured (*right*) radial distributions of velocity at a small vertical distance above the impingement plane in the regime *A* of Fig. 26 (vortex ring reaching to the surface). Measurements with hot-wire anemometer are unable to recognise the change in sign; the computed data were therefore converted to absolute value for easier comparison

engineering point of view, despite the fact that it is topologically similar to configuration A, i.e. with the stationary vortex ring reaching up to the impingement surface.

An important feature of configuration B is its local constriction (i.e. smaller diameter) of the stationary vortex ring at a certain small height above the surface. The fact is that the constriction makes the vortex ring unstable. The computation result shown in Fig. 26 were later reproduced in experiments, with the exception of the configuration B because of its unstable behaviour. It also exhibited complicated effects of hysteresis [16] and bi-stability. When the nozzle height $h$ is increased further, finally a really significant topological change in the characteristics of the flowfield occurred in configuration C. The progressing constriction collapsed into a single point on the jet axis. The part of the vortex ring below the constriction became

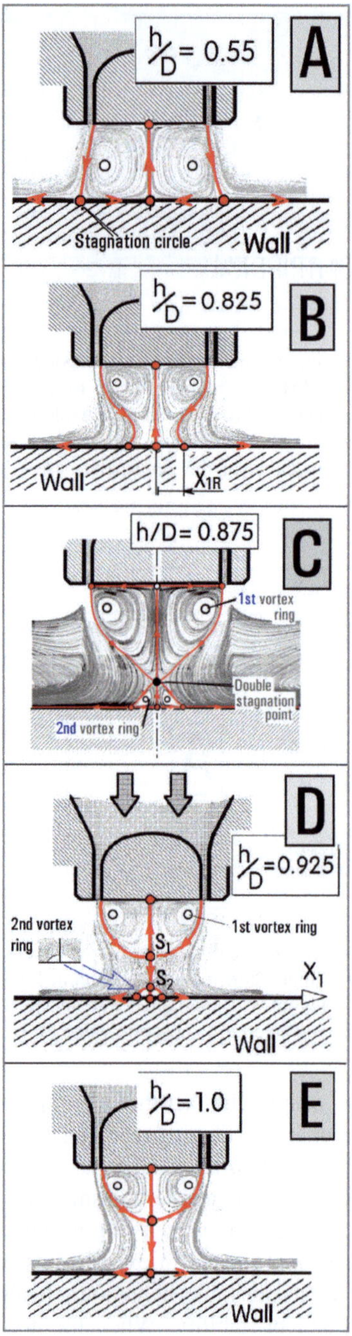
h/D = 0.55
A
Stagnation circle
Wall
h/D = 0.825
B
Wall
X1R
h/D= 0.875
C
1st vortex ring
Double stagnation point
2nd vortex ring
D
h/D = 0.925
2nd vortex ring
1st vortex ring
S1
S2
X1
Wall
h/D = 1.0
E
Wall

◀**Fig. 26** (*Courtesy* of Elsevier B.V.). **Configuration A**: Path-lines computed for the impinging flow from an annular nozzle positioned at a small distance $h$ above the impingement wall. Reynolds number $Re_b = 1045$, nozzle outer diameter $D = 40$ mm, slit width $b = 1$ mm. This is the case of centripetal flow towards the central stagnation point on the jet axis (as in Figs. 24 and 25). **Configuration B**: If the annular nozzle is moved away from the impingement wall then an obvious contraction occurs in the lower (*near-wall*) part of the stationary vortex ring that occupies the space delimited by the stagnation path-lines. As shown in this computed example, below the local constriction the vortex ring expands again. This may be computed but in reality this character of the flow would be highly unstable. This is why this configuration **B**—not actually in a topological sense different from the case **A**—deserves being handled as a separate case. Reynolds number $Re_b = 1045$ as well as the relative width of the annular exit remain the same. The centripetal flowfield configuration A can exist and remain stable only if the distance $h$ (Fig. 22) is shorter than the diameter $D$. If the nozzle distance $h$ is gradually increased beyond this limit, the topological character of the flowfield will change. **Configuration C**: The final stage of the pinching the near-wall part of the stationary vortex in the regime B, caused by to the increase in the distance $h$, is characterised by separation into two co-rotating ring vortices (the *bottom* one much smaller) with very peculiar double stagnation point between them. **Configuration D**: The result of the further movement of the annular nozzle away from the wall. This is a topologically different flow configuration with two stagnation points $S_1$ and $S_2$ the first above the second one, both on the jet axis. As the nozzle continues retreating to higher distances h from the wall, the lower stationary vortex decreases in size. **Configuration E**: The final stage of the changes associated with increasing the nozzle-to-wall height $h$. The flow over the impingement wall becomes "centrifugal" (as opposed to the "centripetal" in previous pictures of this series

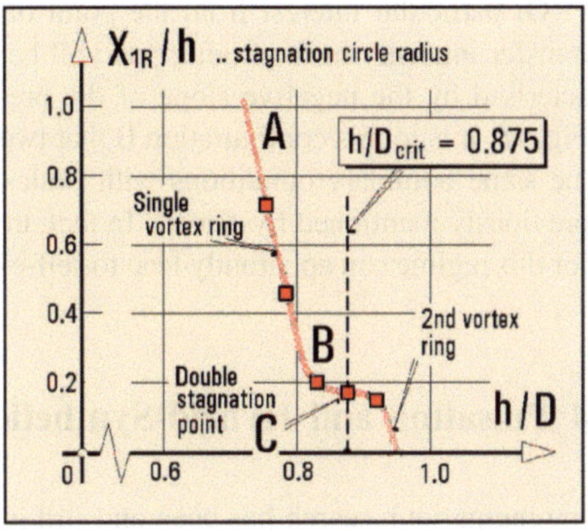


**Fig. 27** Relative size of the stagnation circle on the impingement surface with different relative height $h/D$ of the nozzle (cases A, B, and C in Fig. 26). Evaluated by computations of the case $b/D = 0.025$

separated and formed another co-rotating vortex ring. Further increase in the nozzle height $h$ would lead to configuration D with two different stagnation points on the jet axis. In the upper stagnation point, $S_1$, there is the bottom end of the vortex ring attached to the central body in the nozzle. The very small second vortex ring attached to the impingement surface has the stagnation point, $S_2$, at its top. The size of the bottom ring decreases with further increases in $h$ and finally in configuration E of Fig. 26, it disappears and leaves only the centrifugal stagnation point on the impingement plane. Dimensions of the stagnation circle on the impingement wall are plotted in Fig. 27. In the final configuration E, the part of the flowfield below the

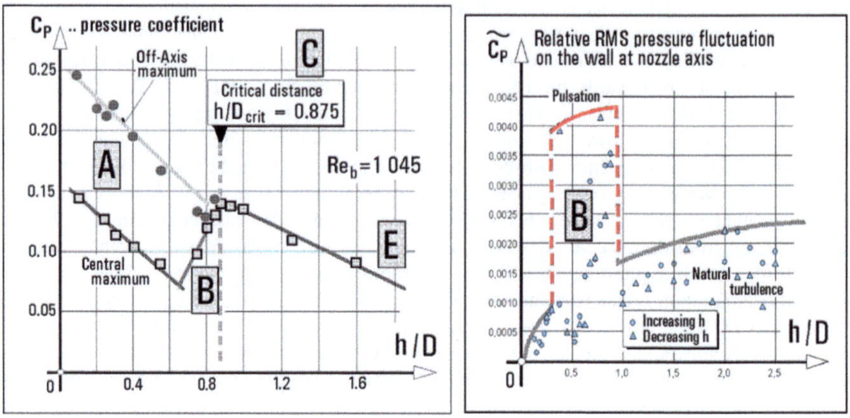


**Fig. 28** *Left* Coefficient $c_P$ of time-mean pressure measured at varying height $h$ in the central stagnation point on the impingement surface—and in regime $A$ also in the off-axis maximum. *Right* Pressure coefficient of self-excited pressure oscillation in the unstable regime $B$

upper vortex ring is practically identical to the impinging jet flowfield generated with a simple round nozzle. In practical applications, of course, this configuration E is not used because it does not bring any advantages over the round jet.

Of particular interest from the point of view of attempts at increasing the heat transfer intensity in impinging jets will be the unstable configuration B. It is characterised by the negative slope of the pressure dependence on nozzle height $h$ in Fig. 28. It is in this configuration B that two different flowfields may be found under the same boundary conditions with (unless Reynolds numbers are very high) the previously mentioned hysteresis. In fact, the hydrodynamic instability characteristic for this regime can apparently lead to self-excited oscillation, documented in Fig. 28.

## 4 Pulsating and Hybrid-Synthetic Impinging Jets

Impinging jet research has been and still is dominated by the idea of increasing the heat and/or mass transfer efficiency. The key problem therefore is the search for ways and means to suppressing the influence of the insulating thermal boundary layer. The most promising solution seems to be pulsation of the supplied fluid flow from which is the jet generated. Because of the economy of such a solution, the method used must be characterised by low investment costs (i.e. the added device must be simple and easy to manufacture) and also low operation costs. It is also generally advisable not to depend upon externally-powered actuators. Low-cost requirements put a restriction on the efficiency of the additional device and identifying the “weak spots” of the insulating layer in which the action has a maximum effect becomes crucial. Ideally, the layer should be removed completely, even though its thinning would also be welcomed (i.e. to decrease the remaining distance over which the heat is to be transported by ineffective conduction). An important fact, known since the first historic analyses of unsteady flows by Stokes [3], is the expression describing the growth of the momentum boundary layer thickness, δ:

$$\delta \propto \sqrt{\nu t}, \tag{12}$$

where $\nu$ [m$^2$/s] is kinematic viscosity of the fluid and $t$ [s] is elapsed time. Because of the existing and well-documented analogy, it may be quite safely assumed that the thickness $\delta_T$ of the thermal boundary layer also grows with time as

$$\delta_T \propto \sqrt{\alpha_T t} \tag{13}$$

If it were possible to remove the layer completely during one-half of the pulsation period, its thickness growth will then be very limited because of the short available time during the next half-period. This idea has been known for some time, but its practical application was found to be by no means easy. Firstly, the mere pulsation in the supplied flow is not likely to remove the insulating layer, which is held on the surface by strong local viscous forces. The layer can absorb significant periodic actions without any degradation. If the pulsation is generated inside the nozzle (or its upstream cavity), it will be strongly damped along the path from the nozzle to the impingement surface. Secondly, there is the problem of costs. Even if an increase in the efficiency of heat transfer is achieved, it may not justify the initial cost of the added mechanism and the operating cost of energy spent on driving this mechanism.

A promising turn came about with the advances in fluidic oscillators [14] and modern methods of their manufacturing, e.g. by photo-etching. A fluidic oscillator —generating self-excited oscillations due to hydrodynamic instabilities—is now quite inexpensive. In principle, it is nothing more than just a special shaping of the fluid inlet cavities. It needs no electric energy, being driven by just a very small percentage extracted from the fluid flow energy supplied to the nozzle. Having no moving or deforming components, fluidic oscillators are reliable, rugged, need no maintenance and their operating lifetime is practically unlimited. The problem that remains is their influences on the thermal-conduction layer. Because of the relatively strong viscous forces that must be overcome, achieving a significant effect calls for relatively large amplitudes of the pulsating component of the flow. Numerous recent publications in available literature recommend to go for the extreme and to use the synthetic jets [17] with no time-mean component at all, as depicted in Fig. 29. The generated convective transport may be effective, but unfortunately only for a short period of time. In any longer operations, the nozzle would inevitably only re-circulate the fluid that is already heated from the cooled object (or already humid by containing the water from a dried object), as shown in Fig. 30. Obviously, some non-zero flow must be supplied into the nozzle, gradually replacing the fluid that in the meantime becomes useless. The resultant phenomenon is called hybrid-synthetic jet (Fig. 21). Its earliest investigation was probably the idea presented in [18], where it combined the piston- type synthetic jet generator with the fluidic rectification ideas as used in the "valve-less" pumps [19]—i.e. with

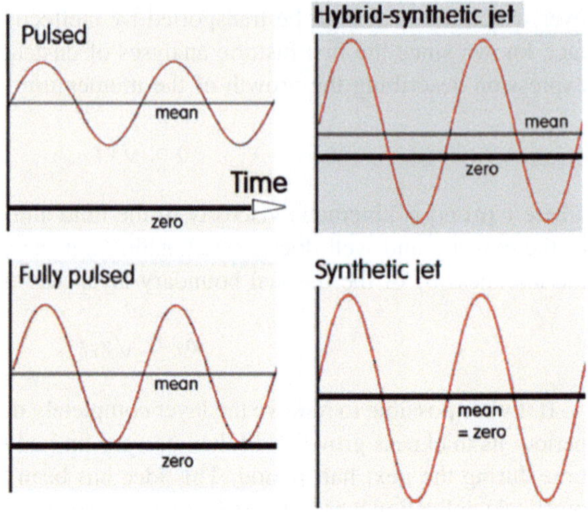


**Fig. 29** Temporal variations of nozzle exit flow rate in combinations of steady and harmonic (sinusoidal) pulsation flow components

fluidic no-moving-part diodes [14]. Their rectification efficiency is not particularly high, but this may be not a real problem in an application where just a measure of fluid replacement may be all that is called for. Of course, the driving by a piston in practical applications is unlikely, being replaced by fluid-displacing deforming membrane or directly by deformed piezoelectric element. The advantage of the generation of the synthetic as well as hybrid-synthetic jets by the displacing mechanism (such as the membrane) is the generation of the jet from fluid available on the spot, with no need of supplying it by tubing that complicates the design. On the other side, the operation of the displacing component (i.e. either a piston or membrane) at the location where the action is needed, necessitates bringing up to this location the electric driving current, which can sometimes also lead to complications. Additionally, with the danger of the piston getting stuck or membrane broken due to fatigue, this movable or deformable solution is less reliable and its life span may be comparatively shorter.

Better from these practical points of view are hybrid-synthetic jet generators based on a different idea—the idea of pure fluidic oscillator as shown in Fig. 31. The advantage of absence of moving parts is obvious: unlimited life, low cost and ruggedness. Early ideas of using fluidic generators are presented in Fig. 32, stemming from an era that preceded the synthetic jets idea. The basic component is the fluidic oscillator, which consists of a fluidic amplifier provided with feedback loop channels. In Fig. 31, it is the jet-deflection type diverter amplifier. Converting the amplifier into an oscillator is achieved by a negative feedback. In case A shown in Fig. 32, as well as the oscillator in Fig. 31, there is the Spyropoulos-type negative feedback loop, connecting the two control nozzles. The return suction into the amplifier through the OFF-side exit is present due to the jet-pump effect entrainment into the supply jet—the one generated by the supply nozzle flow inside the amplifier. The case B in Fig. 32 has a different alternative of the negative feedback, the two-loops version invented by Warren [20].

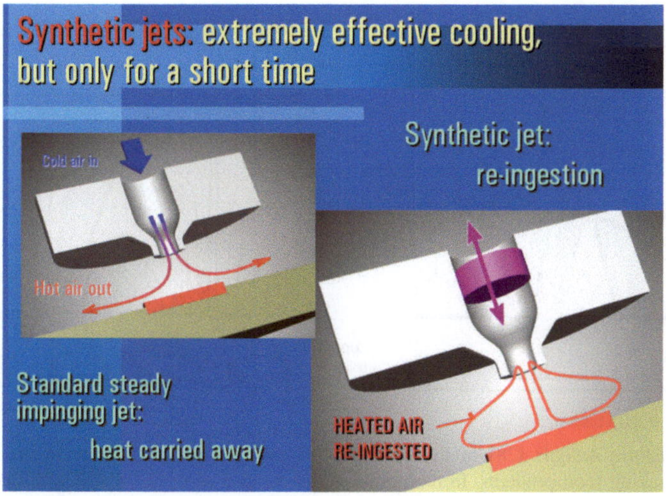


**Fig. 30** Cooling by impinging synthetic jet (at *right*) is currently very popular subject of investigations, especially aimed at keeping cool small electronic components. Obvious shortcoming of synthetic jets in this role is its re-ingestion of the already heated air

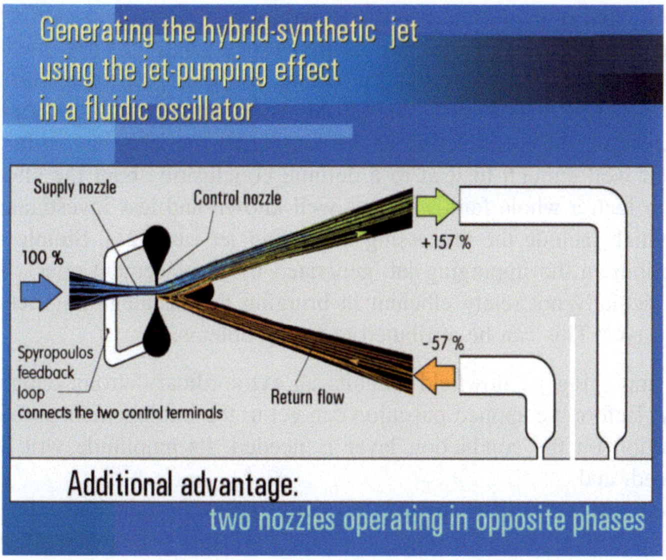


**Fig. 31** Fluidic generator of hybrid synthetic jets [14]. The basic part is a fluidic amplifier with feedback loop that converts it into an oscillator without moving or deformed components. Computed path-lines show in the lower branch the return flow (suction) caused by jet-pumping effect generated in the supply nozzle. *Courtesy* of Elsevier B.V. and Artech House Inc

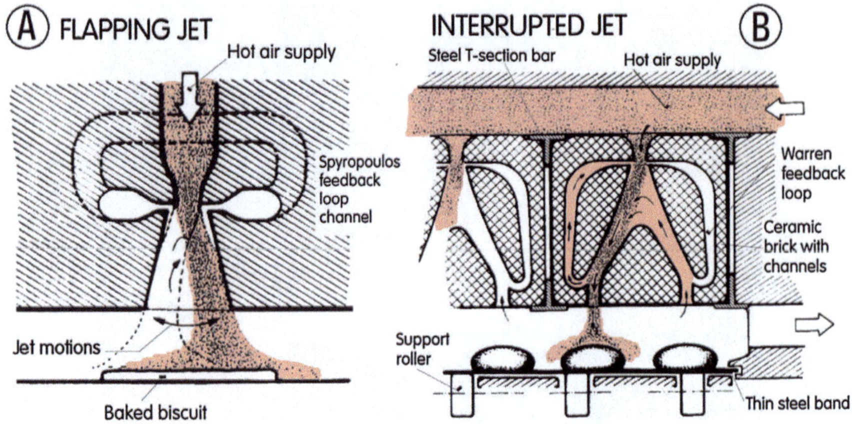


**Fig. 32** Historic designs [28] of fluidic oscillators increasing heat transfer in impinging jets applied to baking food products. In the design **B** the cavities of the oscillator with two feedback loop channels were made by pressing a correspondingly shaped tool into the ceramic brick while it was still deformable prior to sintering

While the advantages offered by fluidics meant that generation of oscillatory excitation in impinging jets became economically feasible, the task of acting upon or downright suppressing the thermal insulation effect of the viscous layer was found to be more difficult than expected. Simply applying a pulsation and hoping for an improvement led to numerous disappointments and thus new methods were constantly searched after. Note that the action on the impingement surface in the case A in Fig. 32 is based not on the pulsating supply flow into the nozzle but on periodic changing of the direction of the impinging jet. This has so far not been investigated well enough to lead to a definite conclusion about the effectiveness. There is, in fact, a whole family of less well known and less investigated configurations which include the precessing impinging jet case [21]. Simple axial pulsatory motions in the impinging jet, generated by displacement effects inside the nozzle, is actually not really efficient in bringing the disintegrating action to the insulating layer. This can be attributed to two problems:

(a) The impinging jet flowfield exhibits an extraordinary strong damping capability. Before the applied pulsation can get to the impingement surface, where its action on the conduction layer is needed, its amplitude will be rapidly reduced, and
(b) The energy of pulsation applied upstream from or inside the nozzle, instead of being effectively used for suppressing the insulating conduction layer, is found to be spent unproductively on the generation of vortical structures in the jet upstream from the impact region.

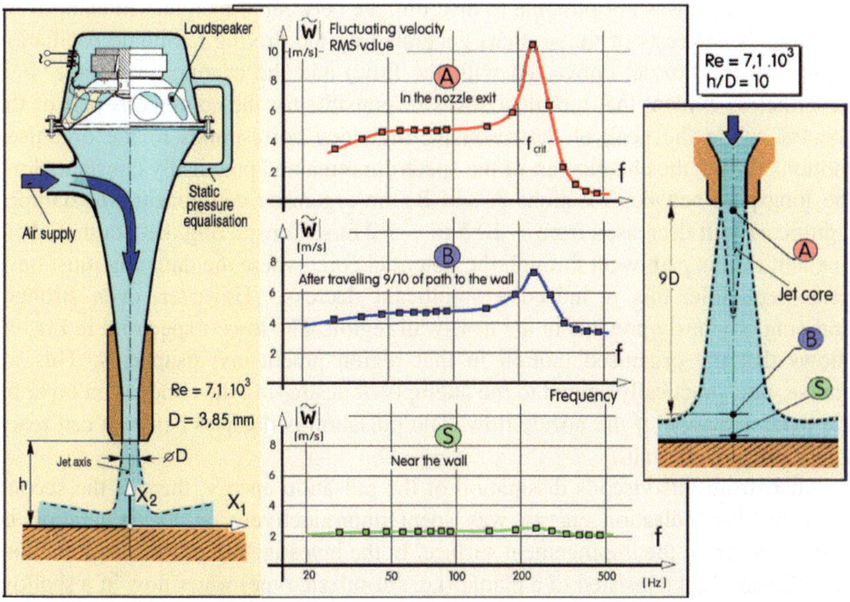


**Fig. 33** Varicose pulsation of impinging jet was investigated using a loudspeaker generator (*left*). Frequency spectra of hot-wire probe signal in the three locations (*A*, *B* and *S*—as shown at *right*) demonstrate the strong damping of pulsation between *B* and *S*

A typical example of these two problems may be demonstrated on the following results of two investigations:

As shown in the drawing of the first setup, in Fig. 33, in this experiment the flow pulsation was generated by a loudspeaker built into the chamber upstream from the nozzle. There were practical reasons for not using fluidics in the experiment. Fluidic oscillators at their present stage of development need themselves a considerable effort to improve their efficiency. Also, they possess a complicating property of the oscillation frequency dependent on the flow rate of supplied fluid. In contrast, keeping the frequency constant in a loudspeaker driven by an electric signal is easy. In the experiment, the driving frequency was set to 235 Hz, which was the natural frequency of nozzle and chamber resonance. The responses to the excitation were measured using hot wire anemometry. The anemometer probe was gradually placed into three locations on the jet axis indicated in the right-hand part of Fig. 33. In each of these positions, large amounts of instant velocity data were measured and accumulated, which made possible computing frequency spectra of unsteady components of absolute velocity (i.e. of the local velocity vector magnitude).

The results obtained are presented in the middle part of Fig. 33, in the order corresponding to the measurement locations, so that it is possible to obtain an idea about the streamwise development of the spectral distributions. In the anemometer output signal, it was not possible to discriminate between the organised component (i.e. mainly the rests of the vortices generated by the driving harmonic oscillation input into the nozzle, convected with the flow) and the chaotic turbulence. It is nevertheless evident that turbulence was responsible for the continuous part of the spectra, while the peak at the resonant frequency corresponds to the organised motions. While the chaotic part of the spectrum remained practically constant along the long path between locations A and B, the organised component was damped significantly. It decreased from $\sim 10.5$ to $\sim 7.2$ m/s. Considering the fact that about one half of this path went through the potential core, where the damping must have been negligible, this is indeed a significant decrease. However, even stronger damping obviously prevails in the near-wall region. The lowest spectrum in Fig. 33 shows that the organised motion in that region practically disappears. This, of course, puts practically an end to the attempts of destroying the conduction layer by the axial pulsation of the nozzle flow. The pulsation is damped before it can reach the impingement wall.

Apart from this viscous dissipation of the pulsation energy, there is the second problem. The pulsation energy was spent unproductively on the generation of vortices far from the impingement surface. In the investigation of this phenomenon, the impinging jet consisted of a planar (i.e. slit-nozzle type) water flow in a shallow cavity between two flat transparent cover plates. Water flow was visualised through dye with suddenly changed dye concentration. Its response to the pulsation was recorded by video camera. The dyed water was released at the start of the video clip, replacing the clear water that was in the cavities before. Of course, for some time the clear water remained in parts of the image, where the flow was slower. On the left of Fig. 34 are two selected images from one of the video clips. Also indicated is their temporal position with respect to the applied harmonic pulsation. The energy of incoming pulsations is evidently used to generate two oppositely rotating vortices on both sides of the jet in each of the flow pulses. From the positions of their formation near the nozzle exit, the vortices were seen in the video clip to move vertically downwards along the jet. Their different sizes recorded at two different phases of the pulsation testifies to their growth during each pulsation cycle. The unpleasant fact is that these vortices are formed above the impingement region B of Fig. 8 so that they do not reach down to the near-wall locations where there is the thin insulating conduction layer that was expected to be influenced. At any rate, the vortices were certainly not found to act upon the layer. On the right of Fig. 34, images taken from another video clip were processed to identify the jet and the vortices by extracting the isophot contour lines. These lines pass through those locations in the flowfield where the displacement of transparent water by the dyed one reached a certain dye concentration density.

As the result of these and similar experimentations it became obvious that if the desirable disintegration of the insulating conductive layer on the heat transfer

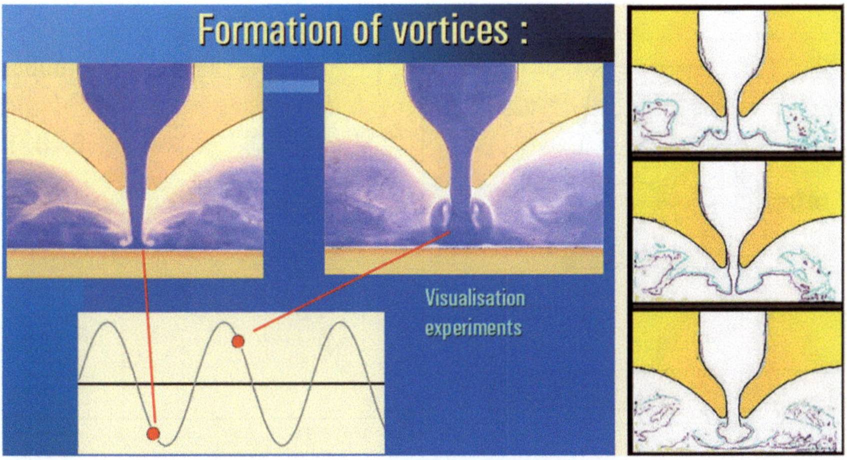


**Fig. 34** Apart from damping the pulsation by viscous dissipation there is another reason limiting an action on the insulating thermal boundary layer by pulsation applied inside the nozzle: instead of acting destructively on the layer, the energy of the pulsation is uselessly spent on generating vortical structures quite high above the impingement wall

surface is to be achieved, the simple approach based on axial pulsation of the fluid in the nozzle—using the mainstream standard round or planar jet—is not the right approach. Either it is necessary to apply the pulsation in a different, more sophisticated way, or this simple approach has to be applied to another impinging flow type, one that is more sensitive towards the pulsation. The possibility of achieving improvement by applying the excitation elsewhere in the flowfield remains a subject of future research. Locating such a place in the flowfield is difficult. Furthermore, it may need a complex geometry—perhaps one with a separate channel to bring the pulsation to more sensitive regions (i.e. perhaps something along the line of the case A in Fig. 32). The idea at any rate contains too many unknown factors. On the other hand, the case of a slightly different impinging flow that has already exhibited the sensitivity to disturbances as a whole came to mind immediately. It is the unstable configuration B of the annular impinging jet in Fig. 26. This idea was tested practically, using an annular nozzle with the same $b/D = 0.025$. The results could be even better if the nozzle were positioned at a distance h with higher sensitivity. Also the driving frequency f = 263 Hz was simply taken over from previous experiments without optimisation.

Measurements were made by naphtalene sublimation technique for investigations of mass transfer [22]. Excitation converted the configuration E, with the centrifugal stagnation point on the impingement wall, to configuration A with the stagnation circle (Fig. 24). The resultant spreading of the mass transfer over a larger

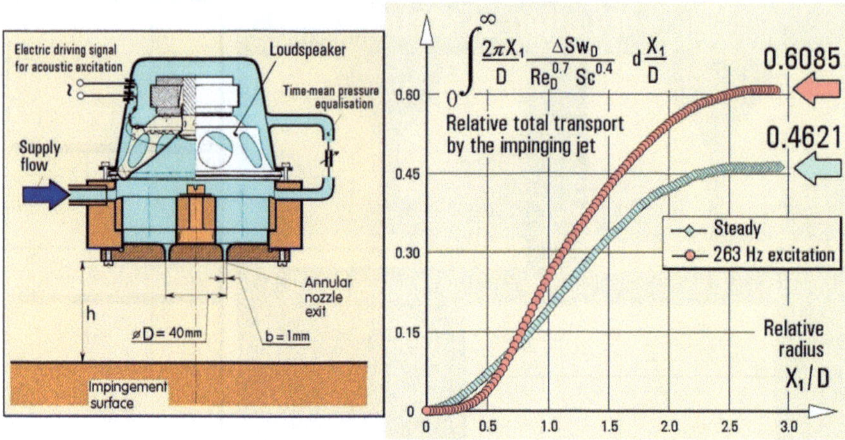


**Fig. 35** Pulsation applied to the annular nozzle was demonstrated to increase the total thermal transport. For these experiments the fluidic oscillator was temporarily replaced by loudspeaker. The increase (red data points, right) was mainly due to the metamorphosis [15] caused by excitation of regime **s** as presented in Fig. 26—adapted from [22]. *Courtesy* of Elsevier B.V

area (due to radial divergence) has led to an improvement of the total integrated value by 32 %—certainly not a negligible improvement, as seen at the right of Figs. 35 and 36.

In another experimental study [23], the setup was designed to correspond more to the envisaged operational layout, with an integral fluidic oscillator and two annular nozzles connected to the oscillator outputs. The two nozzles utilised the fact that the diverter type amplifier in the oscillator has two outlets with alternating outflows from them. The amplifier cavities were laser-cut into a 5 mm thick thermally resistant Teflon plate, as shown at the top left of Fig. 37. The oscillator was placed inside the rectangular metal body seen at the bottom of Figs. 37 and 38. The working fluid (i.e. heated air) was supplied through the inlet terminal S while the control nozzles $X_1$ and $X_2$ were mutually connected by a Teflon tube forming the Spyropoulos-type feedback connecting the two control terminals $X_1$ and $X_2$. Variations in the tube length made adjustment of oscillation frequency possible. The impinging-jet generating nozzles with $b/D = 0.027$ annular slit were located at the ends of the cylindrical extensions.

An increase in the heat transfer efficiency could be clearly demonstrated and was based on monitoring colour changes of thermochromic liquid crystals as shown in the left of Fig. 38. They made possible measurement of the extent of the area $A_{nom}$ on the impingement plane (right-hand side of Fig. 38) where the local temperature was larger than the limit value $T$ = 40 °C, at which with the particular brand of

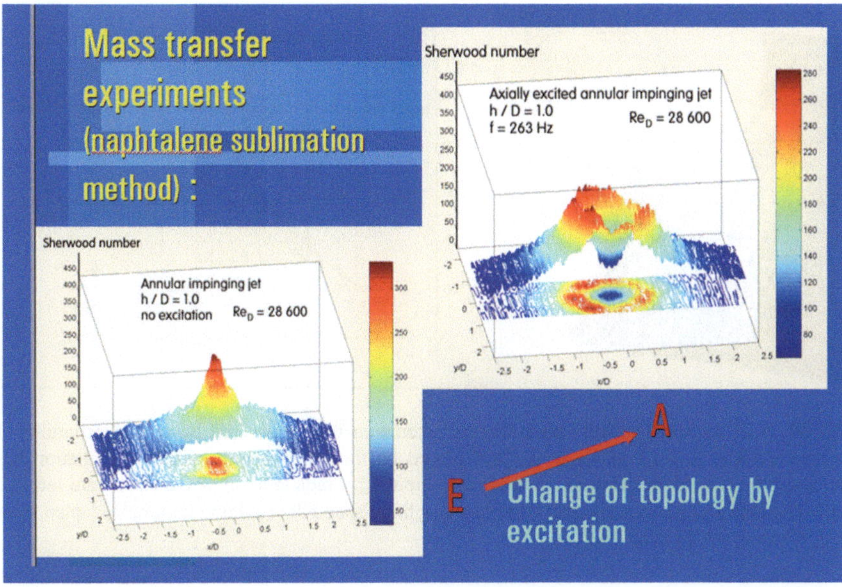


Fig. 36 Computed (*left*) and measured (*right*) velocity along the jet axis in the flow-field configuration *E* (centripetal stagnation point midway between the nozzle and the wall). Co-ordinate axis, $X_1$, oriented in accordance with usage in jet flow studies. Measurements by Prof. Trávníček and Tesař [22]

**Fig. 37** *Top left* Cavities of the fluidic amplifier, laser cut in Teflon (for resistance to heating). It was used in the annular pulsatory impinging jet actuator with two annular nozzles in the heat transfer experiment. Dimensions: nozzle outer diameter D = 30 mm, nozzle slit width b = 0.8 mm. The fluidic amplifier is housed inside the rectangular metal body as seen in Fig. 38. Adapted from [23]. *Courtesy* of Elsevier B. V.

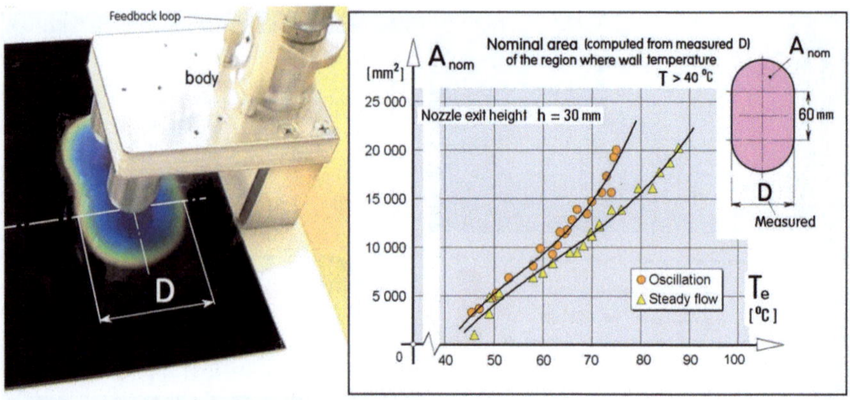


**Fig. 38** Heat transfer changes caused by the excitation were measured by colour variations of thermochoromic liquid crystals in the foil placed on the impingement surface. Oscillation has increased the size of the isothermal coloured region $A_{nom}$ indicating higher transfer rate into the wall. The area $A_{nom}$ was computed from the simplified shape the isotherm contour (adapted from [23]. *Courtesy* of Elsevier B.V.)

crystals the colour changes started. The boundaries of the surface area of temperature $T > 40$ °C were measured first with the steady jet and then with the oscillations activated. In all other respects, the conditions were identical. Because of the comparatory character of the measurements, detailed measurements of the details of the heated area were not necessary and the evaluation was simplified by measuring only a single distance—the width $D$ (as shown in Fig. 38) of the heated region in the location where there the axis of one of the nozzles lies. The areas presented in the diagram were computed from this width assuming the simple geometry (i.e. a rectangle and two half-circles). The measurements were made at a constant component of nozzle exit velocity equal to $w_e = 17$ m/s. As seen in the results, the flow oscillation of the hybrid-synthetic type was demonstrated to increase the heat transfer into the impingement wall by a substantial increment. In some regimes, the heat transfer into the surface was found to be doubled.

Many promising principles of convective heat and mass transfer by impinging jets remain so far untested even though they offer interesting opportunities for research. For example, in spite of all the described advantages of fluidic hybrid-synthetic jet generators, they may find a significant shortcoming in terms of the need to supply compressed air from a central source to the impinging-jet nozzle. The solution shown in Fig. 39 need no such supply with no-moving-part pulsatory driving flow re-circulating the cooling air between the microchip and the heat-removing region provided with cooling ribs. In principle this may be seen as a closed circuit driven by a fluidic pump [19, 24].

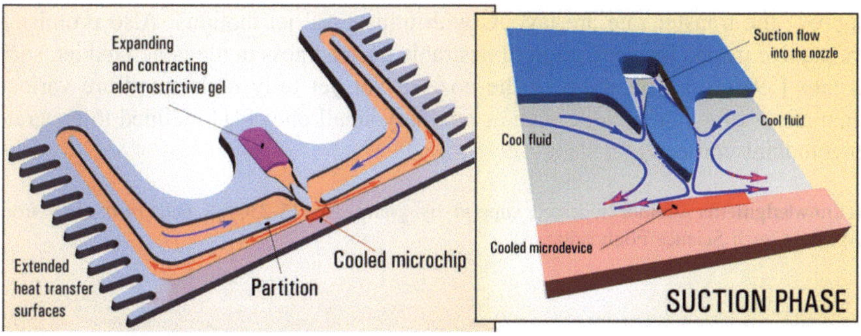


**Fig. 39** Cooling a microchip by hybrid-synthetic jet generated in an actuator which—in contrast to the fluidic oscillators—does not need supply of coolant fluid from an external source. In the loop passing along the surface with ribs the fluid hands over its heat and then becomes entrained by the jet issuing from the cavity with expanding gel. Guiding the jet by the Coanda effect sends coolant towards the microchip in a manner analogous to the forward flow diverter principle in fluidic no-moving-part pumps [24]

## 5 Summary

The *raison d'être* of impinging jets is their capability to act—and act efficiently—on the impingement surface. It may be a force action, but most often this action is heating, cooling, or drying. Impinging jets have the distinction among other convective transfer methods of being capable to generate the highest transfer rate. This places them into the focus of attention mainly for cooling electronic components—microprocessors with their very high localised heat production being the typical example. The flowfield of impinging jet is quite complex and there is no universally valid solution of the governing Navier-Stokes equation. What is available are more or less approximate solutions valid for some parts of the flowfield under certain assumptions, limiting the applicability to specific cases. The main obstacle in increasing the effectiveness of the heat and/or mass transfer is the very thin nearly stagnant insulating thermal conduction sublayer on the impingement surface. Attempts at suppressing its effects concentrated on superposing pulsation on the flow issuing from the nozzle. Most results were disappointing, mainly because the pulsation are effectively damped before they can reach the stagnant sublayer. The energy of the pulsation is also spent on unproductive generation of vortical structures. This is why there has been a search for impinging flows exhibiting hydrodynamic instabilities that the pulsations can excite. A solution was found in an annular jet with stationary vortex under the central body of the nozzle. Under some conditions, the vortex assumes a unstable shape with constriction. Even though the investigations performed so far missed the excitation in the most unstable regime, they have already exhibited an interesting increase in convective transport performance. There are thus significant opportunities for further investigations. Tests with annular jets, on which the attention currently focuses, were so far limited to a few values of the relative exit slit width $b/D$. Other approaches to be studied as means to improve the transfer rate are layouts with transversal jet motions. Also promising are passive means for generation of desirable unsteadiness in the generated jet, such as tabs [39] on the outer lip of the nozzle. Not yet fully understood are various alternatives of synthetic jets, such as auxiliary small ones [41] inclined to generate longitudinal vortices.

**Acknowledgments** Author received support by grants Nr. 13-23046S and 14-08888S from GAČR—Czech Science Foundation.

## References


1. Martin, H. (1977). Heat and mass transfer between impinging gas jets and solid surfaces. *Advances in Heat Transfer, 13*, 1.
2. Garimella, S. V. (2000). Heat transfer and flow fields in confined jet impingement, Chap. 7. In *Annual Review of Heat Transfer* (Vol. XI, p. 413).
3. Stokes, G. G. (1851). On the effect of the internal friction of fluids on the motion of pendulums. In *Transactions of the Cambridge Philosophical Society* (Vol. IX, p. 8).
4. Moore, G. E. (1965). Cramming more components onto integrated circuits. *Electronics, 38*, 114. (Reprinted in *Proceedings of the IEEE, 86*, 82, 1998).
5. Lin, P. T., Chang, C.-J., Huang, H., & Zheng, B. (2011). Design of cooling system for electronic devices using impinging jets. In *Proceedings of COMSOL 2011 Conference*, Boston, U.S.A.
6. Tesař, V. (2005). Time-mean helicity distribution in turbulent swirling jets. *Acta Polytechnica —Journal of Advanced Engineering*, *45*, 9.
7. Tesař, V., Kellnerová, R., Uruba, V., & Něnička, V. (2011). Proper orthogonal decomposition analysis of an air jet with excited helical instabilities. In *Proceedings of Conference on 'Topical problems of Fluid Mechanics'* (p. 121).
8. Tesař, V., & Šarboch, J. (1997). Similarity solution of the axisymmetric turbulent jet using the one-equation model of turbulence. *Acta Polytechnica, 37*, 5.
9. Tesař, V. (2006). Similarity solutions of jet development mixing layers using algebraic and 1-equation turbulence models. *Acta Polytechnica—Journal of Advanced Engineering, 46*, 40.
10. Glauert, M. B. (1956). The wall jet. *Journal of Fluid Mechanics, 1*, 625.
11. Tesař, V., & Kordík, J. (2009). Quasi-similarity model of synthetic jets. *Sensors and Actuators A: Physical, A 149*, 255.
12. Tesař, V., & Kordík, J. (2011). Spectral analysis of a synthetic jet. *Sensors and Actuators, A: Physical, 167*, 213.
13. Trávníček, Z., & Tesař, V. (2003). Annular synthetic jet used for impinging flow mass transfer. *International Journal of Heat and Mass Transfer, 46*, 3291.
14. Tesař, V. (2007). *Pressure-driven microfluidics*. Norwood: Artech House Publishers.
15. Tesař, V., & Trávníček, Z. (2008). Excitational metamorphosis of surface flowfield under an impinging annular jet. *Chemical Engineering Journal, 144*, 312.
16. Trávníček, Z., & Tesař, V. (2013). Hysteresis in annular impinging jets. *Experimental Thermal and Fluid Science, 44*, 565.
17. Trávníček, Z., Hyhlík, T., & Maršík, F. (2005). Synthetic jet impingement heat/mass transfer. In *Proceedings of 16th International Symposium on Transport Phenomena*, Prague.
18. Trávníček, Z., Tesař, V., & Wang, A.-B. (2005). Enhancement of synthetic jets by means of an integrated valve-less pump, part II: Numerical and experimental studies. *Sensors and Actuators A: Physical, 125*, 50.

19. Tesař, V. (2008). Valve-less rectification pumps. In D. Li (Ed.), *Encyclopedia of Microfluidics and Nanofluidics* (pp. 2132–2139). Berlin: Springer.
20. Warren, R. W. (1962). *Negative Feedback Oscillator. US Patent 3,158,166.* filed 7th Aug 1962.
21. Göppert, S., Gürtler Mocikat, H., & Herwig, H. (2004). Heat transfer under a precessing jet: Effects of unsteady jet impingement. *International Journal of Heat and Mass Transfer, 47*, 2795.
22. Trávníček, Z., & Tesař, V. (2004). Annular impinging jet with recirculation zone expanded by acoustic excitation. *International Journal of Heat and Mass Transfer, 47*, 2329.
23. Tesař, V. (2009). Enhancing impinging-jet heat or mass transfer by fluidically generated flow pulsation. *Chemical Engineering Research and Design, 87*, 181.
24. Tesař, V. (2011). Pump for extremely dangerous liquids. *Chemical Engineering Research and Design, 89*, 941.
25. Liu, X., & Lienhard, V. J. H. (1993). The hydraulic jump in circular jet impingement and in other thin liquid films. *Experiments in Fluids, 15*, 108.
26. Tesař, V. (2013). Flow pulsation in the near-wall layer of impinging jets. In *European Physical Journal—Web of Conferences* (Vol. 45) Article number 01129.
27. Popiel, C. O., & Boguslawski, L. (1988). Effect of flow structure on the heat or mass transfer on a flat plate in impinging round jet. In *Proceedings of 2nd UK National Conference on Heat Transfer, University of Strathclyde* (Vol. 1, p. 663).
28. Tesař, V., & Topinka, P. (1987). *Supply of a Hot Gas into the Working Space of a Furnace for Thermal Processing of Food Products.* In Czech, Czechoslovak Certificate of authorship Nr. 263 898, filed December 1987.
29. El Hassan, M., Assoum, H. H., Sobolik, V., Vetel, J., Abed-Meraim, K., Garon, A., & Sakout, A. (2012). Experimental investigation of the wall shear stress and the vortex dynamics in a circular impinging jet. *Experiments in Fluids, 52*, 1489.
30. Trávníček, Z., & Tesař, V. (2013). Hysteresis in annular impinging jets. *Experimental Thermal and Fluid Science, 44*, 565.
31. Tesař, V., Hung, C.-H., & Zimmerman, W. B. J. (2006). No-moving-part hybrid-synthetic jet actuator. *Sensors and Actuators, A: Physical, 125*, 159.
32. Tesař, V., & Trávníček, Z. (2005). Increasing heat and/or mass transfer rates in impinging jets. *Journal of Visualization, 8*, 91.
33. Page, R. H., Chinnock, P. S., & Sezeed-Yagoobi, J. (1996). Self-oscillation enhancement of impinging jet heat transfer. *Journal of Thermophysics and Heat Transfer, 10*, 380.
34. Gardon, R., & Akfirat, J. C. (1965). The role of turbulence in the heat transfer characteristics of impinging jets. *International Journal of Heat and Mass Transfer, 8*, 1261.
35. Chung, Y. M., & Luo, K. H. (2002). Unsteady heat transfer analysis of an impinging jet. *Journal of Heat Transfer, 124*, 1039.
36. Zuckerman, N., & Lior, N. (2006). Jet impingement heat transfer: physics, correlations, and numerical modelling. *Advancers in Heat Transfer, 39*, 565.
37. Sheriff, H. S., & Zumbrunnen, D. A. (1994). Effect of flow pulsations on the cooling effectiveness of an impinging jet. *Journal of Heat Transfer, 116*, 886.
38. Buchlin, J. M., & Laperches, M. (1998). Detailed investigation of aerothermal behaviour of confined impinging jet. In *Proceedings of Conference on QUIRT Quantitative Infrared Thermography*, Lodz, Poland (p. 258).
39. Hayashi, T., et al (2009). Experimental study on control of an impinging jet heat transfer using triangular tabs. *Journal of Fluid Science and Technology, 4*, 292
40. Tesař V., Kordík J., Daněk M. (2008). Lift and separation control on wind turbine blades by vortices having streamwise oriented axes. In *Proceedings of Colloquium FLUID DYNAMICS 2008*, Prague
41. Tamburello D.A., & Amitay M. (2008). Active control of a free jet using a synthetic jet. *International Journal of Heat and Fluid Flow, 29*, 967

# Index





D.T.H. New and S.C.M. Yu (eds.), *Vortex Rings and Jets*,
Fluid Mechanics and Its Applications 111, DOI 10.1007/978-981-287-396-5

MIX
Papier aus verantwortungsvollen Quellen
Paper from responsible sources
FSC® C105338

If you have any concerns about our products,
you can contact us on
**ProductSafety@springernature.com**

In case Publisher is established outside the EU,
the EU authorized representative is:
**Springer Nature Customer Service Center GmbH**
**Europaplatz 3, 69115 Heidelberg, Germany**

Printed by Libri Plureos GmbH
in Hamburg, Germany